The Development of an Aquatic Habitat Classification System for Lakes

Edited by
W.-Dieter N. Busch
U.S. Fish and Wildlife Service
Buffalo, New York
Peter G. Sly
Rawson Academy of Aquatic Science
Ottawa, Ontario

CRC Press
Boca Raton Ann Arbor London Tokyo

Library of Congress Cataloging-in-Publication Data

The Development of an aquatic habitat classification system for lakes : including concepts, the aquatic habitat classification, and background papers from a February, 1988 workshop / edited by W.-D. N. Busch and P. G. Sly.

p. cm.

Includes bibliographical references and index.

ISBN 0-8493-0148-3

1. Aquatic habitats—Classification—Congresses. 2. Fishery management—Congresses. 3. Lakes—Classification—Congresses. 4. Fishes—Habitat—Classification—Congresses. I. Busch, W.-D. N. (W.-Dieter N.) II. Sly, Peter G.

SH329.A66D48 1992

639.9′77—dc20

92-6759

CIP

Direct all inquiries to CRC Press, Inc., 2000 Corporate Blvd., N. W., Boca Raton, Florida, 33431.

International Standard Book Number 0-8493-0148-3

Library of Congress Card Number 92-6759

Printed in the United States 2 3 4 5 6 7 8 9 0

Printed on acid-free paper

DEDICATION

It is with great respect and pleasure that the editors of this book dedicate this work to the memory of Ken Loftus, friend, mentor, and colleague. He was, without doubt, a champion in the understanding of relationships between fish and habitat.

The contributions which comprise this book are a product of a Symposium on the Classification and Inventory of Great Lakes Aquatic Habitats (CIGLAH), the last of a series of Great Lakes Symposia on fisheries and aquatic science and management which have benefited so richly from Ken's participation. Though not blessed with the breadth and depth of his knowledge and insights, like Ken, we know that what we present is only a beginning.

From the start of his career as a fishery biologist with the Ontario Department of Lands and Forests (later the Ontario Ministry of Natural Resources), through his years as the senior fisheries manager in the Province and later science advisor, and through his membership on the Great Lakes Water Quality Board of the International Joint Commission, and especially as a commissioner with the Great Lakes Fishery Commission, Ken Loftus has played a pivotal role in the science and management of fish and aquatic resources. Much of the success of the rehabilitation of the Great Lakes fisheries is directly attributable to Ken's far-sighted ideas and leadership skills. Even though Salmonine fisheries remain heavily dependent on stocking, Ken showed the importance of native species and has surely laid the basis for the eventual recovery of self-sustaining lake trout populations throughout the lakes. Perhaps his greatest contribution, however, has been to foster understanding and linkages between fisheries science and limnology, and the development of management practices which would integrate the knowledge and understanding of specialists in natural science, economics, and sociology.

Ken Loftus was, indeed, a man for this time and we feel privileged to have enjoyed his humor and participation. He recognized how long it takes to achieve a coherent understanding of natural systems. Over the period since the CIGLAH Symposium, we too have learned to appreciate the depths inherent in the topics which are addressed. We hope this dedication will encourage others to carry forward with this work.

W.-Dieter N. Busch and Peter G. Sly

PURPOSE

Approaches to fishery management are changing from a reliance upon regulation of harvest and population enhancement (stocking), to habitat protection and creation, and most recently toward ecosystem management. The ecosystem, of course, includes physical, chemical and biological habitat components. However, applying ecosystem management to large areas of aquatic habitat has been difficult. A basic tool has been missing — an aquatic habitat classification system. This project was undertaken to create such a system and, as far as possible, build on what is already available.

In developing the aquatic habitat classification (AHC) system, considerable effort was directed toward describing and organizing understandings dealing with the functions performed by various components of the aquatic habitat. These functions become particularly important when using a multidimensional approach, as addressed in Chapter 1.

Although no single AHC system has been universally embraced, a number of classification approaches have been developed. These classifications generally use a limited number of physical, chemical, and/or biological variables to produce some form of index (see the detailed reviews later in this publication). Such classification approaches have been and continue to be valuable tools. Many of these classification approaches are summarized in Chapters 3 to 5.

In addition to a general indication of lake productivity, more information is necessary for ecosystem management. Detailed types of information that may be useful are reviewed in Chapters 6 and 7. If the habitat requirements of a fish species at various life stages and seasons of the year are to be included, the information base becomes very complicated. When attempts are made to include factors such as the form and lake bed materials of a basin, nutrient and oxygen levels in the water, water movement, heating rates and light penetration, contaminants, predation, and harvest, the level of complexity becomes so high as to be almost unmanageable.

To implement scientific resource management at the ecosystem level, which includes habitat restoration, enhancement, or protection in addition to fish species management, a wide range of detailed data needs to be available and in a usable format. Furthermore, local or site specific information should be relatable at various scales, from the smallest bays or weed beds to the larger picture, at the basin or the whole-lake scale. This requires standardization of information (language) and some form of structure which provides hierarchical order of the data.

The AHC that has been developed (Chapter 2) provides terms, definitions, and an organizational structure. The level of detail goes from the general to the specific, depending on the needs of the user. A hierarchical approach is used. To accommodate the needs of specific use, a list of modifiers has also been prepared, as have examples of application at several of the levels.

The AHC provides a "language" and a structure that should help standardize field data collection, storage, and use. The AHC does not make "value" judgments. Those must be made by the user and, in the case of the aquatic system, will be dictated by the requirements of the fish species of interest to the manager or researcher.

The AHC is not expected to be all things to all people. Instead, it provides a universal framework. This should allow for data comparisons between sites and at different times, and to identify the intended and potential applicability of information to a site, shoreline, basin, zone, or lake. It is recognized that detailed data collections from large systems are very costly. In addition, such data are often not adequately used because they are not readily available. The approach suggested by the AHC should result in improved comparability of field data. It should encourage greater computer data storage and processing, and, in particular, the latter would allow habitat boundaries to be established by statistical analysis. Statistics can provide not only functional discrimination, but also the degree of separation between habitat types and the groupings by which habitat types may be formed into different associations.

ACKNOWLEDGMENTS AND WORKSHOP PARTICIPANTS

We express appreciation for the financial and moral support provided by the Great Lakes Fishery Commission through its Habitat Advisory Board, the International Joint Commission, and the International Association for Sediment and Water Science. In addition, many federal, provincial and state agencies, and universities permitted their employees to participate at agency expense.

The preparation of detailed background studies was led by the chairpersons of the various task groups. We are very grateful to Walter Duffy, Tom Edsall, Warren Flint, Charles E. Herdendorf, Rex Herron, Pat Hudson, Murray Johnson, Joe Koonce, Joe Margraf, Paul McKee, Ray Oglesby, Jeane Pesendorfer, and Harvey Shear for taking on these tasks.

The program was led by a Steering Committee. Its members were Edward T. LaRoe, Joe Leach, C. Ken Minns, and J. Henry Sather. W. Dieter Busch and Peter G. Sly were co-chairmen.

The editors are extremely grateful for the many detailed and helpful criticisms made by reviewers at several stages of manuscript preparation. Their contributions have undoubtedly added much to the overall value of this volume, both in terms of content and presentation. In particular, we express appreciation to R. Eshenroder, J. Leach, and J. H. Sather for their generous assistance. Finally, a special note of appreciation is owed to Alice Schneider, Marilyn Murphy, Dave Chisolm, and Laura Kracker for their generous help in proofreading and processing the manuscripts.

The following persons participated in the CIGLAH workshop:

Name	**Agency***
Dr. Ray M. Biette	OMNR
Dr. Robert P. Bukata	DOE Canada
Mr. W.-Dieter N. Busch	USFWS
Mr. Victor W. Cairns	DFO
Mr. Gavin Christie	Ontario Hydro (now with GLFC)
Mr. W. Jack Christie	OMNR
Mr. Tom Dahl	USFWS
Dr. Walter Duffy	USFWS
Mr. Tom Edsall	USFWS
Mr. Carlos Fetterolf	GLFC
Dr. R. Warren Flint	SUNY Buffalo
Dr. Tom Fontaine	GLERL
Dr. John Gannon	USFWS
Dr. Mike Gilbertson	IJC
Mrs. Valanne Glooschenko	OMNR
Dr. Lorne Greig	ESSA (Consultants)
Dr. Ronald Griffiths	OME
Dr. Lars Håkanson	University of Uppsala
Dr. Charles E. Herdendorf	Ohio State University
Dr. Rex C. Herron	NMFS
Dr. Brian Hindley	MTRCA
Mr. Pat Hudson	USFWS
Mr. Don Hughes	OMNR
Dr. Murray G. Johnson	DFO
Dr. Mike Jones	ESSA (now with OMNR)

Dr. David Jude	University of Michigan
Dr. Joseph Koonce	Case Western Reserve University
Dr. Edward T. LaRoe	USFWS
Dr. Joe Leach	OMNR
Mr. Ken Loftus	Retired
Dr. Joe Margraf	USFWS
Dr. Paul McKee	BEAK Consultants Ltd.
Dr. C. Ken Minns	DFO
Dr. Ray T. Oglesby	Cornell University
Mr. Bob Pacific	USFWS
Dr. John B. Pearce	NMFS
Ms. Jeane Pesendorfer	OMNR
Dr. Leif Pihl	Institute of Marine Research, Sweden
Dr. C. Nicholas Raphael	Eastern Michigan University
Mr. James Reckahn	OMNR
Dr. Richard A. Ryder	OMNR
Dr. J. Henry Sather	USFWS
Ms. Debbie Schroeder	OMNR
Miss Jean Shaffer	Tahoe Regional Planning Agency
Dr. Harvey Shear	DFO
Dr. Peter Sly	DOE Canada
Mr. Bill Spaulding	USFWS
Dr. Don Stewart	EFS/SUNY Syracuse
Dr. Richard Thomas	DOE Canada
Mr. Thomas J. Wheaton	MacLaren Plansearch Inc.
Dr. Thomas H. Whillans	Trent University

The following persons participated in the CIGLAH activity, although they did not attend the workshop:

Dr. Ted Batterson	Michigan State University
Dr. Stephen Brandt	University of Maryland
Dr. Robert Brock	ESF/SUNY Syracuse
Dr. Mike Brode	USFWS
Mr. James Dawson	BioSonics, Inc.
Mr. Frank Horvath	Michigan Department of Natural Resources
Dr. Eugene Jaworski	Eastern Michigan University
Dr. Charles Liston	USBLR
Dr. William Platts	USFS
Dr. Keith G. Rodgers	DOE Canada
Dr. Paul Sager	University of Wisconsin
Mr. Jim Scurry	USFWS

The following individuals served as chairpersons for groups preparing the background reports and/or workshop facilitators:

Dr. Walter Duffy
Mr. Tom Edsall
Dr. R. Warren Flint
Dr. Lorne Greig
Dr. Charles E. Herdendorf
Dr. Rex C. Herron
Mr. Pat Hudson
Dr. Murray G. Johnson
Dr. Mike Jones
Dr. Joe Koonce

Dr. Joe Margraf	Dr. Ray Oglesby
Dr. Paul McKee	Ms. Jeane Pesendorfer
Dr. C. Ken Minns	Dr. Harvey Shear

*Abbreviations:

DFO	Department of Fisheries and Oceans, Canada
DOE Canada	Department of Environment, Canada
ESF	Environmental Science and Forestry, SUNY Syracuse
GLERL	Great Lakes Environmental Research Laboratory
GLFC	Great Lakes Fishery Commission
IJC	International Joint Commission
MTRCA	Metro Toronto Regional Conservation Authority
NMFS	National Marine Fisheries Service
OME	Ontario Ministry of Environment
OMNR	Ontario Ministry of Natural Resources
SUNY	State University of New York
USBLR	United States Bureau of Land Reclamation
USFS	United States Forest Service
USFWS	United States Fish and Wildlife Service

CONTRIBUTORS

T. R. Batterson
Michigan State University
East Lansing, Michigan

R. H. Brock
College of Environmental Science
and Forestry
State University of New York
Syracuse, New York

R. P. Bukata
Canada Centre for Inland Waters
Burlington, Ontario, Canada

W.-D. N. Busch
U.S. Fish and Wildlife Service
Buffalo, New York

T. E. Dahl
U.S. Fish and Wildlife Service
Arlington, Virginia

J. J. Dawson
BioSonics, Inc.
Seattle, Washington

W. G. Duffy
U.S. Fish and Wildlife Service
Brookings, South Dakota

T. A. Edsall
U.S. Fish and Wildlife Service
Ann Arbor, Michigan

V. Glooschenko
Ontario Ministry of Natural Resources
Toronto, Ontario, Canada

R. W. Griffiths
Aquatic Ecostudies Limited
Kitchener, Ontario, Canada

L. Håkanson
University of Uppsala
Uppsala, Sweden

C. E. Herdendorf
Ohio State University
Columbus, Ohio

R. C. Herron
National Marine Fisheries Service
Bay St. Louis, Missouri

F. J. Horvath
Michigan Department of Natural Resources
Lansing, Michigan

P. L. Hudson
U.S. Fish and Wildlife Service
Ann Arbor, Michigan

E. Jaworski
Eastern Michigan University
Ypsilanti, Michigan

D. J. Jude
University of Michigan
Ann Arbor, Michigan

E. T. LaRoe
U.S. Fish and Wildlife Service
Arlington, Virginia

J. H. Leach
Ontario Ministry of Natural Resources
Wheatley, Ontario, Canada

P. M. McKee
Beak Consultants Limited
Brampton, Ontario, Canada

R. T. Oglesby
Cornell University
Ithaca, New York

J. B. Pearce
National Marine Fisheries Service
Woods Hole, Massachusetts

C. N. Raphael
Eastern Michigan University
Ypsilanti, Michigan

J. A. Reckahn
Ontario Ministry of Natural Resources
Maple, Ontario

P. G. Sly
Rawson Academy of Aquatic Science
Ottawa, Ontario, Canada

T. J. Wheaton
Environmental Advisory Services Limited
Guelph, Ontario, Canada

T. H. Whillans
Trent University
Peterborough, Ontario, Canada

TABLE OF CONTENTS

Chapter 1

INTRODUCTION TO THE PROCESS, PROCEDURE, AND CONCEPTS USED IN THE DEVELOPMENT OF AN AQUATIC HABITAT CLASSIFICATION SYSTEM FOR LAKES

P. G. Sly and W.-D. N. Busch

TABLE OF CONTENTS

ABSTRACT

The approach used to develop an aquatic habitat classification (AHC) system included review and preparation of published habitat criteria reports by six teams, preparation of applied case studies by four teams, and a workshop.

The process included a review of the functional roles of habitat. Further, since the performance of these functions is controlled by the physical, chemical, and biological setting, select descriptors were reviewed by the participants in this project to provide measurement of the aquatic habitat. In addition, it was recognized that the classification system would need to provide a basis for relating information at different scales; at local, regional, and the complete system level of aquatic habitat. A Geographic Information System (GIS) was seen to offer many potential advantages for data handling and processing.

I. INTRODUCTION

Many lake management plans presume that habitat management can be used to sustain and enhance the fishery and that habitat classification provides information useful to estimate fish yield. However, some managers question whether existing means of habitat classification are appropriate for management needs and, further, whether available means for inventory and information processing are being effectively applied. Therefore, a Classification and Inventory of Great Lakes Aquatic Habitat (CIGLAH) project was initiated to address these and related questions.

The CIGLAH project consisted of three units: (1) preparation of background documents providing comprehensive reviews of existing habitat classifications, inventory techniques, and the physical, chemical, and biological variables that need to be considered, (2) preparation of test approaches and their case application, and (3) a workshop (Barrie, Ontario, February 1988).

This report summarizes the concepts and approach used in the development of the AHC system. The major emphasis included an expansion of the "concept" of habitat beyond that provided by the purely physical setting and further evaluation of productivity indices to describe habitat quality. The goal was to address the complex factors that allow specific physically and chemically defined areas to function as habitat(s) for aquatic biota and to formulate an AHC system.

Papers prepared as part of the preworkshop activity are now provided as chapters in this volume. Chapters by Duffy et al., Edsall et al., Herdendorf et al., Hudson et al., Leach and Herron, and McKee et al. review classification systems and inventory techniques for CIGLAH. They indicate that many different approaches to habitat classification are available. However, none of the approaches are completely adequate to meet the needs of large lake limnology.

Conceptual developments in fisheries science and limnology indicate that great benefit may be gained by the integrated management of water quality and fisheries data. The benefits are most visible where both sets of data share a common means of presentation. Any proposed classification system should enable users to understand how existing classification systems interrelate, provide sets of structured information that are common to most user requirements, and allow users to add extra information that may be specific to their needs.

The CIGLAH project addressed the classification of aquatic habitats in large lakes in general, and the Great Lakes in particular. In the Great Lakes region, aquatic habitats include the open lake, nearshore waters, adjacent wetlands, connecting channels, and tributaries. Although most attention was given to relationships between fish and their lake habitat, the participants were also concerned with the habitat in peripheral and connecting areas where society's activities can alter nursery and spawning sites and disrupt access by lake fish populations.

Habitat is described by reference to any variable that defines or delimits the space in which biota live, and covers anything from an entire lake to its smallest part. Habitat is usually described by physical and chemical variables or, to a lesser extent, by biological variables. The latter are most useful when applied as a form of data integration or as a surrogate for physical and chemical characteristics. Classification requires that the variables that best describe aquatic habitat(s) be grouped to allow differentiation among habitat types.

The goal of CIGLAH was to provide an AHC system that could be used by both fishery and water quality agencies, for management decisions, and public discussions. The classification system should be suitable for "problem-oriented" application, be structured in a clear and practical manner that is based on sound science, and allow easy understanding of its uses and limitations. Data for such a system must be compatible with the level of hierarchical resolution required and should be of acceptable quality and quantity. Based on these requirements, the participants decided that a simple hierarchical classification system was best suited to management situations in which data collection and interpretation may be at different scales of resolution.

A hierarchical classification system is rather insensitive to variations in habitat data at the whole-lake level. It is more sensitive to variations in habitat data at the level of smaller component parts of a lake. Hence, much more information is required to construct a total lakewide classification at low levels of hierarchy (e.g., 1 ha sectors) than at the whole-lake level. In creating a comprehensive lakewide classification, one must balance the greater resolution obtainable at low levels of hierarchy against the higher cost and effort required to obtain adequate data. Further, the effects of differing biological conversion efficiencies, patchiness, and entropy that occurs from the use of many habitat variables makes it generally inappropriate to seek statistical correlation between items such as fish yield and lakewide habitat classification, based on summation of detailed information from component parts of a lake. Generic models for specific areas of a lake based on critical habitat features or niche requirements are needed, but represent a task beyond the scope of this project. It is hoped that these topics will be addressed as part of the recommended demonstration project (see Section IV, Recommendations).

Following this introduction is a discussion of the concepts that form the basis for the AHC system. Lastly, limitations to the implementation of the AHC system that are imposed by the use of conventional two-dimensional mapping displays are considered and the need for more advanced techniques of data handling and presentations that would be available through a GIS are discussed.

A short summary of and reflective commentary on the 1984 Symposium on Modeling Habitat Relationships of Terrestrial Vertebrates is provided by J. A. Reckahn in an appendix to this chapter. This unusual contribution points to the significance of scale in terrestrial habitat studies. It also suggests that the CIGLAH approach shares much in common with the lessons learned in terrestrial studies. The authors appreciate the value of this additional material and have provided it as a source of information otherwise not readily available to many readers.

II. CONCEPTS

A. FUNCTIONAL RELATIONSHIPS

One of the underlying goals of the CIGLAH project was to better understand the interactions among major components of habitat and use this to aid in habitat classification. Although habitat is a term that includes both physicochemical and biological components, at the

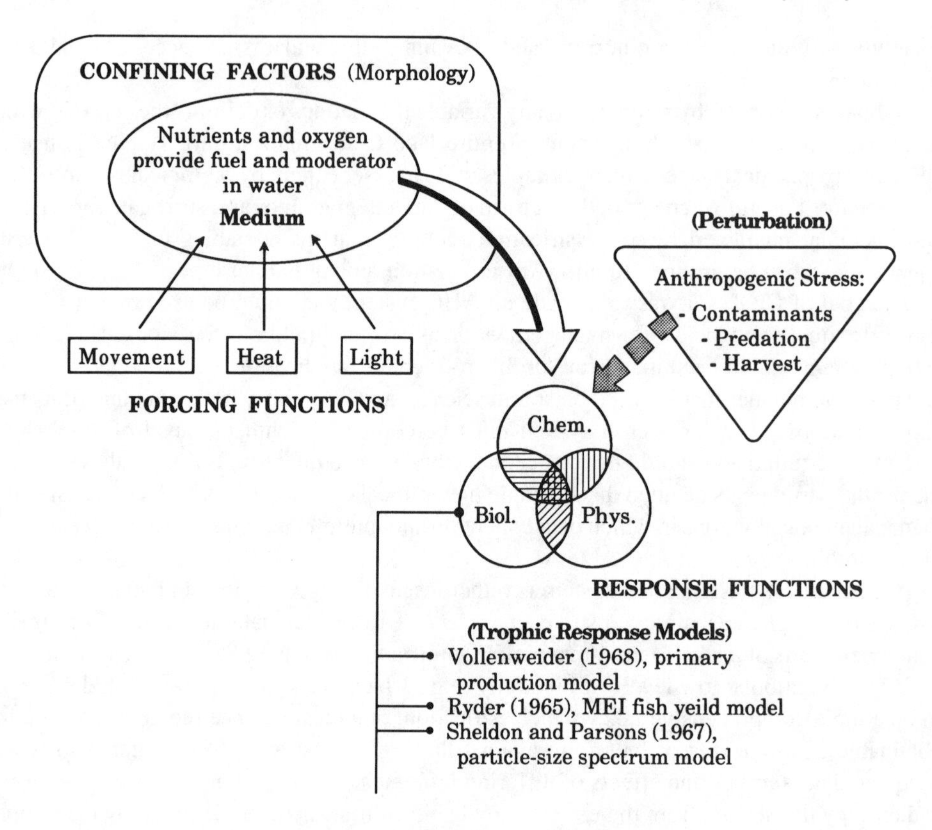

FIGURE 1. Graphic representation of the role of habitat and the connections between confining factors, forcing functions, and response functions; also noted are the effects of perturbations.

macroscale habitat is most often dominated by physicochemical characteristics. The structure of habitat can be expressed functionally in terms of confining factors, forcing functions, and response.

Figure 1 presents such relationships in simple terms. Lake size, morphometry, and nearshore complexity or shoreline development (e.g., the ratio of actual and straight-line distance between shoreline points) characterize the confinement of a body of water. Because morphology is also a product of past geological processes and conditions, earlier conditions may leave an imprint. For example, such imprints may occur as relict substrates or as drowned morphologies; the present is, in part, a product of the past. The forcing functions are energy related and include wind, precipitation, temperature, and light. Forcing functions interact and act differently upon water bodies of different size and shape. Particle-size sorting (including characteristic modifiers such as erosion, transportation, and deposition), turbidity, ice cover, currents, and water residence times are examples of various forms of response to physical forcing.

Water is referred to as a chemical medium, providing both fuel and moderators that drive and control the transfer energy flows through biological pathways. Natural water chemistry is strongly influenced by watershed soils and regional geology (thus, water chemistry could be thought of as another form of response). Although of major importance in determining offshore biological communities, it is usually sufficient to describe water chemistry at the whole-lake or zonal level (see Chapter 2).

Although water and sediment chemistry are recognized as major influences on biotic communities, we have largely excluded chemical attributes from the AHC system to avoid the problems of excessively complex presentation. Where required, water and sediment chemistry, and physical and chemical perturbations of the habitat (nutrient, toxic contaminants, and habitat modification) may be included as data modifiers.

In effect, the environment contains components that act as both a splitter and a buffer through which solar energy directly and indirectly drives physical and chemical processes. It is this energy which is channeled into the biological domain to support life processes at different trophic levels. The complex relations between various parts of the functionally structured habitat classification demonstrate that there is an absolute need for this classification to be flexible. Classification must allow expression of dominant forcing functions that are characteristic of either geographic location or biological requirement. Classifications (e.g., the littoral zone) that implicitly fix response in relation to a single forcing function (such as light) are unworkable in large and complex environments such as the nearshore zone of the Great Lakes where multiple forcing functions interact. Habitat may be variously dominated by wave action, light, temperature, water residence characteristics, and other forcing functions, depending upon the local and regional importance of relations between forcing functions and confining factors. The AHC system is flexible enough to meet these needs, and its hierarchy is tied to temporal and spatial scales of resolution which are otherwise excluded from most classification systems.

The development of the AHC system and its relationships to model concepts draws heavily on a wealth of previously published material, much of which has been summarized by CIGLAH authors in this publication. Further, analogies were also sought in terrestrial ecology where temporal and spatial controls are more easily observed in relation to species population and community. Examples of important ecosystem models that are available include:

Whole lake — The particle-size spectrum model that estimates potential biomass of progressively larger animals in relation to available phytoplankton (Sheldon and Parsons, 1967).

— The same theory may also reflect changes in the spectrum resulting from "top-down" or "bottom-up" stress, termed cascade effects (Dickie et al., 1987; Evans et al., 1987; Leach et al., 1987).

— The Vollenweider trophic state model and its subsequent development (Vollenweider, 1968, 1975; and Vollenweider and Kerekes, 1982), which estimates primary production; the morphoedaphic index (MEI) model (Ryder, 1965), which estimates fish yield and correlations between trophic state and fish yield (Jones and Lee, 1986; Oglesby et al., 1987). Both of these models are based on the use of limited physicochemical variables.

— Within the biological domain, few components are in steady state with each other or, externally, with their habitat. Climax models have their counterpart in aquatic systems, where the rate of change in community structure decreases over time, to be reset at various points between initiation and maturity by events such as climate change (i.e., glaciation and major water level changes over geological time; Karrow and Calkin, 1985); species introductions [i.e., sea lamprey (*Petromyzon marinus*), rainbow smelt (*Osmerus mordax*), and white perch (*Morone americanus*); Smith and Tibbles, 1980; Emery, 1985]; and cyclical changes of population [e.g., lake whitefish (*Coregonus clupeaformis*); Reckahn, 1986].

Part lake — At a level of much greater detail, the habitat suitability index (HSI; Aggus and Bivin, 1982; McMahon et al., 1984) provides another empirical approach to modeling relations between functional groups (guilds), species, and some fish communities and their habitat; for example, cool- and coldwater fishes (Aggus and Bivin, 1982), walleye (*Stizostedion vitreum*) (McMahon et al., 1984), and lake trout (*Salvelinus namaycush*) (Marcus et al., 1984). This approach can be useful as a means of locally quantifying ways that maintain no net loss of habitat.

Descriptors of processes in both whole- and part-lake aquatic habitats were assessed, along with habitat requirements for both communities and single species. The merging of diverse variables and their integration, often by surrogates, constitutes an important development in conceptual thinking and model building.

Habitat classification is most effective when biotic elements are appropriately related to physicochemical functions (confining, forcing, and response). Different users may have similar concepts of important functional processes and may select similar physical, chemical, and biological variables for use in models, but have very different needs for the information that is derived. Classification systems whose structures do not depend upon specific end-use requirements (of any one user group) are most applicable for multipurpose use.

The application of biological measures was considered in relation to classification and measurement of fish habitats of large lakes. Traditionally, many biotic classification systems have been of the index or indicator types and indices have been developed for fish that require either community structure or composition to format input data. The most valuable hierarchical approach includes biotic measures at organism, population, community, and ecosystem levels. The niche approach was explored at the population level and the guild approach at the community level. Aspects such as particle-size and food-web structure were considered key elements at the ecosystem level.

It is agreed that fine tuning will improve predictive abilities and the resolution of top hierarchy models. However, the underlying variabilities in such data mean that they must be expressed by some form of integration such as averaging, or as biological surrogates which are also an expression of integration. We are uncertain just how far it may be useful to attempt to refine top hierarchy models, but proposed improvements to the early MEI model of Ryder (1965), such as the introduction of a "climate" variable, provide a useful guide (Hanson and Leggett, 1982; Jenkins, 1982; Leach et al., 1987; Ryder, 1982; Schlesinger and Regier, 1982; Youngs and Heimbuch, 1982).

B. HIERARCHICAL SCALES

The following discussion of a generic Atlantic salmon (*Salmo salar*) riverine habitat illustrates differences in scales of data derived at various levels within an AHC system. The total northwest Atlantic salmon population is assumed to be composed of stocks A, B, C, D, …, each of which spawns in a different watershed. Productive capacity is used to describe the yield of smolts. In Figure 2, some form of empirical model is used to estimate the productive capacity of aquatic habitat; examples of possible variables indicate the types of information that may be required. For one selected stock, type "L" data describe large-scale but low-resolution characteristics. The gradient, for example, may be used as a crude expression that naturally integrates substrate type, water residence time, and flow regime (the latter being somewhat more refined by use of further characterization of the November to April precipitation over the watershed). Most of these type "L" data are inherently stable.

Within each watershed, each river or stream can be further divided into numerous smaller site-specific habitat units. Based on a detailed knowledge of habitat requirements, such as used to establish the HSI (Aggus and Bivin, 1982), the river and stream units could be functionally grouped as spawning, nursery, juvenile, and other forms, and summed to produce total habitat area or volume. These high-resolution and small-scale (site-specific) pieces of information are characterized by type "S" data. Their lack of inherent stability is illustrated in the plots of a generic variable "x" (Figure 3).

At first glance, it may seem possible to refine estimates of habitat productivity based on type "L" data by carefully collecting large quantities of site-specific type "S" data for each watershed. However, the summation of site-specific habitat areas and volume rarely determines the actual productive capacity of a habitat because many of the forcing functions that

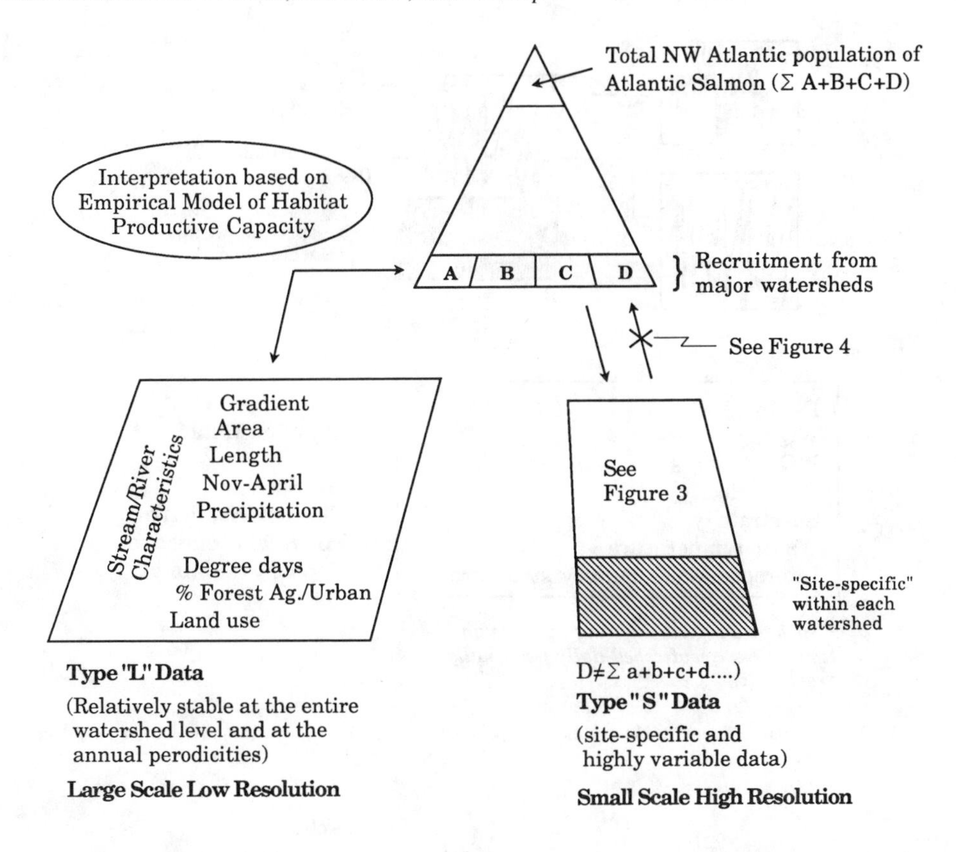

FIGURE 2. Habitat productive capacity: large-scale-low-resolution ("L") data compared to small-scale-high-resolution ("S") data.

control habitat use and availability are probabilistic (not specifically predictable). Also, increasing variability of forcing functions may result in greater species diversity but decreased productivity by individual species. For example, good spawning habitat may be limited by the effects of winter ice after egg deposition, or changes in flow regime may result in juvenile crowding within summer pool habitats. These constraints are not predictable in detail, but collectively, they restrict the availability of habitat. The amount of habitat that is actually available is termed "critical habitat". Critical habitat, for example, includes connectedness which allows fish to move between spawning and nursery habitats. The productive capacity of habitat is most closely related to critical habitat [assuming that stocks are not otherwise limited by excessive harvest or by conditions at sea (Watt, 1989)]. Also, it should be noted that density dependence may further complicate relations between community structure, populations, and habitat (Morris, 1988, 1989).

Type "L" data are essential for regional planning and setting harvest estimates. Type "S" data are an essential tool for site-specific habitat management, including the identification of limiting conditions that may restrict critical habitat and the expansion of its time-space volume (enhancement of productive capacity). Both sets of information are complimentary.

Figure 4 illustrates how the same point lying between data sets at high and low levels of hierarchy may be seen as one of integration or probability. Analysis by reduction passes through various levels of natural integration, from large to small scale, to determine cause and effect relationships with increasing resolution. Synthesis moves in the opposite direction. Usually, it is easier to derive interpretations based upon reduction of data; it is harder to do the reverse, which requires large quantities of data, if based on actual measurement instead of statistical simulation.

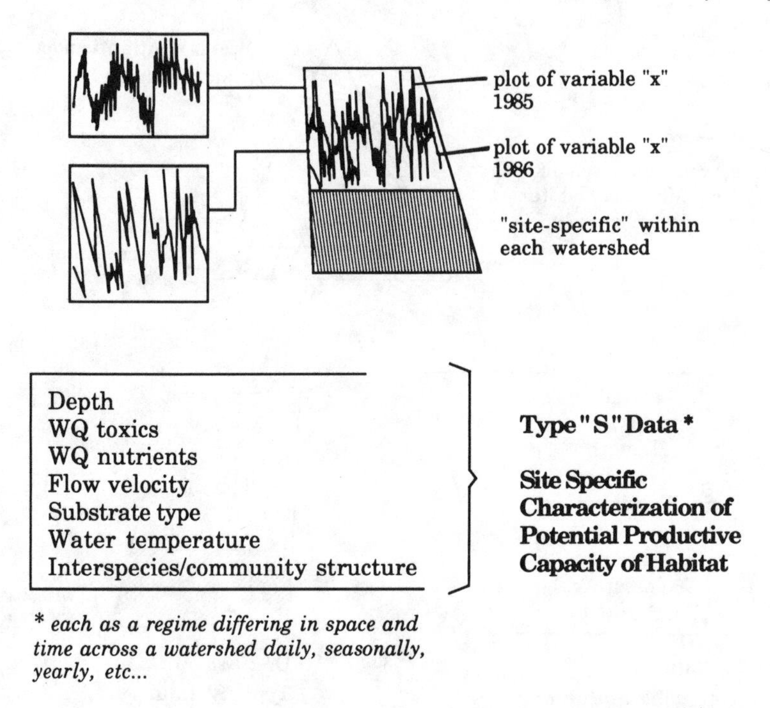

FIGURE 3. Graphic demonstration of the inherent instability of site-specific (high-resolution "S") data.

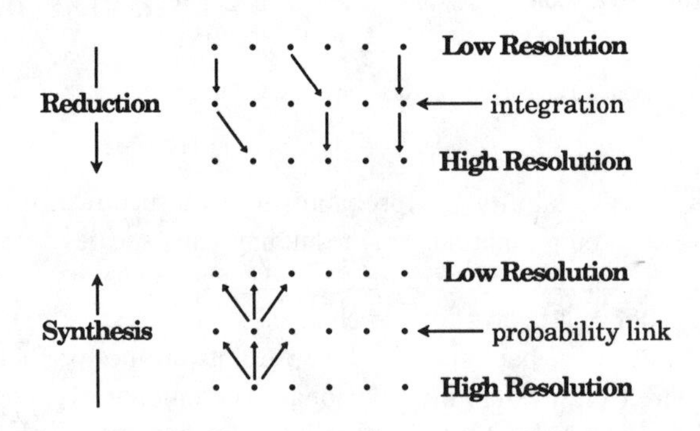

FIGURE 4. Integration and probability in relation to data reduction and synthesis, and scalar hierarchy.

It is also possible to hybridize data from different levels of hierarchy. For example, site-specific data can be collected from type locations and, based on the amount of each type available by locality, total quantities of each type of habitat can be estimated. We are unsure how much this approach will improve estimates of productive capacity because it does not take into account the temporal and spatial connectedness between type localities. Almost certainly, however, the approach must be watershed- and species-specific.

III. THE NEED FOR A GEOGRAPHIC INFORMATION SYSTEM (GIS)

Whereas it is possible to present a limited number of habitat characteristics on two-dimensional map displays, it is impractical to present the great number of combinations of an n-dimensional habitat array, as shown in Figure 5. Even the use of multiple map sheets will

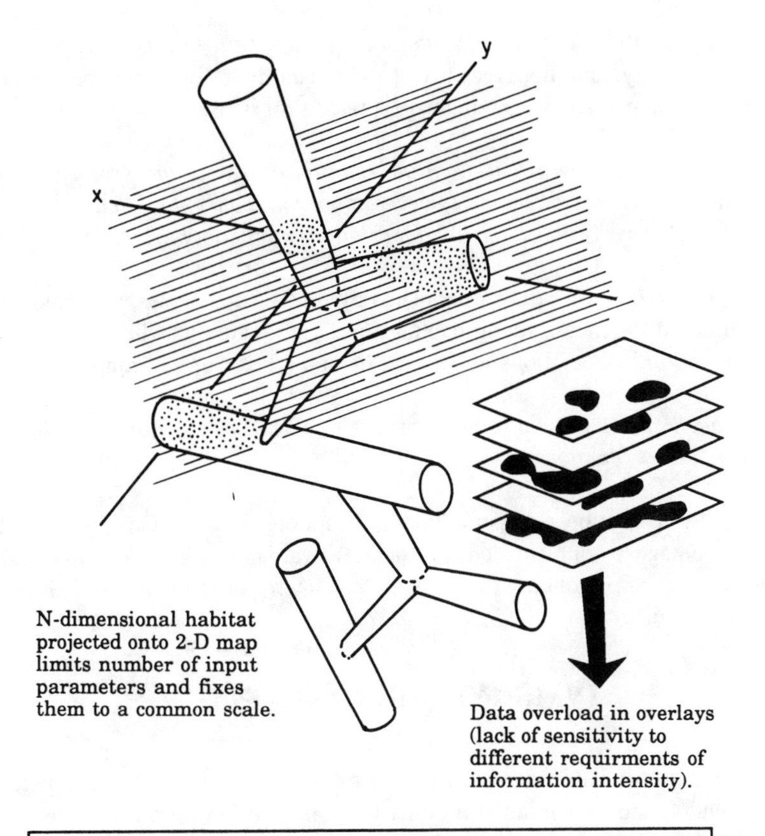

Data Interaction Envelope (examples of time/space dynamics)
Stable repeat... e.g. Spawning shoals.
Semi-stable repeat... e.g. Metalimnion or thermal interface
Unstable repeat... e.g. Phytoplankton blooms.
Events/catastrophy... e.g. Community restructuring.

FIGURE 5. The complexity of multidimensional habitat presentations are limited in a two-dimensional display, but can be accommodated by a computerized GIS.

not clearly show all important relations among the various habitat units at levels below the subdivision (defined in Chapter 2).

Several options are potentially available:

1. Plot key data (such as the distributions of critical temperature and dissolved oxygen values) on existing base maps (such as bathymetric charts), and then relate biotic community data to them.

   This approach is limited by the lack of available charts, differences in chart scales, or the visual overload that occurs when large quantities of information are displayed.

2. Statistical processing of data by computers can be used to reconstruct distribution data to any common scale although at different values of probability (reflecting temporal and spatial precision).

   This, to some extent, overcomes problems of presentation at a common scale. It does not overcome problems of information overload in display.

3. Existing classification techniques (categorizing habitat types according to concepts in classical limnology, and more recently, geology and hydrology) may be combined and plotted at common scale by computer-generated programs.

   This overcomes, to some extent, both information overload and problems associated with the need for a common scale. However, it imposes conceptual limits in our analysis of habitat, so that many niche and habitat relations might not be adequately considered.

4. If habitat boundaries were to be established by statistical analyses, we could determine the functional differences among habitat types, the degree to which habitat types are separated, and the associations into which habitat types can be grouped.

   This approach allows habitat to be categorized without constraint, and information to be presented at a common scale.

Until recently, it has not been possible to pursue options 3 and 4. However, the development of GIS and linkage to sophisticated statistical software now allows us to assess the value of these options as a measure of classification and display for different associations of habitat and niche characteristics.

IV. RECOMMENDATIONS

The authors recommend that a demonstration project be sponsored to compare outputs from options 3 and 4, and to determine if either is applicable for use with habitat inventory, water quality, and fisheries management issues. The demonstration project should test if concepts such as the HSI can provide widely applicable models of specific habitat and niche requirements. Also, statistical techniques should be used to establish precision at different levels of the hierarchy. Statistical techniques could be also used to link data sets derived at different levels of hierarchy to form hybrid models of habitat.

The usefulness of high level, low level, and hybrid hierarchical models of habitat should be tested against various types of user needs. The classification system, as presented, should be reviewed and modified after demonstrations and actual field applications. A further detailed review is suggested within a few years of the start of general applications based on the AHC system.

APPENDIX

WILDLIFE 2000: MODELING HABITAT RELATIONSHIPS OF TERRESTRIAL VERTEBRATES

This 1984 symposium in California produced a large volume of modeling approaches to wildlife and habitat simulations (Verner et al., 1986). The 60 papers had a combined literature section of 795 references, creating a valuable entrance to the literature of this topic. The most interesting apsect of the collected symposium presentations is that each of the six sections has two summaries, one from a manager's and one from a researcher's viewpoint. The similarities and differences are informative in themselves. A brief summary of their conclusions may provide some additional insight into potential similarities to be expected in Great Lakes studies.

Shugart and Urban (1986) reported that both wildlife and terrestrial modeling approaches sustained a significant stimulus from the International Biological Program (IBP). They also emphasized the diversity of approaches in this new area of research, but suggested two major avenues: (1) describing functional relationships between the environment and animals and (2) simulating the dynamics of the environment. Again, two levels of study were put forward: (1) large-scale approaches in which the most work has been done on territorial breeding males of nongame birds on spatial scales of 40 ha (100 acres). For large game animals a greater sense of multiple habitat use existed on scales of 100 ha. Second, fine-scale approaches examined presence (sometimes abundance) vs. absence on grids of about 0.1 ha. These studies used multivariate correlative statistics to determine significant relationships. This topic was examined previously in a separate symposium (Capen, 1981).

It was pointed out that model building is a dual approach that uses both inductive and deductive processes as the model is tested and improved (Stormer and Johnston, 1986). Their statement, "MODELS ARE ALWAYS TENTATIVE AND NEVER ABSOLUTELY PROVEN", is capitalized here for added emphasis. Noon (1986), a biometrician, pointed out the multiple stages in the modeling process and they are reiterated here because of their clarity and appropriateness:

1. An initial observation period in which the system is roughly defined and a set of questions posed
2. An exploratory stage using statistical techniques (i.e., multivariate analysis) designed to find patterns in the data
3. A period of empirical model building still largely based on data gathered from observational studies
4. A period in which the predictions of empirical models are statistically tested with observational data, the model refined, retested, etc.
5. A phase in which the results of the empirical model testing are synthesized into a theoretical model that makes statements (predictions) about the relationships between population processes and habitat components
6. The final stage, at which the predictions of the theoretical model are tested by confirmatory statistical methods. This final stage lends itself most readily to experimental studies, or to creatively designed observational studies in which confounding covariates are controlled statistically.

The theoretical model may be rejected at this time and the research forced to go back to an earlier stage and the process begun again. A quick survey of the wildlife-habitat literature suggests that most wildlife ecologists are at the third stage in this process (Noon, 1986). Aquatic ecologists are at a similar stage in the process, and the CIGLAH workshop can be viewed as an important step in this process for Great Lakes aquatic habitats.

Throughout the WILDLIFE 2000 symposium both managers and researchers called for greater cooperation between their groups to provide more usable models that have been thoroughly tested. Model validation is one aspect that everyone stressed, but it is apparently not happening rapidly enough. Salwasser (1986) made this sobering comment, "It is unsettling that 50 years into the business of wildlife management, people still argue about what habitat is", which indicates that aquatic biologists are not likely to reach consensus on the basis of a single workshop. The highly active controversies during CIGLAH may thus be a healthy symptom of the diverse viewpoints that characterize any newly opened avenue of research (Shugart and Urban, 1986).

J. A. Reckahn, Research Scientist
Fisheries Branch
Ontario Ministry of Natural Resources
P.O. Box 5000
Maple, Ontario L6A 1S9 Canada

January 23, 1989

REFERENCES

Aggus, L. R. and W. M. Bivin. 1982. Habitat Suitability Index Models: Regression Models Based on Harvest of Coolwater and Coldwater Fishes in Reservoirs. FWS/OBS–82/10.25, U.S. Fish and Wildlife Service, Washington, D.C.

Dickie, L. M., S. R. Kerr, and P. Schwinghamer. 1987. An ecological approach to fisheries assessment. *Can. J. Fish. Aquat. Sci.* 44(Suppl. 2):68–74.

Duffy, W. G., R. T. Oglesby, and J. A. Reckahn. Biological measures having applications to classifications of aquatic habitat in the Laurentian Great Lakes. This volume.

Edsall, T. A., J. J. Dawson, R. P. Bukata, R. H. Brock, I. J. Horvath, and W. J. Christie. Inventory techniques applicable to habitat classification of lakes. This volume.

Emery, L. 1985. Review of Fish Species Introduced into the Great Lakes, 1819–1974. Tech. Rep. 45, Great Lakes Fisheries Commission, Ann Arbor, MI.

Evans, D. O., B. A. Henderson, N. J. Bax, T. R. Marshall, R. T. Oglesby, and W. J. Christie. 1987. Concepts and methods of community ecology applied to freshwater fisheries management. *Can. J. Fish. Aquat. Sci.* 44(Suppl. 2):448–470.

Hanson, J. M. and W. C. Leggett. 1982. Empirical prediction of fish biomass and yield. *Can. J. Fish. Aquat. Sci.* 39:257–263.

Herdendorf, C. E., L. Håkanson, D. J. Jude, and P. G. Sly. A review of the physical and chemical components of the Great Lakes: a basis for classification and inventory of aquatic habitats. This volume.

Hudson, P. L., R. W. Griffiths, and T. J. Wheaton. A review of habitat classification schemes appropriate to streams, rivers, and connecting channels in the Great Lakes drainage basin. This volume.

Jenkins, R. M. 1982. The morphoedaphic index and reservoir fish production. *Trans. Am. Fish. Soc.* 111:133–140.

Jones, R. A. and J. F. Lee. 1986. Eutrophication of the lake for water quality management: an update of the Vollenweider OECD model. *WHOI — Water Qual. Bull.* 11:67–119.

Karrow, P. F. and P. E. Calkin, Eds. 1985. Quaternary Evolution of the Great Lakes. Spec. Pap. No. 30. Geological Association of Canada. Memorial University, Newfoundland, Canada.

Leach, J. H. and R. Herron. A review of lake habitat classification. This volume.

Leach, J. H., L. M. Dickie, B. J. Shuter, U. Borgmann, J. Hyman, and W. Lysack. 1987. A review of methods for prediction of potential fish production with application to the Great Lakes and Lake Winnipeg. *Can. J. Fish. Aquat. Sci.* 44(Suppl.2):471–485.

Marcus, M. D., W. A. Hubert, and S. H. Anderson. 1984. Habitat Suitability Index Models: Lake Trout (Exclusive of the Great Lakes). FWS/OBS-82/10.84, U.S. Fish and Wildlife Service. Washington, D.C.

McKee, P. M., T. R. Batterson, T. E. Dahl, V. Glooschenko, E. Jaworski, J. B. Pearce, C. N. Raphael, and T. H. Whillans. Great Lakes aquatic habitat classification based on wetland classification systems. This volume.

McMahon, T. E., J. W. Terrell, and P. C. Nelson. 1984. Habitat Suitability Information: Walleye. FWS/OBS-82/10.56, U.S. Fish and Wildlife Service. Washington, D.C.

Morris, D. W. 1988. Habitat-dependent population regulation and community structure. *Evol. Ecol.* 2:253–269.

Morris, D. W. 1989. Density-dependent habitat selection: testing the theory with fitness data. *Evol. Ecol.* 3:80–94.

Oglesby, R. T., J. H. Leach, and J. Forney. 1987. Potential (*Stizostedion*) yield as a function of chlorophyll concentration with special reference to Lake Erie. *Can. J. Fish Aquat. Sci.* 44(Suppl.2):166–170.

Reckahn, J. A. 1986. Long-term ecological trends in growth of lake whitefish in South Bay, Lake Huron. *Trans. Am. Fish. Soc.* 115:787–804.

Ryder, R. A. 1965. A method for estimating the potential fish production of north-temperate lakes. *Trans. Am. Fish. Soc.* 94:214–218.

Ryder, R. A. 1982. The morphoedaphic index — use, abuse and fundamental concepts. *Trans. Am. Fish. Soc.* 111:154–164.

Ryder, R. A. and S. R. Kerr. 1989. Environmental priorities: placing habitat in hierarchic perspective. *Can. Spec. Publ. Fish. Aquat. Sci.* 105:2–12.

Schlesinger, D. and H. A. Regier. 1982. Climatic and morphoedaphic indices of fish yields from natural lakes. *Trans. Am. Fish. Soc.* 111:141–150.

Sheldon, R. W. and T. R. Parsons. 1967. A continuous size spectrum for particulate matter in the sea. *J. Fish. Res. Bd. Can.* 24:909–926.

Smith, P. R. and J. J. Tibbles. 1980. Sea lamprey (*Petromyzon marinus*) in Lake Huron, Michigan, and Lake Superior: history of invasion and control 1936–1978. *Can. J. Fish. Aquat. Sci.* 37:1780–1801.

Vollenweider, R. A. 1968. Scientific Fundamentals of the Eutrophication of Lakes and Flowing Waters, with Particular Reference to Nitrogen and Phosphorus as Factors in Eutrophication. Rep. DAS/SCI/68.27. U.N. Organization for Economic and Cultural Development (OECD). Paris.

Vollenweider, R. A. 1975. Input/output models with special reference to the phosphorus loading concept in limnology. *Schweiz. Arch. Hydrol.* 37:53–84.

Vollenweider, R. A. and J. J. Kerekes. 1982. Eutrophication of Waters: Monitoring, Assessment, and Control. U.N. Organization for Economic and Cultural Development. Paris.

Watt, W. D. 1989. The impact of habitat damage on Atlantic salmon (*Salmo salar*) catches. *Can. Spec. Publ. Fish. Aquat. Sci.* 105: 154–163.

Youngs, W. D. and D. G. Heimbuch. 1982. Another consideration of the morphoedaphic index. *Trans. Am. Fish. Soc.* 111:151–153.

APPENDIX REFERENCES

Capen, D. E., Ed. 1981. The use of Multivariate Statistics in Studies of Wildlife Habitat. Gen. Tech. Rep. RM-87, U.S. Department of Agriculture, Fort Collins, CO.

Noon, B. R. 1986. Summary: biometric approaches to modeling — the researcher's viewpoint, in *WILDLIFE 2000: Modeling Habitat Relationships of Terrestrial Vertebrates,* J. Verner, M. L. Morrison, and C. J. Ralph, Eds., University of Wisconsin Press, Madison, 197.

Salwasser, H. 1986. Modeling habitat relationships of terrestrial vertebrates — The manager's viewpoint, in *WILDLIFE 2000: Modeling Habitat Relationships of Terrestrial Vertebrates,* J. Verner, M. L. Morrison, and C. J. Ralph, Eds., University of Wisconsin Press, Madison, 419.

Shugart, H. H., Jr. and D. L. Urban. 1986. Modeling habitat relationships of terrestrial vertebrates — the researcher's viewpoint, in *WILDLIFE 2000: Modeling Habitat Relationships of Terrestrial Vertebrates,* J. Verner, M. L. Morrison, and C. J. Ralph, Eds., University of Wisconsin Press, Madison, 425.

Stormer, F. A. and D. H. Johnston. 1986. Introduction: biometric approaches to modeling, in *WILDLIFE 2000: Modeling Habitat Relationships of Terrestrial Vertebrates,* J. Verner, M. L. Morrison, and C. J. Ralph, Eds., University of Wisconsin Press, Madison, 159.

Verner, J., M. L. Morrison, and C. J. Ralph, Eds. 1986. *WILDLIFE 2000: Modeling Habitat Relationships of Terrestrial Vertebrates,* University of Wisconsin Press, Madison.

Chapter 2

A SYSTEM FOR AQUATIC HABITAT CLASSIFICATION OF LAKES

P.G. Sly and W.-D. N. Busch

TABLE OF CONTENTS

ABSTRACT

An aquatic habitat classification (AHC) system is described that defines, organizes, and presents variables used to describe the abiotic and biotic characteristics of a body of water. The framework of the classification is based largely on the physical characteristics of habitat, but can be modified to incorporate biological and chemical characteristics.

The AHC system is hierarchical and provides differentiation at various levels between a whole lake and its component parts (e.g., bay, cove, or weed bed). Certain types of data are essential for fisheries interests, but a wide range of additional information, such as water quality data, can be incorporated. The system shares some conceptual similarities with the U.S. Fish and Wildlife Service wetland classification and may be interfaced directly with it. However, because the two systems are dominated by different structural elements of habitat, their frameworks are different. The AHC system does not name any of the possible associations of habitat variables but leaves this choice open to the user.

Selected features of aquatic habitat can be plotted on base maps or charts, but the maps must have a common scale and can be difficult to interpret if too much information is displayed. A geographic information system (GIS) allows habitat data to be processed statistically, grouped without constraint, and displayed at any scale. Therefore, computerized GIS is needed to make full use of the AHC system.

I. INTRODUCTION

Background concepts addressed in the development of the AHC system have been outlined in Chapter 1. Although this chapter presents the AHC system, an understanding of the background concepts is essential in order to make full use of this classification system.

Habitat descriptors include any variable that defines or delimits the space in which biota live, and covers anything from an entire lake to its smallest part. Classification requires that the variables that best describe aquatic habitat(s) be grouped to allow differentiation among habitat types. The authors decided that a simple hierarchical classification system was best suited to management situations in which data collection and interpretation may be at different scales of resolution. As discussed in Chapter 1, a hierarchical classification system is rather insensitive to variations in habitat data at the whole-lake level. It is more sensitive to variations in habitat data at the level of smaller component parts of a lake. Therefore, in order to have a classification system that can be used lakewide or at sites, the system must be flexible. This allows users to balance the greater resolution obtainable at low levels of hierarchy against the higher cost and effort required to obtain adequate data.

The system for large lake AHC is presented here and explained by means of example applications.

II. CLASSIFICATION

The AHC system is not intended to be "all things to all people", but it is intended to provide a means of structuring data that cover a wide range of information. Essential data provide a minimum framework of information for habitat comparisons between sites, or over time at the same site. More data (in the form of modifiers) will probably be needed by most users to meet specific habitat management tasks.

The AHC system (Figure 1 and Tables 1 and 2) defines and organizes the variables used to describe both the abiotic and biotic characteristics of a body of water. A major effort was made to make the classification system logical and to employ easily defined terms.

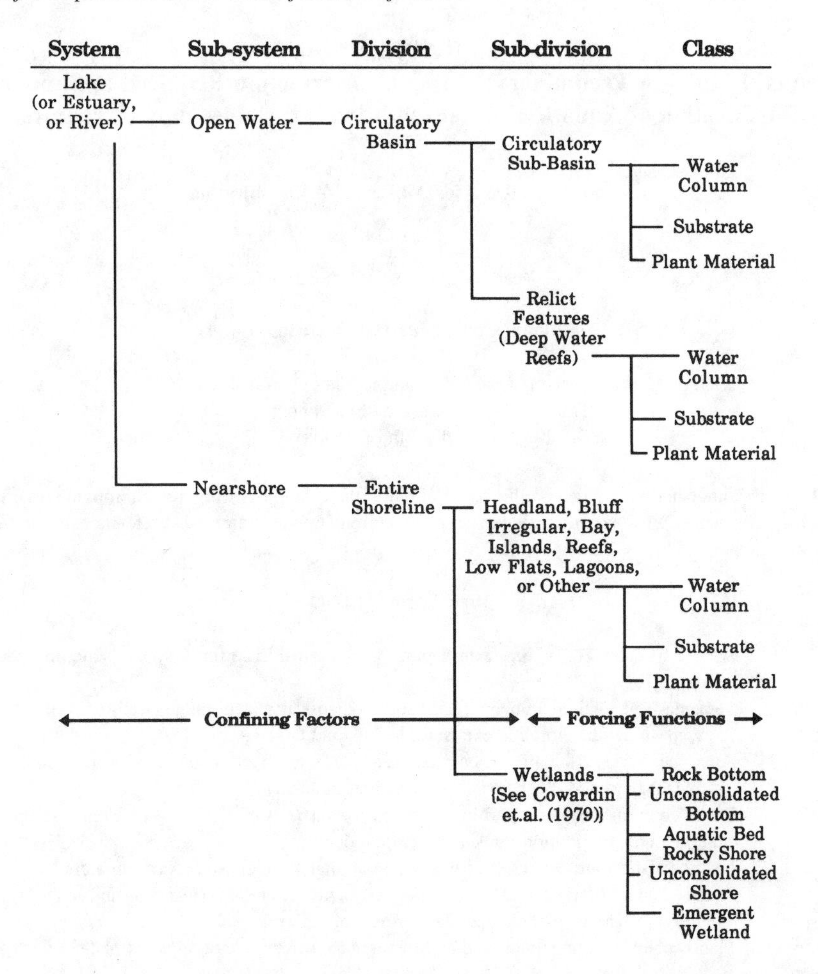

FIGURE 1. The AHC, showing system, sub-system, division, sub-division, and class levels, and the interface with the Cowardin et al. (1979) wetland classification system.

The hierarchy of the AHC system consists of the system, sub-system, division, sub-division, and class (Figure 1), and each is briefly described in the following sections. The AHC goes from general to specific and can be addressed at any level required to meet the needs of users. Also, each level of the hierarchy is related to confining factors and forcing and response functions. Quite literally, these characterize confinement of aquatic habitat, the degree of forcing (e.g., by heat, wind, and light), and the forms of response in an aquatic habitat (such as the growth of aquatic vegetation and sediment sorting). These factors and functions are discussed in Chapter 1 as part of the conceptual basis of a classification system.

Each level of the classification hierarchy requires a basic set of data, termed "essential information", to adequately describe habitat (Table 1). Basic data sets can be expanded or modified at any level of classification through the use of modifiers (examples in Table 2).

A. SYSTEM

The system is determined by the principal structure of aquatic habitat to be classified, such as a river, a connecting channel, a lake, or an estuary. Classification into one of these categories is based largely on flow characteristics and residence time. Estuaries generally represent the zone of fluctuation at the freshwater-saltwater interface, but estuarine-like conditions can also occur where rivers enter very large lakes.

TABLE 1
Essential Data — A Preliminary Listing of Information Required to Support the Classification of Aquatic Habitat (additional variables may be required)

Coordinate Data for All Levels of Classification

Space	Midpoint coordinates and geographic name
Time	Year and time period (month or season)

Additional Coordinate Data for Class Level

Space	Specific sample point coordinates
Time	Day (Julian date) and hour
Depth	Actual depth of sample

Note: Although latitude and longitude coordinates are used in the examples of Figures 2 through 5, many data sets use UTM coordinates. Either system may be adopted; current trends indicate a preference for the Universal Transverse Mercator (UTM) system.

Other Forms of Data

System	Area, dimensions (x, y, z), mean depth, hydraulic residence time, trophic state, morphoedaphic index (MEI)
Sub-system	Percentage of total area, mean depth, and median water level range (wetlands only)
Division	Area, mean depth, shoreline exposure, length, and development
Sub-division	Area, mean depth, feature dimensions (x, y, z), feature type, openness, local hydraulic residence time (for bays and other forms of partial entrainment)
Class (point samples)	Water column: mean epi- and hypolimnion (or surface and bottom water) temperatures, depth to thermocline, epilimnion transparency, total depth
	Lake bed: substrate type, mean particle size of total sediment sample, mean particle size exclusi of coarse material (gravel and cobbles); cone-penetrometer data (Håkanson and Jansson, 1983) may be used instead of particle-size data or as a modifier
	Plants: plant type (emergent and/or submergent), percentage of adjacent plant cover

TABLE 2
Modifiers

		System	Subsystem (and lower rank)		
	Modifier	**Lacustrine**	**Open water**	**Nearshore and shoreline**	**Wetland**
1	Thermal stability		x	x	
2	Temperature		x	x	x
3	Nepheloid layer		x		
4	Up-/downwelling and thermal bars		x	x	
5	Currents		x	x	
6	Groundwater intrusion			x	x
7	Water levels			x	x
8	Residence time	x		x	x
9	Shoreline exposure			x	
10	Transport/accumulating/erosional substrate		x	x	x
11	Contiguity/patchiness		x	x	x
12	Light penetration		x	x	x

TABLE 2 (Continued)
Modifiers

	System	Subsystem (and lower rank)		
Modifier	Lacustrine	Open water	Nearshore and shoreline	Wetland
13 Water color		x	x	x
14 Oxygen		x	x	x
15 pH		x	x	x
16 Toxics	x	x	x	x
17 Nutrients	x	x	x	x
18 Other anthropogenic stress			x	x
19 Cladophora			x	x
20 Vegetation (emergent and submergent)			x	x
21 Invertebrates			x	x
22 Fish species	x	x	x	x

Note: An infinite number of modifiers are possible. As a guide to information selection, the examples above are linked to general levels of ranking within the AHC system.

B. SUB-SYSTEM

In the sub-system, lakes are divided into zones based on the extent to which the shoreline and lakebed characteristics influence aquatic habitat. This influence may be thought of as an edge effect. Terms such as limnetic (pertaining to or inhabiting the pelagic region) and littoral (the biogeographic zone between high water mark and that depth to which 1% of light penetrates) carry a strict scientific meaning and should not be distorted in their use. Therefore, terms such as "open water" and "nearshore and shoreline" zones are preferred because they more reasonably characterize the concept of edge effect. In the AHC system, the "nearshore/open water" boundary is defined by either the sand/silt sediment boundary or the boundary formed by attached vegetation, depending upon whichever reaches the greater depth. In the main lake, the nearshore/open water boundary is usually found at a point at which the turbulent shear caused by surface wave motion begins to affect the lake bed. In sheltered bays, the boundary generally forms at the point at which light penetration is sufficient to allow growth of attached vegetation.

Wetlands form a transition between aquatic and terrestrial ecosystems (Jaworski and Raphael, 1976). They are dominated by edge effects that shift in response to water level fluctuations. The authors have included the wetland classification of Cowardin et al. (1979) in this hierarchy without change. The wetland component of this classification, which extends to a maximum depth of 2 m, could be adopted in its entirety for Great Lakes wetlands.

C. DIVISION AND SUB-DIVISION

Habitats are categorized at the division and sub-division levels, based on the size and complexity of physical features (including bottom configuration). The degree to which each level of complexity is useful, however, depends largely upon the extent to which features or groups of features remain distinct, one from another. For example, along an extensive stretch of shoreline, indentation and headland development may occur only as minor features. In this case, the stretch of shoreline (which might be tens of kilometers in length) would be categorized at the division level and without need for further resolution at the sub-division level. On the other hand, the presence of well-developed headlands, bays, lagoons, bars, reefs, shoals, and islands would require resolution at the sub-division level. Islands and shallow shoals in midlake areas should be described as highly localized nearshore environments, even though geographically remote from the main shoreline.

In open water areas, divisions can be established on the basis of distinct circulatory basins or sub-basins, as defined by the distributions of fine lake bed materials (silt and clay size) and large-scale water motions. Deepwater reefs in midlake areas should be described as geologically relict features of the open water environment. Deepwater reefs will most often be identified at the sub-division level and then further characterized at the class level.

D. CLASS

At the class level, separate information is required for each major component of habitat, including the water column, substrate, and plant material, if present. These components respond to various forcing functions and each component is considered separately in terms of its response. Classes provide the most detailed level of information required for the AHC system.

At the class level, essential data are drawn from all three components and a limited number of variables (including temperature, particle size, and plant type) can be used to describe a wide range of habitats.

1. Water Column

Motion in the water column may be influenced by current flow or wind and wave turbulence. Density structures and winter ice cover form in response to various types of thermal regimes which are modified by water motions. Photochemical processes and light-sensitive behavior of biota are controlled by levels of light intensity and water clarity. Quantities of suspended particulates may be regulated by physical, chemical, and biological in-lake processes, and also by processes external to a lake.

2. Substrate

The particle-size distribution of contemporary bottom sediments is often strongly related to hydraulic motion in overlying waters, as a result of an integration of current flows and wave turbulence. However, particle-size distribution in relict materials usually reflects site conditions prevalent at the time of formation rather than present conditions. Formative conditions for relict habitats may differ significantly from those of the present because of the effects of changing lake levels or glaciation. Both substrate type and water motion strongly influence the distribution and type of benthic biota.

3. Plant Material

Plant growth is largely influenced by water turbulence, water quality, heat, and light penetration.

E. COMBINATIONS OF VARIABLES

In the AHC system, no attempt is made to provide formal terminology for any of the combinations of variables that may be used to define habitat. This is because of the great number of combinations that are possible. Rather, we emphasize obtaining essential data that are needed to use this system of classification, and we provide the option of including additional (modifier) information that is specific to a wide range of individual user needs. Further, computer treatment of data can be used to categorize the same data in many different ways and to discriminate between different data sets (within or between habitat sites). As a means of limiting "noise" in class level data, most variables should be selected on the basis of temporal stability, time integration, and sessile behavior (plants). Clearly, accuracy and precision are also essential attributes to be assured in all data entered into the AHC system.

F. EXAMPLE APPLICATIONS

Figures 2 through 5 provide examples of how the AHC system might appear if applied to Lake Ontario. In general, only essential data are presented at the upper levels of the hierarchy. Somewhat greater detail is provided at the class level in which both area (summary form) and point-source data are shown. Greater amounts of data are available and most can be stored, retrieved, and used within the proposed framework.

The examples of Figures 2 through 5 demonstrate the application of a hierarchical approach to habitat classification. Some examples relate to very large forms of habitat (e.g., whole lakes), and others refer to disaggregated or much smaller forms of habitat. The examples provide structure data that can be used to show relationships between biotic and abiotic components of habitat. Each habitat (at different levels in the classification) is characterized by variables that can be distinguished, measured, or mapped.

The examples demonstrate a standardized approach to the definition, identification, measurement, and analysis of habitat units. The structure of the habitat classification system allows for aggregation or disaggregation of habitat units depending upon user needs. The system identifies and defines useful information at different levels of hierarchy. The system also incorporates elements of both classical limnology and the developments of more recent concepts in geology and hydrology (Sly, 1978) to provide integrated data sets that reflect the significance of subtle as well as strongly defined differences in habitat type.

G. MODIFIERS

Modifiers provide a source of information that acts as an adjunct to the essential data of the AHC system. Modifiers can be used, for example, to increase the number of variables and thus discriminate between different types of habitat, as qualitative descriptors of habitat conditions, or as records of specific events or regimes that affect habitat. Modifiers may be used in conjunction with any of the various levels of the AHC system. Example modifiers are given in Table 2 and others may be added as appropriate. Insofar as it may be possible, the measured values of selected modifiers should be compatible with the need for precision at each level of hierarchy (Ryder and Kerr, 1989); these requirements should be addressed as part of the proposed demonstration project (see Chapter 1, Section IV, Recommendations). Explanations and definitions of the listed modifiers are as follows:

1. The stability of a thermal layer can be defined by examination of temperature gradients. Two designations may be useful: stable and unstable. Stability is also influenced by wind effects and density flows.
2. Some temperature information is considered essential data, but actual temperatures and values such as the mean, maximum, minimum, range, and rates of warming or cooling also may be used as modifiers. Water temperature and thermal stability are key factors that not only influence the distribution of fish species, as a result of physiological control, but also the structure of the food-web (in particular, primary production and rates of microbial degradation).
3. The presence or absence of a nepheloid layer at the bottom of a hypolimnion unit may be of significance, although there is no agreement on the persistence of such features. Nepheloid layers may be associated with suspended sediment focusing and nutrient exchange at the sediment-water interface and, through this, to characteristics of the benthic food chain.

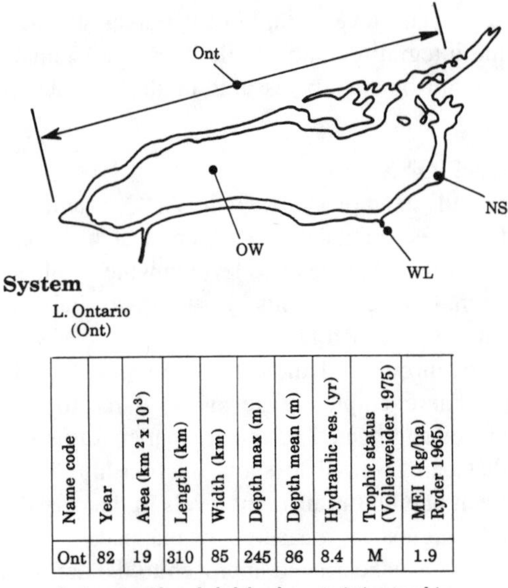

System

L. Ontario
(Ont)

Name code	Year	Area (km² x 10³)	Length (km)	Width (km)	Depth max (m)	Depth mean (m)	Hydraulic res. (yr)	Trophic status (Vollenweider 1975)	MEI (kg/ha) Ryder 1965)
Ont	82	19	310	85	245	86	8.4	M	1.9

System provides whole lake characteristics, trophic state and potential productivity.

Sub-System

Open water
(OW)

Code	% of total area	Depth mean (m)
OW	83	100

Nearshore
(NS)

Code	% of total area	Depth mean (m)
NS	16	10

Wetland
(WL)

Code	% of total area	Depth mean (m)	Water level mean range (m)
WL	1	0.8	0.5

Sub-system apportions to cold/deep and warm/shallow species and to wetlands

FIGURE 2. Example application of the AHC at the system and sub-system levels: Lake Ontario.

4. Areas with recurrent up- or downwellings may be defined as well as the approximate seasonal location of thermal bars.
5. The direction, average, and range of current velocities in thermal layers are considered to be of great significance. The metalimnion is generally a shear zone. Organic matter is frequently entrained along shear zones and thermal bars, and may provide an important site for fish feeding on insect larvae and other entrained materials.
6. Groundwater intrusions may be defined by substantial temperature or water quality differences. Intrusions may be important as sites of overwinter temperature refuge or egg incubation (particularly in streams).
7. Water level fluctuations may be included, for example, as a range, mean, standard deviation, maximum/minimum, frequency, and duration. Water levels are key to successful spawning and use of nursery areas by many species of fish.
8. The balance between outflow and surface inputs, as runoff, largely regulates residence time in a basin. Residence time affects long-term nutrient concentrations and thus the primary production of different bodies of water.
9. Exposure may be delineated by the categories of Jaworski and Raphael (1976) or by means of Håkanson-type calculations (Håkanson and Jansson, 1983), e.g., mean slope (percent), topographic openness of shoreline, and immunity to ice scour. Exposure has a major impact on the community structure of different habitats.

Division

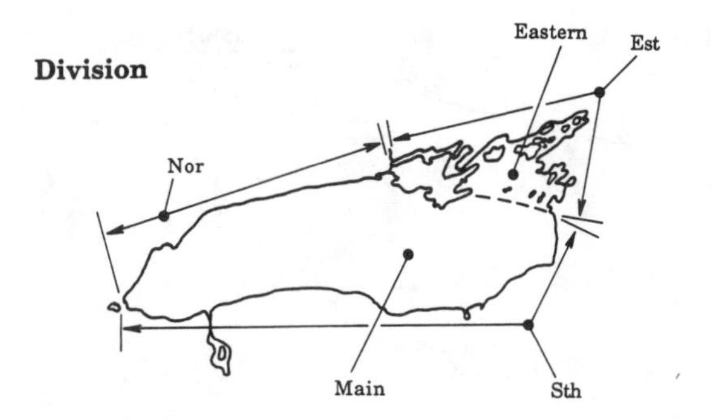

Basin (BA)

Code	Name	Area (km^2 x 10^3)	Depth mean (m)	Depth max (m)	Mid-point co-ordinates	
BA	Eastern	1.5	28	55	N 44 00	W 76 35
BA	Main	17.5	99	245	N 43 40	W 78 00

Shore (SH)

Code	Name	Length (km)	General exposure (8 points)	Shoreline development	Mid-point co-ordinates	
SH	Nor	190	S	1.2	N 43 50	W 78 50
SH	Est	180	W	3.7	N 44 05	W 76 45
SH	Sth	315	N	1.3	N 43 20	W 77 45

Geographic apportioning: circulation of main circulatory cells defines major basins, shore stretch selection largely reflects geological process controls.

FIGURE 3. Example application of the AHC at the division level: Lake Ontario (basin and shore).

10. Substrate data, when characterized according to the scheme of transport/accumulation/erosion (Håkanson and Jansson, 1983), provide a surrogate for the combined effects of wave turbulence and flow. Also, substrate may be characterized by origin and geologic history. Substrate data provide useful information about probable benthic community composition and the potential for specialized use by fish for spawning, nursery, and refuge.
11. If habitat mapping is used to record distributions of certain key biotic guilds, a need may arise to indicate the contiguity or patchiness of certain habitat types.
12. Light penetration may be characterized by photic depth, optical transmission, secchi depths, etc.
13. Water color may be recorded by comparison with color standards or spectral absorption. Light penetration, water clarity, and color are major factors affecting the growth of aquatic plants, phytoplankton production and the distribution of sight-feeding species of fish. These habitat characteristics are particularly significant in shallow nearshore waters.
14. Oxygen is considered to be a component of essential data, but it may also be used as a modifier (e.g., as stand-alone data or to qualify a thermal unit). Useful long-term exposure criteria are >2 mg/l for survival of benthic macroinvertebrates, and >5 mg/l for coolwater fish. The rate of dissolved oxygen depletion is of primary concern in the

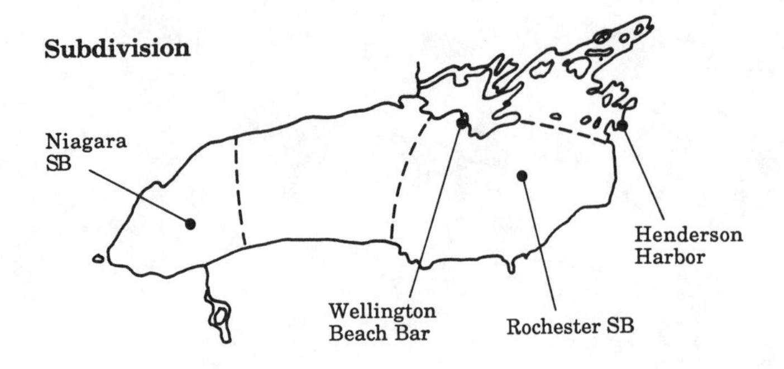

Sub-Basin examples (SB)

Code	Name	Area (km^2)	Depth mean (m)	Depth max (m)	Mid-point co-ordinates	
SB	Rochester	7.0	118	245	N 43 35	W 77 00
SB	Niagara	2.5	68	137	N 43 35	W 79 10

Geographic apportioning: circulation in less well-defined cells characterizes Sub-basins, features are generally defined on the basis of the largest individual form.

Features Examples (FE)
Dimensions (km)

Code	Name	Parallel to shore	Perpendicular to shore	Type	Openess	Hydraulic res. time (d)	Depth mean (m) if <50% open to main lake	Mid-point co-ordinates
FE	Henderson Harbour	10	4	Bay	30	5	5.5	N 43 53 W 76 10
FE	Wellington Beach	6.5	0	Beach Bar	100	0	—	N 43 56 W 77 19

FIGURE 4. Example application of the AHC at the sub-division level: Lake Ontario (two sub-basins and two features).

hypolimnion, but metalimnion problems are known and localized oxygen depletion can occur in the epilimnion.

15. Measurements of pH may be used to characterize habitat. High pH values are usually associated with high primary production in the water column. Sediment acidity is usually associated with decomposition of organic matter. pH is also significant in terms of the toxicities of metals and other chemical compounds such as ammonia.
16. Toxic contaminants may be indicated by total, speciated, or method-specific fractionations, expressed as concentrations (in water, sediments, or biota), or possibly by forms of bioindex, e.g., presence/absence of species/communities or other evidence of health effects. Under extreme conditions, contaminants (including extremes of pH) may limit *in situ* species survival, feeding, and reproduction. For the most part, however, contaminants are of concern because of their potential effects on human health (as a result of fish consumption).
17. Nutrient conditions may be indicated by plankton counts, chlorophyll *a* concentrations, chemical concentration values (of P, N, C, Si, in particular), or by a trophic state, i.e.,

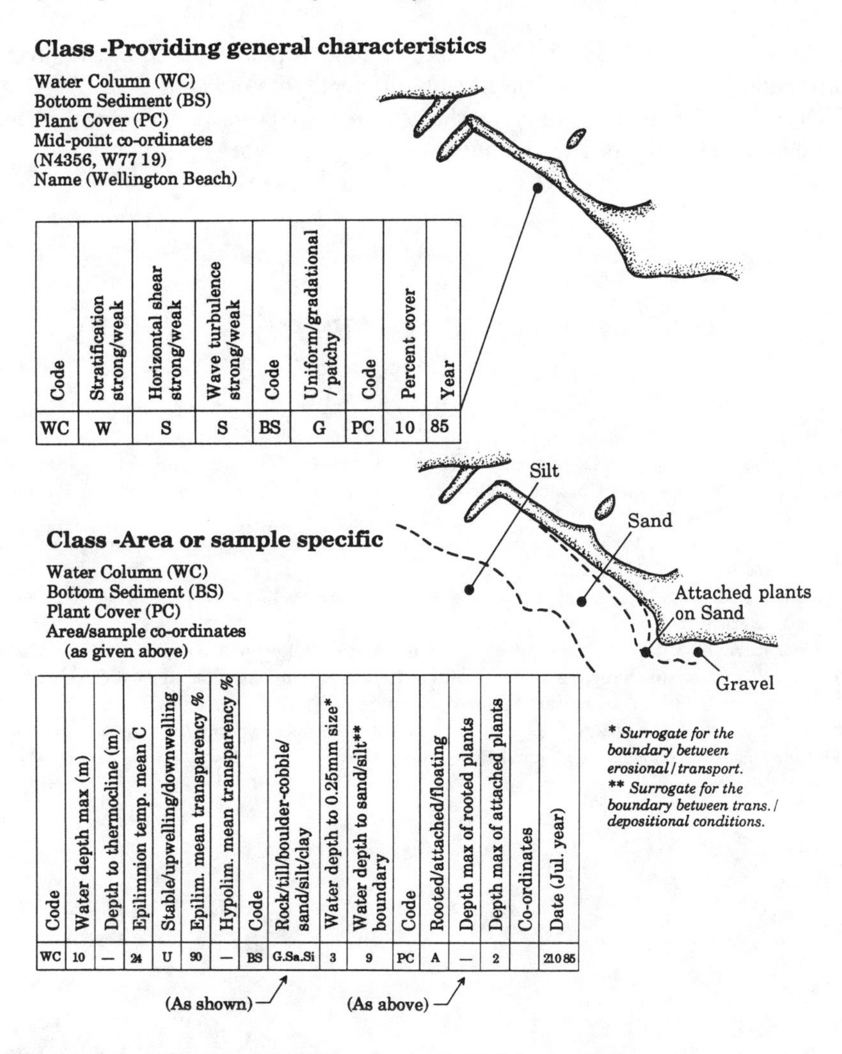

FIGURE 5. Example application of the AHC at the class level: Lake Ontario (general, area, or sample specific).

oligotrophic, mesotrophic, or eutrophic (Vollenweider, 1968; Vollenweider and Kerekes, 1982). Trophic state is indicative of potential food-web and community structures, and production efficiencies of natural systems are often higher than those responding to cultural enrichment.

18. Point sources of anthropogenic stress may include culvert/outfall, floodgate/lock, contaminant spill, marina/port facility, etc., and can be noted by type and geographic location. Additional information could provide loading estimates and timing of events at specific sites. Point sources can be important both as sites of toxic impact and increased productivity.
19. The presence, absence, or change in abundance of *Cladophora* may be used to quantify physicochemical habitat changes. During summer, *Cladophora* may enhance the nearshore habitat for some fish species that utilize associated benthic invertebrates. However, seasonal degradation of this alga may limit egg survival of some fall spawning species.
20. Vegetation types (emergent, subemergent, floating) by proportion, species composition, and percentage cover may be used to characterize nearshore zones (primarily controlled by the integrated effects of light penetration and water motion). The type and cover of aquatic vegetation is a major factor controlling the communities of fish, waterfowl, and some mammalian species in nearshore and wetland habitats.

21. Indices may be used to indicate the species composition, abundance, and diversity of invertebrate populations associated with different types of habitat.
22. Indices may be used to similarly characterize the species composition, abundance, diversity, and health of fish communities in different types of habitat.

REFERENCES

Cowardin, L. M., V. Carter, F. C. Jolet, and E. T. LaRoe. 1979. Classification of Wetlands and Deepwater Habitats of the United States. FWS/OBS-79/31, U. S. Fish and Wildlife Service, Washington, D.C.

Håkanson, L. and M. Jansson. 1983. Principles of Lake Sedimentology. Springer-Verlag, Heidelberg.

Jaworski, E. and C. N. Raphael. 1976. Modification of coastal wetlands in northeastern Michigan and management alternatives. *Mich. Acad.* 8(3):303-317.

Ryder, R. A. and S. R. Kerr. 1989. Environmental priorities: placing habitat in hierarchic perspective. *Can. Spec. Publ. Fish. Aquat. Sci.* 105:2-12.

Sly, P. G. 1978. Sedimentary processes in lakes, in *Lakes: Chemistry, Geology and Physics,* A. Lerman, Ed., Springer-Verlag, New York, p. 65.

Vollenweider, R. A. 1968. Scientific Fundamentals of the Eutrophication of Lakes and Flowing Waters, with Particular Reference to Nitrogen and Phosphorus as Factors in Eutrophication. Rep, OECD DAS/SCI/68.27. U.N. Organization for Economic and Cultural Development, Paris.

Vollenweider, R. A. and J. J. Kerekes. 1982. Eutrophication of Waters: Monitoring, Assessment, and Control, U.N. Organization for Economic and Cultural Development, Paris.

Chapter 3

A REVIEW OF LAKE HABITAT CLASSIFICATION

J. H. Leach and R. C. Herron

TABLE OF CONTENTS

ABSTRACT

Lake classification systems have been reviewed to provide background information for the development of classification and inventory techniques for Great Lakes aquatic habitats. Characteristics of lakes and their drainage systems used in classification are numerous and approaches range from single variables such as surface area or total phosphorus to complex multivariate models. Almost all abiotic and biotic factors that influence lake metabolism have been used in classification schemes. Of the 32 categories reviewed, 24 were trophic classifications, reflecting the recent interest in quantifying the eutrophication process. Habitat is a major factor in determining fish community structure and, in turn, influences fish production and yield. In general, some habitat variables that characterize lakes on the basis of fish associations are also useful for empirical prediction of fish production and yield. We suggest that there may be advantages of simplicity, availability of data, and relevance in including these variables in habitat classification systems. Because of the local nature of climate, basin morphometry, and edaphic inputs, lakes classified by fish communities may also be classified on a regional basis. For fisheries agencies responsible for large numbers of lakes, regional classification may be an expedient first ordering of fisheries habitat.

I. INTRODUCTION

Lake classification grew out of the desire and need of early limnologists to reduce and simplify an ever-increasing amount of descriptive information. In his history of limnology, Elster (1974) described the early attempts to group lakes according to geological origin, geography, morphometry, hydromechanical processes, temperature regimes, edaphic inputs, etc. The trophic nature of lakes was introduced into typologies by Naumann and Thienemann in the early part of this century (Naumann, 1932; Thienemann, 1925). In 1918, Thienemann characterized lakes as oligotrophic and eutrophic depending on oxygen depletion in the hypolimnion in summer and the types of benthic organisms associated with oxygen-rich and oxygen-poor sediments (Hutchinson, 1973). At about the same time, Naumann classified Swedish lakes as oligotrophic or eutrophic depending on plankton biomass. Since this early work, much interest has been generated in the development of indices that indicate trophic levels. The other major lake classification system still in common use today is the thermal mixing classification of Hutchinson and Loffler (1956).

Elster (1974) considered three main reasons that comparative science attempted to order its findings: (1) to gain an overall view, (2) to simplify understanding of complex systems by characterizing a few common factors, and (3) to predict properties or relationships of parts of systems from other measured properties. Rigler (1975) considered that the predictive capabilities of classification schemes were not readily apparent until stated by Elster (1958). Rigler suggested that the lack of predictive application was due to the failure of early limnlogists to understand the practical need for predictions. Fisheries biologists were the first to recognize the value of predicting fish harvests from lake characteristics (Rawson, 1955; Ryder, 1964).

This chapter reviews indices and systems developed for lake habitat classification to provide background information. A general catalog of classification schemes and typologies is proposed in Table l. Rawson (1939) proposed a schematic presentation of the interrelationships of abiotic factors that influence lake metabolism. His figure was modified by Steedman and Regier (1987) to include the influence of human factors (Figure 1). Classification schemes have been proposed for virtually all of the factors involved in lake metabolism as indicated in Figure 1.

We have attempted a brief description of classification types with reference to examples of each in the literature. We do not suggest that the review is comprehensive of all known classification systems. The advantages and disadvantages and suitability of the systems in characterizing Great Lakes habitat are included in the discussion.

II. REVIEW OF LAKE CLASSIFICATION SCHEMES AND TYPOLOGIES

A. ORIGIN, SHAPE, AND LOCATION

1. Origin

A classic review and synthesis of origins of lake basins was written by Hutchinson (1957) and is based on global literature, much of which was written in the 19th century and early decades of the 20th century. In a formal summary, Hutchinson differentiates 76 lake types based on geomorphology of basins grouped under 11 processes or events (Table 2). In general, the processes have been responsible for the building, excavating, and damming of lake depressions. The processes of origin constitute a system of classification that may also be geographical. For example, the Rift Valley lakes in Africa are of tectonic origin and most of the lakes in central North America are due to glaciation.

2. Morphometry and Morphology of Lake Basins

The importance of morphometric parameters of a lake basin to physical, chemical, and biological processes was recognized early in the study of lakes. The most common parameters that are measured and used are maximum length, maximum width, surface area, drainage basin area, volume, maximum depth, relative depth, shoreline length, and shoreline development. Several classifications or inventories of lakes based on one or more morphometric parameter or ratios of parameters have been published, the most recent being that of Herdendorf (1982). The importance of morphometry and (especially) depth as factors in lake productivity has been recognized since the early work of Thienemann (Rawson, 1952, 1955; Hayes and Anthony, 1964; Ryder, 1964).

The morphology of a lake basin is largely determined by its origin and history. Hutchinson (1957) has proposed a classification based on basin shape. Most of the categories can be related to the processes responsible for lake origin, as listed in Table 2.

3. Location

Lake classification based on geographic location of lakes was considered by early limnologists who proposed two fundamental types of lakes, caledonian-subalpine and baltic, accord-

TABLE 1
A Selection of Lake Classification Schemes and Typologies

Basis of classification	References
Origin, shape, and location	-
Origin	Hutchinson, 1957; Odum, 1971; Wetzel, 1975; Zumberge, 1952
Morphometry and morphology	Herdendorf, 1982; Hutchinson, 1957; Rawson, 1955; Zimmermman et al., 1983
Location	Brylinsky and Mann, 1973; Herdendorf, 1982; Schneider, 1975
Physical	
Thermal properties (mixing)	Hutchinson, 1957; Hutchinson and Loffler, 1956; Lewis, 1983, Patterson et al., 1984 Walker and Likens, 1975; Wetzel, 1975, Whipple, 1898
Optical properties	Baker and Smith, 1982; Smith and Baker, 1978
Chemical	
Edaphic — major ions	Hutchinson, 1957; Kemp, 1971; Moyle, 1945a, b, 1946; Schneider, 1975
Water quality	Michalski and Conroy, 1972; Newton and Fetterolf, 1966; Pitblado et al., 1980; Warren, 1971; Welch, 1978
Trophic status	
Single parameter	
Dissolved oxygen	Eberly, 1975, Hutchinson, 1938, 1957; Strom, 1931; Thienemann, 1928; Walker, 1979
Primary production	Chapra and Dobson, 1981; Rodhe, 1958; Winberg, 1961
Total phosphorus	Chapra and Dobson, 1981; Dillon and Rigler, 1975; Forsberg and Ryding, 1980
Total nitrogen	Forsberg and Ryding, 1980
Chlorophyll *a*	Chapra and Dobson, 1981; Dillon and Rigler, 1975; Dobson et al., 1974; Forsberg and Ryding, 1980
Transparency (Secchi disc)	Chapra and Dobson, 1981; Forsberg and Ryding, 1980
Organic matter in sediments	Entz, 1977
Composite indices	
Morphoedaphic index	Ryder, 1965; Ryder et al., 1974
Productivity index	Hayes and Anthony, 1964
Quality index	Hayes, 1957
Trophic state index	Aizaki et al., 1981; Carlson, 1977; Kratzer and Brezonik, 1981; Rao and Berkel, 1978; Shannon and Brezonik, 1972
Absolute ranking	Lueschow et al., 1970
Trophic index numbers	U.S. EPA, 1974
Lake condition index	Uttormark and Wall, 1975
Bioproduction number	Håkanson, 1984

TABLE 1 (Continued)
A Selection of Lake Classification Schemes and Typologies

Basis of classification	References
Naumann and Thienemann trophic scales	Chapra and Dobson, 1981
MSS-related models	Boland, 1976; Boland and Blackwell, 1977; Witzig and Whitehurst, 1981
Zafar's taxonomy	Zafar, 1959
Biological indicators	
Bacteria	Aizaki, 1985; Rai and Hill, 1980
Phytoplankton	Beeton, 1966; Hornstrom, 1981; Nygaard, 1949; Rawson, 1956; Rott, 1984; Round, 1958; Stockner, 1971
Macrophytes	Canfield et al., 1983; Jensen, 1979
Zooplankton	Beeton, 1966; Brooks, 1969; Sprules, 1980
Benthos	Alm, 1922; Berg, 1938; Brundin, 1942, 1949, 1956, 1958; Decksbach, 1929; Deevey, 1941; Jarnefelt, 1925, 1953; Lundbeck, 1926, 1936; Miyadi, 1933; Prat, 1978; Reynoldson, 1978; Saether, 1975; Thienemann, 1925; Valle, 1927
Regional typology	
N. Scandinavia	Jarnefelt, 1958
N. Saskatchewan	Rawson, 1961
N. Colorado	Pennak, 1958
N. Spain	Margalef, 1958, 1975
Victoria, Australia	Williams, 1964
Connecticut	Deevey, 1941
Minnesota	Moyle, 1956
Wisconsin	Uttormark and Wall, 1975
British Columbia	Larkin and Northcote, 1958
Fish assemblage and habitat	Anonymous, 1978; Harvey, 1975, 1978, 1981; Johnson et al., 1977; Marshall and Ryan, 1987; Schneider, 1981; Tonn and Magnuson, 1982; Tonn et al., 1983

ing to types of midge larvae present (Thienemann, 1909). However, Theinemann later found typical examples of subalpine and baltic lakes situated on either side of a ridge in Germany (Rodhe, 1975).

Cole (1977) considered that the geographic concept was still useful but perhaps limited to regional applications. He suggested, for example, that it could be used to differentiate lake types in different parts of New England. On the other hand, Brylinsky and Mann (1973), in a study of International Biological Program (IBP) lakes, found that on a global scale, latitude alone explained about 57% of the variability in production. On the same scale, altitude explained only 1% of the variability, but became the most important variable determining productivity in lakes within a narrow range of latitude.

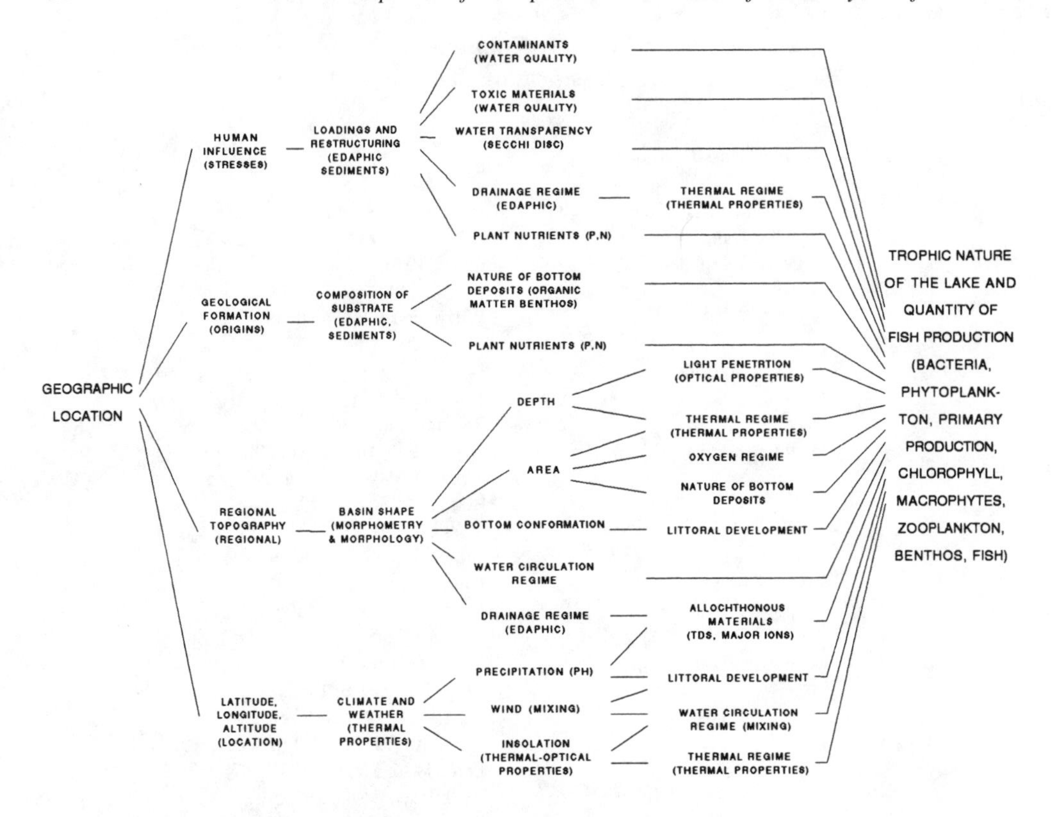

FIGURE 1. Interrelations of abiotic and human factors that influence lake metabolism. Lake classification schemes and typologies are added in parentheses. (Modified from Rawson, 1939 by Steedman and Regier, 1987.)

TABLE 2
Summary of Geomorphological Processes Responsible for Origins of Lake Basins (Hutchinson, 1957; Strahler and Strahler, 1973)

Process	No. of lake types	Examples
Tectonic	9	Basins formed by movement of deeper parts of the Earth's crust, e.g., Lake Baikal
Volcanic	10	Craters of extinct or dormant volcanoes, e.g., Crater Lake, OR
Landslides	3	Valleys blocked by landslides. e.g., Clear Lake, CA
Glacial	3	Lakes held by ice or moraines of existing ice sheets — glacier lakes
	4	Glacial rock basins — Great Lakes
	4	Moraine and outwash dams — Finger Lakes
	9	Drift basins — Kettle Lakes
Chemical weathering	5	Limestone solution lakes, e.g., Deep Lake, FL
Fluviatile action	12	Oxbow Lakes, lateral lakes, plunge-pool lakes
Wind action	4	Basins formed by wind scour or blocked by dunes

TABLE 2 (Continued)
Summary of Geomorphological Processes Responsible for Origins of Lake Basins (Hutchinson, 1957; Strahler and Strahler, 1973)

Process	No. of lake types	Examples
Shoreline processes	5	Basins formed behind barrier beaches and bars or coral reefs
Organic accumulation	3	Basins formed by accumulation of plant matter
Animal activity	3	Beaver ponds, reservoirs
Meteorite impact	2	Chubb Lake, Quebec

Zafar (1959) applied a plant geography system of climatic zonation to his taxonomy of lakes. In categorizing lakes in terms of fish communities, Johnson et al. (1977) and Schneider (1975) found that climate, as indicated by latitude, was indeed an important factor. Herdendorf (1982) inventoried the large lakes of the world according to geographic distribution.

B. PHYSICAL PROPERTIES OF LAKES

1. Thermal Mixing

Forel (1892) was the first to propose a classification of lakes based on their thermal conditions. His simple terminology (temperate, tropical, polar) was modified and expanded by others, including Whipple (1898) and Birge (1915). Hutchinson and Loffler (1956) presented a scheme which, because of its simplicity, was generally well received and is still in use today. Walker and Likens (1975) reviewed meromixis and suggested a revised typology of lake circulation patterns.

Lewis (1983) has proposed a revision of the Hutchinson-Loffler classification (Table 3) to take into account deficiencies noted by him and others. Lewis summarized the limitations of the Hutchinson-Loffler classification as exclusion of shallow lakes, unsatisfactory relationship between meromixis and the six basic lake types, excessively complex treatment of tropical lakes, and difficulties in the classification of cold lakes. As in the original, the revised classification separates lakes by three criteria: ice cover, mixing, and direct stratification. It also provides for the above limitations, particularly the meromixis-holomixis dichotomy with seasonal mixing types (shown schematically in Figure 2). Lewis attempted to show the approximate distributions of lake types in relation to latitude, altitude, and depth (Figure 3).

2. Optical

A method for optically classifying natural waters on the basis of total chlorophyll-like pigment concentration was proposed by Smith and Baker (1978). In a later paper, Baker and Smith (1982) added a dissolved organic material component to the classification model. These models were designed for remote sensing applications in ocean waters. Remote sensing technology is also applied to lakes and the optical classifications of Baker and Smith (1982) may have relevance to trophic classification and production estimation.

C. CHEMICAL PROPERTIES OF LAKES

1. Edaphic Inputs

The relationship between chemical composition in surface waters and edaphic and atmospheric inputs of major ions has been used to follow the historic influence of forest clearing, development of agriculture, urbanization, and industrialization. For example, Beeton (1969)

TABLE 3
Comparison of the Hutchinson-Loffler Classification of Lakes Based on Mixing and Lewis' Revised Classification (Lewis, 1983)

Hutchinson-Loffler classification	Lewis' revised classification
Amictic: always ice covered	Amictic: always ice covered
Cold monomictic: ice covered most of the year and warming sufficiently to thaw but not to exceed 4°C	Cold monomictic: ice covered most of the year, ice-free during the warm season, but not warming above 4°C
Dimictic: ice covered part of the year, stably stratified part of year, and mixing in spring and fall	Continuous cold polymictic: ice covered part of the year, ice-free above 4°C during the warm season, and stratified at most on a daily basis during the warm season
Warm monomictic: never ice covered, mixing once per year, and stably stratified the rest of the year	Discontinuous cold polymictic: ice covered part of the year, ice-free above 4°C, and stratified during the warm season for periods of several days to weeks, but with irregular interruption by mixing
Oligomictic: never ice covered, stratified most of the time, but cooling enough to mix at irregular intervals longer than 1 year	Dimictic: ice covered part of the year and stably stratified part of the year with mixing at the transitions between these two states
Polymictic: never ice covered, stratified, but with several episodes of complete mixing per year	Warm monomictic: no seasonal ice cover, stably stratified part of the year, and mixing once each year
	Discontinuous warm polymictic: no seasonal ice cover, stratifying for days or weeks at a time, but mixing more than once per year
	Continuous warm polymictic: no seasonal ice cover, stratifying at most for a few hours at a time.

has shown historic changes in concentrations of major ions and total dissolved solids (TDS) in the Great Lakes.

The classification of lake waters on the basis of chemical composition has been utilized by limnologists for many years. In his review of inorganic chemistry of surface waters, Hutchinson (1957) discussed water types on the basis of anions and cations and various ratios of these. Moyle (1946) proposed a classification of lake waters in Minnesota on the basis of hardness. He categorized the waters from very soft to alkalai on the basis of carbonate and sulfate content and related the categories to productivity of lakes in terms of plants and fish. Hooper (1956) and Schneider (1975) have also classified Michigan lakes on the basis of carbonate alkalinity, and Schneider used this and other characteristics in a typology of Michigan lakes in terms of fisheries potential.

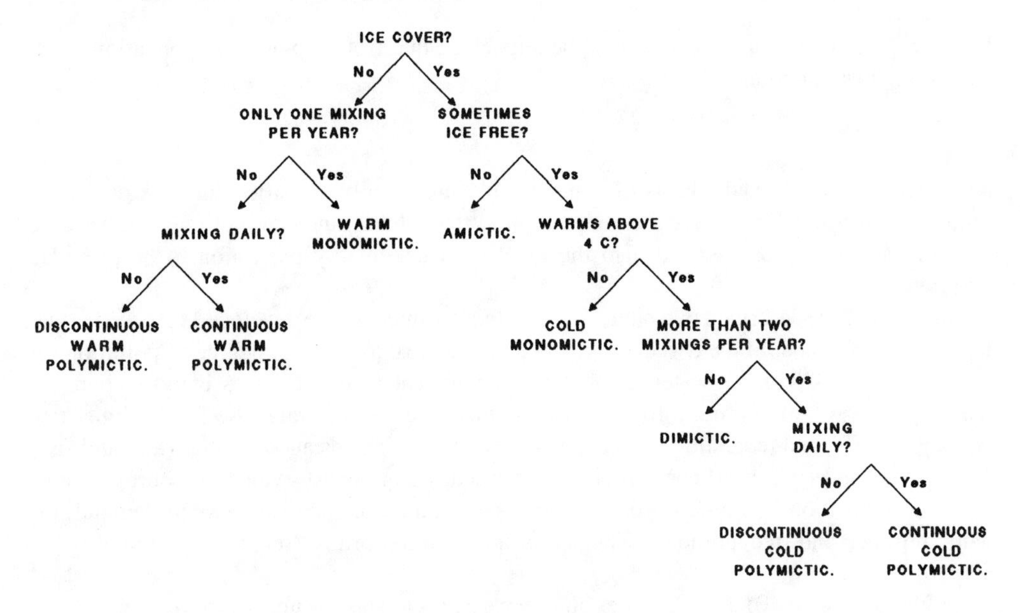

FIGURE 2. A revised classification of lakes based on mixing. (From Lewis, W. M., *Can. J. Fish. Aquat. Sci.*, 40, 1779, 1983. With permission.)

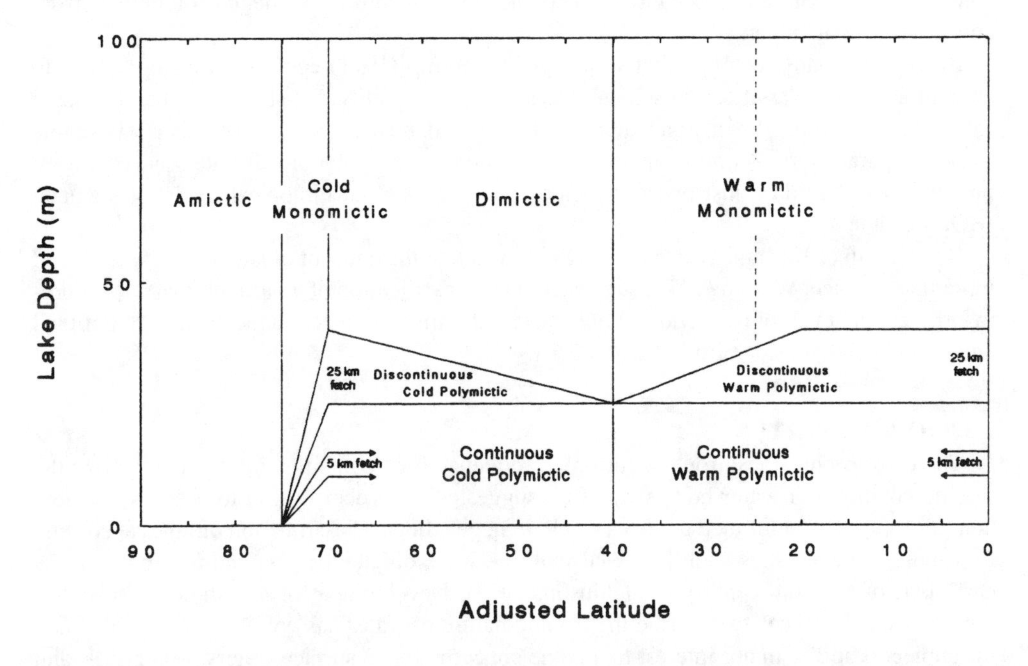

FIGURE 3. Estimated distribution of the eight lake types in Lewis' revised mixing classification in relation to latitude (adjusted for altitude) and water depth. Latitude has been adjusted by adding degrees of latitude equivalent to the elevation of the lake. (From Lewis, W. M., *Can. J. Fish. Aquat. Sci.*, 40, 1779, 1983. With permission.)

Pearsall (1923) related a basic ratio [(Na + K)/(Ca + Mg)] to algal periodicity in lakes of the English lake district. Zafar (1959) incorporated this cation ratio into his taxonomy of lakes on the basis that the ratio is <0.2 in eutrophic waters and >2.0 in oligotrophic waters. Shannon and Brezonik (1972) also included Pearsall's cation ratio as a factor in their multivariate approach to trophic conditions of lakes. Although Kemp (1971) recognized the importance of ionic proportions in classifying waters, he suggested that from a practical application, TDS was of greater importance.

2. Water Quality

Waters classified on the basis of water quality are usually classified for uses or for the protection of users. Therefore, a classification system should indicate not only the specific type of use to be protected but also the level of water quality protection to be provided (Warren, 1971).

Shapiro (1975) described two national water quality indices, one developed for Sweden and the other for Canada. In the Swedish scheme, waters are classed as to degree of pollution for three purposes — bathing, water supply, and fishing. The Canadian index is in two parts — the first rates pollution criteria for their importance in affecting water use for drinking, fish and aquatic life, and recreation; the second deals with toxic substances in the water and fish.

Newton and Fetterolf (1966) computed "chemical water quality evaluation scores" from a regression equation involving orthophosphate concentration, chemical oxygen demand, organic nitrogen and free ammonia. The scores are used to portray "relative water quality" in a set of lakes.

Pitblado et al. (1980) used 23 water quality variables to classify northeastern Ontario lakes that were impacted by acid precipitation. Chemical variability within the lakes was related to four components: nutrient status, buffering status, atmospheric deposition status, and sodium chloride status. The authors then grouped the lakes according to degree of impact from atmospheric contaminants.

Because so many factors affect water quality, Kemp (1971) considered it impractical to attempt a general classification scheme based on water quality. He suggested that a scheme should be based more directly on water use and favored the incorporation of only a few simple chemical parameters. As an example, he proposed a classification of waters according to probable uses based on just two parameters, conductivity and biological oxidation demand (BOD) (Table 4).

Both Kemp (1971) and Warren (1971) believed that the concept of water quality involved more than science. Warren (1971) suggested that "classification of a particular water, according to uses and levels of protection, should be based primarily on economic, technical, political and other social considerations."

D. TROPHIC STATUS

The oligotrophic–mesotrophic–eutrophic sequence used so commonly to characterize the trophic condition of water bodies was first suggested by Weber (1907) to describe nutrient conditions which determined the flora of German peat bogs. Classification of lakes according to trophic characteristics soon followed with the work of Thienemann and Naumann in the early part of the 20th century (Hutchinson, 1973). Development of terminology, schemes, typologies, and indices to describe the trophic nature of water has continued to this day and the subject is still of major interest to anyone concerned with surface waters, water uses, and aquatic life. Of all lake classification schemes, trophic classification has the widest acceptance and use at the present time. The literature on the subject is voluminous and we have attempted only a partial and brief review below.

1. Single Parameter

a. Dissolved Oxygen

Following the earlier work of Thienemann and Strom, Hutchinson (1938) proposed a trophic classification based on the areal hypolimnetic oxygen deficit. Hutchinson and Mortimer

TABLE 4
Classification of Waters According to Their Probable Utility (Kemp, 1971)

Conductivity	BOD: Low (at least 95% of the time <4 ppm)	BOD: High (>5% of the time above 4 ppm)
Low: at least 95% of the time below 750 μmho cm^{-1}	Class 1	Class 4
Intermediate	Class 2	
High: >95% of the time above 2250 μmho cm^{-1}	Class 3	Class 5[a]

Class 1 Probably suitable as a source of municipal water supply and for most other uses.

Class 2 Probably suitable as a source of municipal water supply provided it is abstracted by means of a suitably designed impoundment dam. Probably suitable for drinking by private consumers and probably for most other uses, but not for irrigation except in special circumstances.

Class 3 Not suitable as a source of municipal water supply nor for industrial use nor ordinarily for irrigation, but in many instances suitable for drinking by private consumers and for watering cattle if the conductivity is not excessive.

Class 4 Probably suitable for irrigation, but not for drinking, stock watering, or industrial purposes.

Class 5 Unsuitable for almost every use except perhaps irrigation under special circumstances.

Note: A water described as probably suitable for some specific use must not, in fact, be accepted for that use until further details of the relevant specification have been studied and other matters considered.

[a] All toxic waters are also to be included here.

(Hutchinson 1957) proposed the following ranges for hypolimnetic oxygen loss expressed as mg cm^{-2} $month^{-1}$:

	Hutchinson	Mortimer
Oligotrophic	<0.5	<0.75
Eutrophic	>1.0	>1.65

Hutchinson (1957) suggested that Mortimer's limits may be more convenient. Problems are inherent in using only the areal oxygen depletion rate for comparative purposes because of the variability in hypolimnion thickness between lakes due to morphology and between years due to climatic conditions. For example, Burns (1976) recorded 1970 areal oxygen depletion rates of 1.29 and 2.61 mg cm^{-2} $month^{-1}$ for the central basin (CB) and eastern basin (EB) of Lake Erie, respectively. These data would class the CB as mesotrophic and EB as eutrophic under Mortimer's classification. On the other hand, the volumetric oxygen depletion rate for the CB is twice that of the EB and, therefore, on a volumetric basis the trophic status of the basins would be reversed. Burns (1976) suggested that both areal and volumetric depletion rates were necessary to fully describe hypolimnion oxygen depletion.

Use of the areal hypolimnetic oxygen deficit is restricted primarily to large, deep stratified lakes such as the Great Lakes. Chapra and Dobson (1981) stated that "oxygen deficit is a fundamental measure of lake trophic state". They have included the concept in a quantification of the Thienemann typology which will be discussed below under Section II.D.2, Composite Indices.

b. Primary Production

Rodhe (1958) suggested that lakes be classified as oligotrophic or eutrophic on the basis of rate of primary production. For European waters, he suggested that lakes with annual primary production rates of <25 and >75 gC m^{-2} be classed as oligotrophic and eutrophic, respectively. Rodhe (1969) later indicated that the approximate range of production might be from 7 to 700 gC m^{-2} $year^{-1}$. Other workers have prepared various estimates of production for the trophic classes but within Rodhe's overall range:

Oligotrophic	Mesotrophic	Eutrophic	References
<25	25–75	>75	Rodhe, 1958
<30	30–200	>200	Winberg, 1961
<100	100–200	>200	Vollenweider et al., 1974
<145	145–240	>240	Chapra and Dobson, 1981

The estimates of Chapra and Dobson (1981) are suggested for offshore waters of the Great Lakes.

One obvious advantage of using primary production as a basis for trophic classification is that it represents the first stage in the food web and reflects production at other levels. Disadvantages are difficulties in measuring the parameter accurately and the paucity of dependable data. Problems with the measurement of photosynthesis in the Great Lakes have been reviewed by Vollenweider et al. (1974).

c. Total Phosphorus

Because of its importance in plant nutrition, phosphorus is used in the empirical prediction of algal biomass and as an index of trophic status of lake waters. Examples of ranges of ambient concentrations (milligrams per cubic meter) of total phosphorus for three trophic classes are

Oligotrophic	Mesotrophic	Eutrophic	References
< 8	8–23	>23	Bachmann, 1980
<10	10–20	>20	Chapra and Robertson, 1977
<11	11–21.7	>21.7	Chapra and Dobson, 1981
<15	15–25	>25	Forsberg and Ryding, 1980
<15	15–30	>30	Welch and Lindell, 1978

The ranges of Chapra and Robertson (1977) and Chapra and Dobson (1981) were suggested for use in classifying the Great Lakes.

The obvious advantage of using total phosphorus in trophic classification is its ease of measurement and availability of data. A single determination of phosphorus concentration in the spring is sufficient to indicate trophic conditions in some lakes. For example, springtime assimilable phosphorus concentrations >10 mg m^{-3} indicates eutrophication potential

(Vollenweider, 1968). However, the universal application of a simple phosphorus index is not always appropriate due to other factors of production such as morphometry, flushing rate, optical properties, climate, and other nutrients.

d. Total Nitrogen

Although not in common use, total nitrogen has been proposed as a simple indicator of trophic status. For example, the following ranges of total nitrogen concentrations (milligrams per cubic meter) have been suggested:

Oligotrophic	Mesotrophic	Eutrophic	References
<300	300–650	>650	Vollenweider, 1968
<400	400–600	>600	Forsberg and Ryding, 1980

A eutrophication problem is indicated when the springtime concentration of assimilable nitrogen is >300 mg m^{-3} (Vollenweider, 1968).

The main disadvantage of using nitrogen as a trophic state indicator is that phosphorus is usually the limiting nutrient. For example, despite a steady increase in concentrations of inorganic nitrogen in the Great Lakes during the 1980s, a reverse trend has occurred in phosphorus concentrations and plant production. Moreover, since some blue-green algal species are capable of fixing atmospheric nitrogen, ambient concentration in the water may not be a valid indicator of trophic potential.

e. Chlorophyll a

The pigment, chlorophyll *a*, has long been used as a surrogate of algal biomass, and therefore it is logical to expect its concentration in lake waters to reflect trophic conditions. A few ranges of the pigment (milligrams per cubic meter) in relation to trophic classes selected from the literature are recorded below:

Oligotrophic	Mesotrophic	Eutrophic	References
<2.9	2.9–5.6	>5.6	Chapra and Dobson, 1981
<3	3–7	>7	Forsberg and Ryding, 1980
<3.7	3.7–10	>10	Welch and Lindell, 1978
<4.3	4.3–8.8	>8.8	Dobson et al., 1974

The ranges of Chapra and Dobson (1981) and Dobson et al. (1974) are suggested for classification of waters in the Great Lakes.

The principal advantage of chlorophyll *a* as a trophic indicator is ease of measurement and therefore availability of data. However, in the Great Lakes chlorophyll *a* data are variable due to temporal and spatial differences in measurement. Moreover, Nicholls and Dillon (1978) have pointed out several physiological and ecological considerations associated with using chlorophyll *a* as an indicator of algal biomass.

f. Secchi Disc Transparency

Because light extinction in a water column is a reflection of density of algal particles, Secchi disc transparency has been proposed as an index of trophic state. Several examples of ranges of Secchi depth (m) in relation to trophic class are listed below:

Oligotrophic	Mesotrophic	Eutrophic	References
>4.0	2.5–4.0	<2.5	Forsberg and Ryding, 1980
>5	5–3	<3	Chapra and Dobson, 1981
>6	6–3	<3	Vallentyne et al., 1969

The simplicity of the Secchi disc transparency is overshadowed by the problems associated with its use in dystrophic lakes or where inorganic particles substantially contribute to extinction of light. Hence, use of the Secchi disc as an indicator of trophic conditions may be adequate for offshore waters of the Great Lakes, but not in nearshore areas or embayments.

g. Organic Matter in Sediments

Entz (1977) proposed the use of organic content of sediments (expressed as a percentage of dry weight) as a descriptor of trophic classes of lakes. From his experience with central European lakes, he found that sediments with percent organic matter <17 were oligotrophic and >30 eutrophic with the range in between indicating mesotrophy.

2. Composite Indices

a. Morphoedaphic Index (MEI)

Although originally developed by Ryder (1965) for fish yield prediction in north-temperate lakes, the MEI has had global application (Ryder, 1982). In a comparison with seven other empirically derived yield models, Leach et al. (1987) considered the MEI to be the best overall estimator of fish yield in the Great Lakes, especially for the upper Great Lakes where it predicted yields within 19% of Matuszek's (1978) estimates of maximum sustained yield.

Ryder (1965) suggested that a MEI of about 6 to 7 (metric scale) indicated mesotrophy and that lakes with indices below and above that range were oligotrophic and eutrophic, respectively. In a multivariate analysis of habitat and fish associations in Ontario lakes, Johnson et al. (1977) found that the MEI was a strong factor in delimiting lake types. For example, an MEI of 6 clearly separated lake trout from nonlake trout lakes. The importance and application of the MEI as a descriptor of fisheries habitat as well as a yield estimator have been discussed by Ryder et al. (1974), Cole (1977), and Ryder (1982).

b. Productivity Index and Quality Index

For 150 north temperate lakes, Hayes (1957) calculated a productivity index (PI) by summing the weights of harvests of fish species according to length of food chains and dividing the weights by Carlander's (1955) factors. The PI is an index of fish production, and hence trophic status, and may be used to compare lakes. Hayes (1957) also calculated a quality index (QI) from the data to facilitate comparison of lakes of different depth. The QI is equal to PI × $\sqrt{m/5}$ where m = depth in meters. The QI attempts to modify the effect of depth on productivity in favor of inherent productivity of the drainage basin. The PI, QI, and MEI for the Great Lakes are compared below:

	PI	QI	MEI
Superior	0.17	0.90	0.98
Michigan	0.35	1.40	1.82
Huron	0.31	1.05	1.92
Erie	1.57	2.92	3.70
Ontario	0.21	0.82	1.91

Hayes and Anthony (1964) later modified the PI with multiple regressions of lake dimensions and water chemistry. In an analysis of the original data set, 67% of the variability of the PI was accounted for by depth (29%), area (20%), and alkalinity (18%), and 33% was due to unidentified residual factors. As Ryder et al. (1974) have pointed out, mean depth and area were correlated, which tends to diminish the significance of the analysis.

c. Trophic State Index (TSI)

A TSI was developed for use in evaluating Florida lakes by Shannon and Brezonik (1972). Secchi disc transparency (SD), specific conductance (COND), total organic nitrogen (TON), total phosphorus (TP), primary productivity (PP), chlorophyll *a* (CHA), and Pearson's cation ratio (CR) were the environmental parameters used as annual averages in multivariate expressions for clear lakes (a) and colored lakes (b):

$$\text{(a) TSI} = 0.936\left(\frac{1}{\text{SD}}\right) + 0.827(\text{COND}) + 0.907(\text{TON}) + 0.748(\text{TP}) + 0.938(\text{PP}) + 0.892(\text{CHA}) + 0.579\left(\frac{1}{\text{CR}}\right) + 4.76$$

$$\text{(b) TSI} = 0.848\left(\frac{1}{\text{SD}}\right) + 0.809(\text{COND}) + 0.887(\text{TON}) + 0.768(\text{TP}) + 0.930(\text{PP}) + 0.780(\text{CHA}) + 0.893\left(\frac{1}{\text{CR}}\right) + 9.33$$

The index was developed to rank sets of lakes in a logical and objective sequence. For a group of 55 Florida lakes, TSIs of 3 and 7 bracketed mesotrophy. The index is recommended by the authors for regional or local use only because interpretation of results may be difficult if the data set is too diverse. Furthermore, information needs are high and several of the factors may be meaningless for some lakes.

Carlson (1977) proposed a simpler TSI based on any one of several parameters such as SD, chlorophyll (CHL), and TP. He derived the following empirical relationships:

$$\text{TSI}_{(\text{SD})} = 10\left(6 - \frac{\ln \text{SD}}{\ln 2}\right)$$

$$\text{TSI}_{(\text{CHL})} = 10\left(6 - \frac{2.04 - 0.68 \ln \text{CHL}}{\ln 2}\right)$$

$$\text{TSI}_{(\text{TP})} = 10\left(6 - \frac{\ln \frac{48}{\text{TP}}}{\ln 2}\right)$$

Carlson assumed that the indicators used in these equations are correlates of algal biomass. His TSI runs from 0 to 100 and each division of 10 represents a doubling of algal biomass (Table 5). The selection of trophic indicator may vary seasonally, according to lake conditions or from lake to lake. Advantages include ease of obtaining data, TSI values are numbers rather than names, TSI values are derived from a variety of lakes and therefore are applicable to

TABLE 5
Trophic State Index and Associated Parameters (Carlson, 1977)

TSI	Secchi disc (m)	Surface phosphorus (mg m^{-3})	Surface chlorophyll (mg m^{-3})
0	64	0.75	0.04
10	32	1.5	0.12
20	16	3	0.34
30	8	6	0.94
40	4	12	2.6
50	2	24	6.4
60	1	48	20
70	0.5	96	56
80	0.25	192	154
90	0.12	384	427
100	0.062	768	1183

many lakes, and there is no loss of information by combining unrelated parameters. Furthermore, the TSI has predictive capabilities and can be used for regional classification of all surface waters.

On the other hand, Aizaki et al. (1981) considered that Carlson's TSI based on SD had definite limitations in waters in which the extinction coefficient of nonalgal substances is large. They proposed a modification of Carlson's TSI based on particulate organic carbon (POC) and particulate organic nitrogen (PON) concentrations which they used to classify Japanese lakes.

Rao and Berkel (1978) proposed two bacterial index ratios (BIR) to classify water bodies and assist in establishing the zone of influence of point source pollution. BIR is the ratio between hetero- or oligotrophic bacterial populations from plate counts and total bacteria from epifluorescence microscopic counts.

d. Absolute Ranking

Lueschow et al. (1970) classified a set of Wisconsin lakes simply by ranking them for each of five parameters which indicate trophic status: hypolimnetic oxygen depletion, SD, organic nitrogen, total inorganic nitrogen, and net plankton. For each lake a composite ranking was calculated by summing the ranking for each parameter. The lake with the lowest composite score was considered most oligotrophic and the one with the highest, most eutrophic.

Rawson (1960) used a similar approach to compare lakes in Saskatchewan. Two rankings were made by Rawson, one using five physical parameters and one based on three biological parameters.

The major disadvantage of these indices is that they are relative. The positions of any lake in a series are dependent on the position of the other lakes.

e. Trophic Index Number (TIN)

The U.S. Environmental Protection Agency (EPA) (1974) modified the Lueschow et al. (1970) methodology to classify over 200 lakes for the National Eutrophication Survey. Percentile rather than absolute rankings were used for each of six parameters (median levels of total phosphorous, inorganic nitrogen, and dissolved phosphorous, mean levels of chlorophyll *a*, SD, and minimum dissolved oxygen). Composite ranking values (TINs) are determined by summing each of the six percentile ranks. Highest TIN values indicate the most oligotrophic lakes and strongest eutrophic lakes. The TIN ranking is also relative and therefore limited for comparing lakes in a data set.

f. Lake Condition Index (LCI)

Uttormark and Wall (1975) recognized the paucity of good chemical and biological data for most lakes and developed a classification system that uses more readily available indicators of trophic condition. Four parameters were selected and assigned a range of penalty points to cover lake conditions from desirable to undesirable:

Parameter	**Points**
Dissolved oxygen	0 – 6
Transparency	0 – 4
Fishkills	0 – 4
Use impairment	0 – 9
Total	0 – 23

Composite lake ratings (LCIs) were calculated by summing the points assigned to each parameter. The LCI values are not synonymous with trophic status. However, the authors found the following relationship between trophic classification and LCI for over 1100 Wisconsin lakes: <4.9, oligotrophic; 5 to 9.9, mesotrophic; >10, eutrophic. The LCI was found to be a useful classification system for fisheries managers in Wisconsin.

g. Bioproduction Number (BPN)

Håkanson (1984) considered lake sediments as a bank of environmental information and therefore useful as an indicator of lake trophic level. Sediment nitrogen (N), phosphorus (P), carbon (C), and loss on ignition (IG) have been suggested as indicators of trophic level. In a study of Swedish lakes, Håkanson found that the slope of the regression line between N and IG, which he termed the BPN, provides accurate information about lake trophic status. However, the BPN cannot be used as a trophic indicator when the organic content (IG) is larger than 20% of dry sediments. Lakes with IG between 20 and 30% of dry sediments are considered by Håkanson as transitional and cannot be accurately classified by the BPN (Figure 4). He has also developed a classification system for dystrophic lakes when IG is >30% based on the ratio IG/N. Mesohumic lakes are bounded by IG/N values of 20 and 25.

There are limits to the application of Håkanson's method (e.g., basin morphology and currents influence the deposition of sediments), but it appears to provide reasonable agreement with water-phase trophic indicators in some Swedish lakes. The method requires further refinement and application to other sets of lake data.

h. Naumann Trophic Scale and Thienemann Trophic Scale

Chapra and Dobson (1981) have quantified the lake typologies of Naumann and Thienemann with application to Great Lakes data. They derived two separate but related trophic schemes for the Great Lakes based on Naumann's surface water quality concept and Thienemann's hypolimnetic oxygen depletion concept (Figure 5). They developed scales for each scheme and made them comparable by expressing each with a lower boundary of zero and a mesotrophic range from five to ten. The trophic scales for the Great Lakes are compared in Figure 6.

The upper Great Lakes are classed as oligotrophic on both scales and eastern Lake Erie is mesotrophic. The Thienemann indices from Lakes Superior, Huron, and Ontario are lower than the Naumann indices, probably because of the large oxygen reserves in the hypolimnia of those lakes. The much larger Thienemann index in central Lake Erie reflects the hypolimnetic oxygen problem there. Lake Ontario is mesotrophic by the Naumann scale, but very oligotrophic on the Thienemann scale.

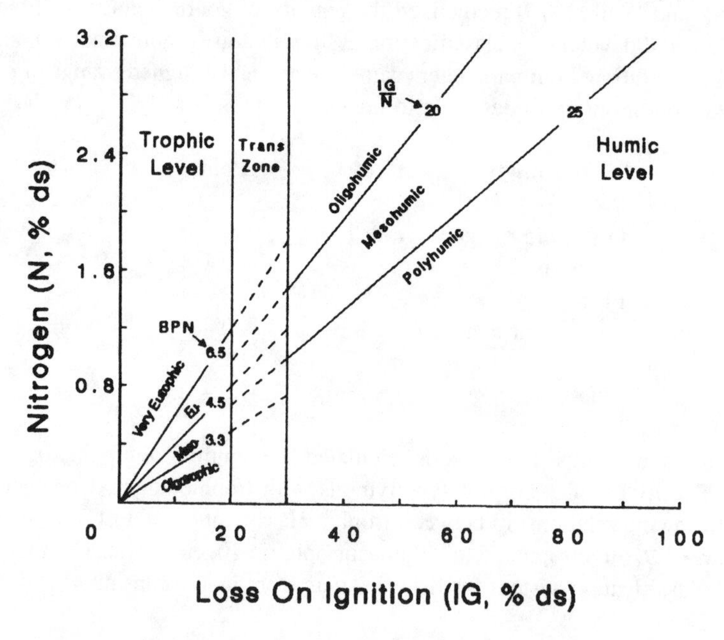

FIGURE 4. The relationships between nitrogen concentration and loss on ignition of surficial sediments (0 to 1 cm) relative to lake trophic level and lake humic level. (From Håkanson, L., *Water Res.*, 18, 303, 1984. With permission.)

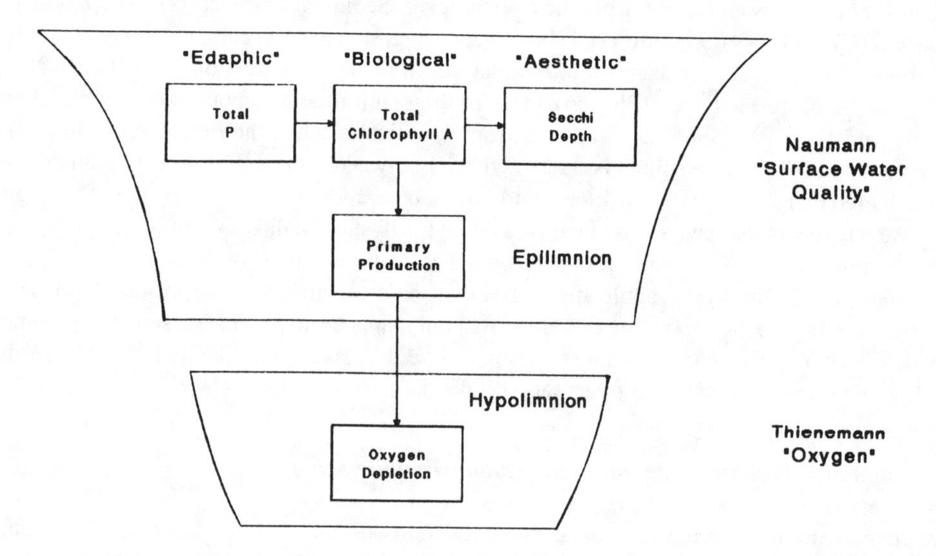

FIGURE 5. Interrelations among variables in Naumann's and Thienemann's concepts of eutrophication. (From Chapra, S. C. and Dobson, H. F. H., *J. Great Lakes Res.*, 7, 182, 1981. With permission.)

The authors suggest that both scales be considered when classifying lakes and that the maximum values be used when determining remedial measures. For example, Lake Ontario should be managed to improve surface water quality, and the rehabilitation of central Lake Erie should be directed toward alleviating the oxygen problem.

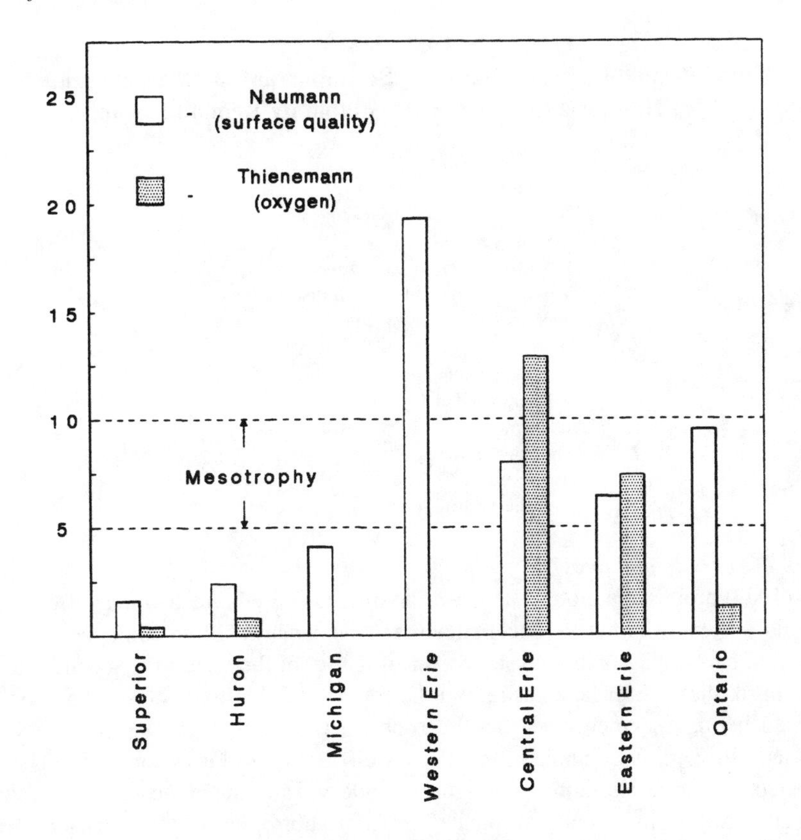

FIGURE 6. Comparison of Naumann and Thienemann indices for the Great Lakes. Thienemann indices are not included for Lake Michigan due to lack of data and for western Lake Erie due to nonstratified conditions. (From Chapra, S. C. and Dobson, H. F. H., *J. Great Lakes Res.*, 7, 182, 1981. With permission.)

i. Multispectral Scanner (MSS) Indices

MSS data acquired from LANDSAT and aircraft have been used to estimate the trophic status of lakes in several states (Boland, 1976; Boland and Blackwell, 1977). In a study of Colorado lakes (Boland and Blackwell, 1977) contact-sensed data for chlorophyll *a*, SD, conductivity, TP, TON, and algal assay yield were used to develop a TSI. Relationships between this index, several trophic indicators, and digitally processed MSS data were examined using regression and MSS classification techniques. The authors found significant correlations between spectral bands, the TSI and chlorophyll *a*, SD, and TON. They then generated color-coded photomaps to depict spectral aspects of trophic state.

The authors concluded that the methodology had practical value in lake monitoring and survey programs, including large numbers of lakes within a single MSS frame. Incorporation of MSS color ratios into regression models yielded good predictions of SD. Predicted values of chlorophyll *a* were considered to be a crude index. Models for both indicators do not give valid predictions when used with MSS data from different dates. Ground truth and MSS data must be collected concurrently for most accurate model development.

Remote sensing studies (Witzig and Whitehurst, 1981) have demonstrated several features or processes which might be useful in characterizing fish habitat in pelagic areas, especially in marine environments (Table 6). Many of these habitat parameters are dynamic and affect fish abundance and probably production. Satellite remote sensing offers a relatively easy and inexpensive methodology to study and quantify these features.

TABLE 6
Some Pelagic Features that May Be Important in Characterizing Fish Habitat and that are Detectable by Remote Sensing

Thermal fronts
Interfaces of oceanic water masses
Upwelling zones
Divergent and convergent zones
Two-dimensional chlorophyll distributions
River and/or turbidity plumes
Eddies and/or gyres
  Warm and cold core
  Cyclonic and anticyclonic
Streamers
Prevailing currents
Salinity fronts
Intrusions (warm or cold)

j. Zafar's Taxonomy of Lakes

Zafar (1959) proposed a global classification of lakes based on a naming system similar to the binomial system of Linnaeus, except that the name consists of four components rather than two. The first part indicates the physicochemical nature of the water; the second, the climate zone; the third, the type of lake basin; and the fourth, the dominant class of phytoplankton. The physicochemical segment describes the trophic nature of the water and considers calcium, phosphorus, nitrogen, and humus content as well as Pearsall's cation ratio. The climate component of the name takes into account the latitude and altitude of the lake according to four zones on either side of the equator: tropical (0° to 20°), subtropical (20° to 40°), temperate (40° to 60°), and arctic (60° to 80°). Subdivisions of the four zones are made to allow for elevation. For the basin component Zafar (1959) uses the volume development classification of Welch (1948) simplified to Greek letters. The fourth part of the name is provided by the dominant class of phytoplankton.

Zafar's taxonomy is an admirable attempt to provide a global classification system for lakes. Because it tries to cover all bases, it is descriptive rather than quantitative. It is cumbersome at the local scale and we doubt that it has application to the Great Lakes.

3. Biological Indicators

a. Bacteria

Aizaki (1985) proposed the use of total number of bacteria as a TSI after he found very significant relationships between direct counts of bacteria and several environmental parameters in 25 lakes and 4 streams in Japan. Highest correlations occurred with numbers of bacteria and POC, PON, chlorophyll *a*, SD, and chemical oxidation demand (COD). Aizaki did not attempt to quantify the relationships into an index which could be used to evaluate other lakes. Bacteria are not routinely counted in most lake surveys because of lack of expertise, and hence, it is unlikely that a TSI based on bacteria numbers will become commonplace.

b. Phytoplankton

Few attempts have been made to use algae in lake classification. The use of algae as indicators of trophy is difficult because of the range of habitat that genera or even individual species will tolerate. However, Round (1958) considered that lake classification from an algal standpoint was possible on the basis of dominant species or dominant communities. For example, he suggested that lakes in the U.K. could be classed as eutrophic or oligotrophic on the basis of dominant diatom species in littoral sediments.

Rawson (1956) considered that in addition to the ecological dominants, the number of species present was important in characterizing lakes. He pointed out that several species commonly considered to be indicative of eutrophy were common in large oligotrophic lakes in Canada. Beeton (1965) found a similar pattern in the Great Lakes. However, on the basis of plankton abundance as well as presence of Rawson's trophic type dominant algal species, he considered that Lakes Superior, Huron, and Michigan would be oligotrophic and Lake Erie eutrophic.

In recognition of weaknesses associated with the use of single-organism trophic indicators, Nygaard (1949) and others have suggested the use of various phytoplankton quotients. Stockner (1971) has also proposed a trophic classification based on ratios of diatom frustules from two groups — Araphidineae and Centrales (A/C) — in surface sediments. For temperate lakes he suggested the following values for A/C: oligotrophic 0 to 1.0, mesotrophic 1.0 to 2.0, eutrophic <2.0. All of these systems require a considerable amount of expertise to accurately determine the ratio. Furthermore, Shapiro (1975) considers that they provide little insight into the condition of the lake.

c. *Macrophytes*

Most trophic classification systems have been developed for open water and do not take into account the contributions of macrophytes to lake conditions. Canfield et al. (1983) recognized that trophic state assessment based on the usual suite of limnological indicators may be in error when applied to macrophyte-dominated lakes. For example, nutrient and chlorophyll concentrations may be low and SD high in lakes rich in macrophytes, and therefore trophic status may be underestimated. Canfield et al. (1983) suggested that nutrients in the macrophytes be added to those in the water column to determine a potential water column nutrient concentration which is then used with existing trophic indices. Problems associated with the determination of plant nutrients are recognized by the authors and they recommend the approach as a first-order approximation of potential input of macrophytes on lake trophic state. We feel that this approach may have relevance in embayments of the Great Lakes and is worthy of consideration.

d. *Zooplankton*

There are problems associated with attempting to use zooplankton in lake classification schemes. Many of the general effects of enrichment on zooplankton can be masked by predation by planktivores. For example, an apparent increase in zooplankton abundance in response to an increase in food supply in a lake undergoing enrichment, may not show up in a sampling schedule if predation by fish also increases. Changes in species composition may be a more obvious effect of enrichment, but can also be caused by size-selective predation (Brooks and Dodson, 1965). In the Great Lakes and elsewhere, there has been a shift in dominance by *Eubosmina coregoni* to the smaller *Bosmina longirostris*; however, Brooks (1969) urges caution in interpretation of shifts in these species in the water column and sediments because of predation.

Sprules (1980) suggested that size-feeding ecology structure of zooplankton communities might be used to group lakes into major types with distinct pathways of energy flow. He considers that this approach could lead to the separation of lakes into three types on the basis of zooplankton size and feeding behavior.

e. *Benthos*

Of all the biological indicators of trophic status, benthos has received the most attention. The history and recent status of the use of benthic organisms in lake classification have received comprehensive coverage in an excellent review by Brinkhurst (1974). Many benthic typologies have been proposed since the early work of Thienemann, at first using the

Chironomidae and later an array of benthic organisms. Most were developed for European lakes, but problems with the taxonomy of groups of organisms (particularly Chironomidae) have hampered efforts to relate them to lakes in North America. Saether (1975) reviewed North American efforts to use chironomids as indicators of lake typology and compared indicator communities with those in Europe. Often, closely related indicator species of the Nearctic and Palaearctic shared similar niches on both continents. He concluded that species composition of chironomid communities can assist the evaluation of lake types and cultural stresses. However, he cautioned that much remains to be done with North American chironomids regarding taxonomy, autecology, and zoogeography.

Very little attention has been paid to the application of Great Lakes benthos to lake typology. Cook and Johnson (1974) reviewed benthos research in the Great Lakes back to 1870 and outlined changes in bottom fauna due to cultural stresses. They considered the upper Great Lakes as a major subdivision with average proportions of taxa in summer as follows: *Pontoporeia affinis* 60 to 80%, oligochaetes 10 to 20%, sphaeriids 5 to 15%, and chironomids 0 to 5%. Oligochaetes increased in abundance in Lake Ontario (30% oligochaetes, 50% *P. affinis*) and were dominant in eastern Lake Erie (34% oligochaetes, 27% *P. affinis*) (in the whole of Lake Erie the proportion of oligochaetes declined from 86% in the west basin (WB) to 55% in the CB to 34% in the EB). Cook and Johnson found that the gradient of oligochaete density from the upper to lower lakes correlated with those of carbonate alkalinity and particulate organic matter. They also concluded that eutrophic and mesotrophic associations of oligochaetes were generally similar in both North America and Europe, but that dominants in the oligotrophic profundal were different. Problems with sampling and sorting methodology and taxonomy are drawbacks to widespread application of benthos as an indicator of water quality and lake type.

4. Regional Typology

Lakes have been classified on a regional basis in many areas (see Table 1 for a partial list). This approach is especially appealing to managers responsible for large numbers of lakes.

The importance of latitude and altitude in explaining variability in production of International Biological Program (IBP) lakes distributed locally has been indicated by Brylinsky and Mann (1973). On a finer scale the importance of local climate to aquatic production has been recognized by limnologists and fisheries ecologists (Zafar, 1959; Schlesinger and Regier, 1982; Schlesinger and McCombie, 1983; Colby and Nepszy, 1981). In addition, other factors of production (e.g., basin morphometry, nutrient inputs) are usually distributed on a local or regional basis. Therefore, there are usually one or more habitat characteristics that are uniform on a regional scale and are useful in classification models.

Cole (1977) considers the best approach in North America to be a regional classification of lakes. He advocates the development by state and provincial natural resources agencies of habitat classificatons that can be applied regionally even though they may be less useful elsewhere.

E. CLASSIFICATION OF LAKES BY ASSEMBLAGES OF FISH SPECIES AND FISH HABITAT

The assumption that lakes with similar habitats will support similar fish communities has led to the consideration of "type" lakes and a community approach to management. This approach is particularly appealing to fisheries managers responsible for large numbers of lakes. Despite their potential usefulness, classification schemes that link habitat directly to fish community structure and productivity are comparatively recent and not numerous. This is probably because the comparative approach in these schemes requires extensive multivariate analyses with demands for large computer capacity.

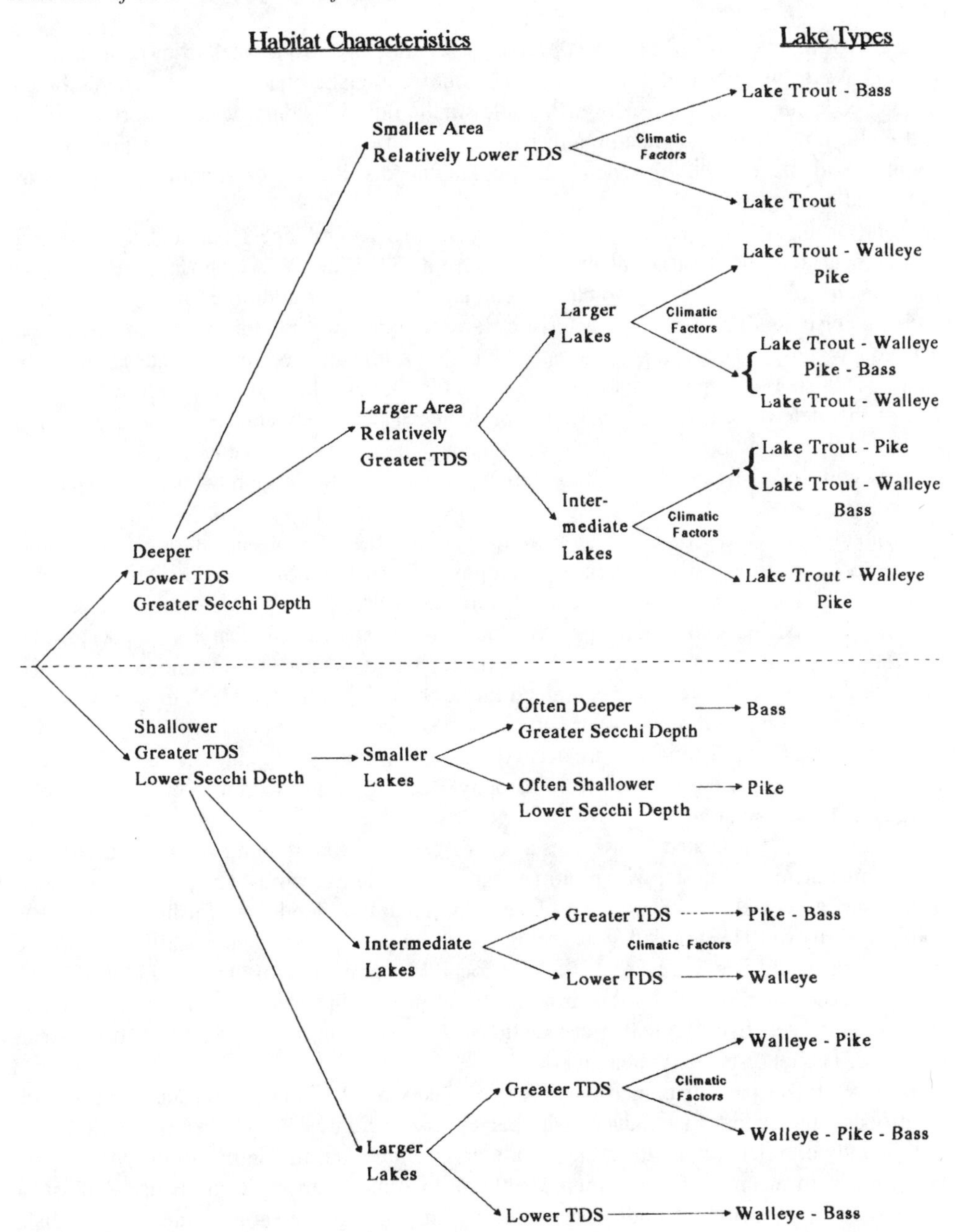

FIGURE 7. Schema of 15 lake types in Ontario classified by fish associations in relation to habitat characteristics or habitat factors. (From Johnson, M. G. et al., *J. Fish. Res. Bd. Can.*, 34, 1592, 1977. With permission.)

Johnson et al. (1977) related 7 limnological characteristics to 15 associations of 4 fish species in 2496 Ontario lakes using principal component analysis (summarized in Figure 7). The most common lake types were walleye-northern pike (22%), northern pike only (19%), lake trout only (16%), and smallmouth bass only (10%). Habitat characteristics most significant in distinguishing lake types were depth and area, but Secchi disc transparency and TDS also contributed to lake characterization. Mean depth was >6 m in most lake trout lakes. TDS of 50 mg l^{-1} could also be used to separate lake trout and nonlake trout lakes, although some overlap occurred. Larger lakes contained more species, probably because they have greater diversity of habitats. An MEI of 6 appeared to separate lake trout lakes from nonlake trout lakes. Lake trout only and lake trout-smallmouth bass lakes had hypolimnion dissolved oxygen concentrations of 6 to 7 mg l^{-1}, while smallmouth bass only, walleye-northern pike, and walleye-northern pike-smallmouth bass lakes had dissolved oxygen concentrations of about 2.9 to 4.2 mg l^{-1}. In general the geographic distribution of smallmouth bass was explained by climatic factors.

Tonn et al. (1983) distinguished two discrete fish assemblage types and their habitat requirements from multivariate analysis of data from 18 northern Wisconsin lakes. From these relationships they accurately predicted assemblage types in 11 additional lakes in the same region using data for 5 habitat characteristics (area, maximum depth, pH, drainage basin area, and conductivity). Tonn et al. suggested that the application of multivariate community analyses to fisheries management should occur in two stages. The first would delineate habitat factors that determine fish assemblages in particular lakes. The second phase would use the results of the community analysis for several management activities, including the classifying or predicting of fish assemblage types in other lakes in the region with similar habitat characteristics.

Schneider (1981) used factor analysis to examine the relationship among five habitat factors (area, maximum depth, climate, macrophyte density, and SD) and the distribution and abundance of fish species in 126 Michigan warmwater lakes. The analysis indicated general groupings of species with certain environmental parameters. Cluster analysis was performed on 221 lakes to determine associations of species based on correlation among abundance indices (Table 7). The lakes were examined further by discriminant analysis to assess fishing quality in relation to community structure, species size structure, and growth and habitat characteristics. Good fishing was associated with good quality habitat, i.e., lakes that were deeper and clearer, without excessive macrophyte densities, and with a layer of cool, well-oxygenated water in summer.

Harvey (1978) associated fish species assemblages in 51 Manitoulin Island, Ontario, lakes with habitat factors using stepwise multiple regression and multivariate techniques. A number of fish species correlated strongly with lake area, maximum depth, and flushing period. By contrast, Harvey (1975) found that these three physical factors correlated weakly with species number in the acid-stressed lakes of the La Cloche Mountain area of Ontario. In those lakes pH alone accounted for 40% of the variance. By clustering the Manitoulin Island lakes data, Harvey delineated five major fish associations. The communities were related to physical aspects of habitat, especially area and depth.

Harvey (1981) later examined data from 57 lakes on the Bruce Peninsula, Ontario, to identify habitat factors that induce fish species associations. The number of species was significantly correlated with area, lake width, and volume, but in contrast to the Manitoulin lakes, not with mean depth, maximum depth, and flushing period. In the Bruce Peninsula lakes, six physical habitat variables accounted for only 51% of the variance in number of fish species, compared to 77% for three habitat factors in the Manitoulin lakes. As in the La Cloche and Manitoulin lakes, pH was the only chemical factor that contributed significantly to species number variance in the Bruce lakes. Clustering techniques indicated four major fish communities and these induced seven main lake groups.

The Ontario Ministry of Natural Resources has used fish communities, defined as any combination of six major species [brook trout (*Salvelinus fontinalis*), lake trout, lake whitefish (*Coregonus clupeaformis*), northern pike, smallmouth bass, and walleye], as a basis for identification of lake types (Anonymous, 1978). The species are top predators or are of sport and/or commercial importance. The 63 major communities occurred in 4999 of 6455 lakes listed in Ontario's Lake Inventory Data Base. Twenty-four communities were selected on the basis of prevalence and importance as type communities from which candidate lakes were

TABLE 7
Associations among Habitat Factors and Species Abundance as Determined by Factor Analysis and Associations among Species as Determined by Cluster Analysis in Michigan Warmwater Lakes (Schneider, 1981)

	Factor analysis (126 lakes) Scaled factor loadings Factors				
Variable	**1**	**2**	**3**	**4**	**Cluster analysis (221 lakes)**
Environmental characteristic					
Area	0.0	0.8	–0.2	0.1	
Maximum depth	–0.2	–0.1	0.2	0.0	
Macrophyte index	0.4	0.0	0.0	0.3	
Secchi index	–0.1	–0.2	0.3	0.1	
Climate index	0.4	–0.3	–0.2	–0.3	
Abundance					
Bluegill	0.3	–0.2	0.2	0.6	
Pumpkinseed	0.4	–0.1	0.3	0.3	
Black crappie	0.1	0.1	–0.1	0.2	
Northern pike	0.2	0.2	0.1	0.6	
Bullheads	0.5	0.2	0.2	–0.1	
Grass pickerel	0.6	–0.2	0.1	0.0	
Minnows	0.6	0.1	0.2	–0.3	
Lake chubsucker	0.8	0.1	0.2	–0.3	Warmwater
Warmouth	0.5	0.1	0.0	–0.2	species
Bowfin	0.4	0.6	–0.1	0.0	
Largemouth bass	0.1	–0.3	0.1	0.4	
Carp	0.0	0.0	–0.2	–0.2	
Yellow perch	–0.0	0.0	–0.5	–0.1	
Rock bass	–0.2	0.3	0.7	0.0	
Smallmouth bass	–0.3	0.2	0.7	–0.2	Coolwater
Walleye	–0.1	0.7	–0.2	0.0	species
White sucker	–0.1	0.1	0.2	–0.1	
Percent variance (cumulative)	13	22	30	37	

chosen. These communities represented 4618 or 92% of the 4999 lakes referred to above. A subsample of 911 candidate lakes was selected after consideration of frequency distribution in regions, representation of fish community stresses (exploitation, eutrophication, acidification, water level fluctuations, physical alterations, and introductions), and inherent properties (area, mean depth, oxygen/temperature, TDS). A final selection of 172 lakes was made, and these were grouped for logistic reasons into 23 assessment units. A major role of assessment units is to evaluate responses of fish communities to various stresses by monitoring the community, the habitat, the resource harvest, and the resource users. From the long-term monitoring of type lakes trend-through-time data are developed that can be used in the management of other lakes of the same type.

Marshall and Ryan (1987) used direct gradient analysis to determine patterns of associations of eight key species [walleye, northern pike, white sucker (*Catostomus commersoni*), yellow perch (*Perca flavescens*), lake trout, lake whitefish, lake herring (*Coregonus artedi*), and burbot (*Lota lota*)] along clines of lake area, mean depth, SD, and the MEI in 75 lakes in northwestern Ontario. Based on their responses to various habitat conditions, the eight species of fish were separated into two main assemblages. Walleye, yellow perch, northern pike, and white sucker responded similarly to environmental gradients in the analysis. These species constitute the main core of Ryder and Kerr's (1990) harmonic percid community. The remaining species — lake trout, lake whitefish, lake herring, and burbot — are the four main species occurring in Ontario's lake trout lakes and may be considered as a harmonic salmonid community. Marshall and Ryan (1987) found that an SD of 2 m separated percid and salmonid communities. They also found that the percid assemblage was predominant in lakes with an MEI >4.0 and that the salmonid group contributed the greatest biomass in lakes with an MEI <4.0. In general, the results of Marshall and Ryan (1987) supported those of Johnson et al. (1977) in distinguishing fish communities on the basis of depth, area, SD, and TDS.

III. DISCUSSION

As we reviewed existing classification schemes and typologies, we tried to keep fish habitat in mind. We have reviewed 32 categories of classification systems or indices based on factors that are considered to affect lake metabolism. As indicated in Figure 1, all of these ultimately affect the quality and quantity of fish production. The habitat factors in Figure 1 form a hierarchy that increases in complexity toward fish production, but their usefulness as indicators of satisfactory fish habitat does not necessarily follow a parallel order. For example, components of the MEI are quite remote from fish output in the hierarchy.

Many of the classification systems we reviewed were developed by limnologists who were not directly concerned with their specific application to fish habitat or production. Current interest in lake classification is focused on those of a trophic nature, although Lewis (1983) indicated that interest in mixing classifications persists. Of the 32 categories reviewed, 24 pertain to trophic classification. Interest in this area was spurred by a need to quantify the eutrophication process which has been accelerated in recent decades by cultural influences.

Some of the components of trophic classifications have been used in empirical relationships to estimate fish production and yield (see Leach et al., 1987 for a review). Since fish production at both species and total levels is closely related to habitat, it is not surprising that these variables are useful as first-order predictors of yields. Models using surface area, TP, primary production rate, chlorophyll *a*, benthos biomass, and the MEI have potential use in at least some of the Great Lakes as yield estimators, but generally require refinement and improved input information. The MEI, which describes physical and chemical characteristics that control production at the bottom of the food web, was the best overall yield predictor for those lakes least influenced culturally. Food-web components, when fitted to biomass-size spectrum models, were also good predictors of production and yield. For example, the Borgmann model has been modified to predict potential fish production and yield in two size categories from production estimates of zooplankton and macrobenthos (Leach et al., 1987).

Fish production is also influenced by community structure, which is determined partly by habitat. In our consideration of lake classification by fish species association we found several common habitat parameters. Lake surface area was used in six of the multivariate approaches we examined, and depth (either maximum or mean) and SD were included in five. TDS (or correlates, conductivity and alkalinity) and pH were used in four. In all of the analyses, surface area and depth were the most significant habitat characteristics determining lake types based on fish communities. TDS (or correlates) and SD also contributed to community characterization. In acid-stressed areas, pH was the principal habitat indicator. It is interesting to note that all of these habitat factors are simple and that information on them is easy and inexpensive to obtain. Data on these factors have been used in some of the above analyses because they were readily available from lake survey files. There may be other factors equally useful, at least in specific situations, which have yet to be tested with multivariate analyses.

Lake classification by fish species association will likely be delineated on a regional basis because of morphometric, edaphic, and local climatic considerations. We agree with Cole (1977) that a regional classification of lakes may be the best first-order approach to fisheries habitat inventory and classificaton by fisheries agencies.

Moreover, Ryder (1982) suggested a hierarchy of approaches to fish yield estimation which we feel has relevance to fisheries habitat classification. The hierarchic levels and prime determining variables are

1. Global — Temperature and area (interrelated through total caloric content)
2. Regional — Air temperature is a constant
   — Area reduced to constant units to compare lakes
   — Nutrient input variations and lake morphometry are more important (MEI = $N/\bar{z}$)
3. Infraregional — Either nutrient inputs or morphometry ($\bar{z}$) may be constant
   — Yield may be predicted from $\bar{z}$ in the first instance and from nutrient levels in the second

We are not suggesting that habitat criteria be limited to Ryder's variables, but we recognize that there is merit in considering aquatic habitat in those terms that adequately predict fish production and yield.

From the viewpoint of fisheries managers, many of the lake classification schemes and typologies reviewed above can be criticized for lack of direct application to fish and fisheries. However, as indicated in Figure 1, the linkages exist. We feel that the variables that adequately characterize fish communities and predict production are the most suitable for habitat classification purposes. Inherent in the former are criteria that characterize habitat suitable for early life history stages.

ACKNOWLEDGMENTS

This chapter benefitted from discussions with K. H. Loftus, R. A. Ryder, and R. T. Oglesby. We are grateful to Dieter Busch, Peter Sly, and Henry Sather for review comments. The diligence and patience of Debbie Warner and Nancy Robson in processing numerous drafts of the manuscript are greatly appreciated. Jeff Wright and Doug Lee redrafted the figures.This is Contribution No. 91-08 of the Ontario Ministry of Natural Resources, Research Section, Fisheries Branch, Box 5000, Maple, Ontario L6A 1S9 Canada.

REFERENCES

Aizaki, M. 1985. Total number of bacteria as a trophic state index. *Verh. Int. Ver. Limnol.* 22:2732–2738.

Aizaki, M., A. Otsuki, T. Fukushima, M. Hosomi, and K. Muraoka. 1981. Application of Carlson's trophic state index to Japanese lakes and relationships between the index and other parameters. *Verh. Int. Ver. Limnol.* 21:675–681.

Alm, G. 1922. Bottenfauna och fiskens biologi i Yxtasjon samt jamforande studier over bottenfauna och fiskavkastning i vara sjoar. *Medd. Landbruksstyr.* 236:1–186.

Anonymous. 1978. Designation of Assessment Units. Rep. SPOF Working Group No. 1. Ontario Ministry of Natural Resources, Toronto.

Bachmann, R. W. 1980. The role of agricultural sediments and chemicals in eutrophication. *J. Water Pollut. Control Fed.* 52:2425–2432.

Baker, K. S. and R. C. Smith. 1982. Bio-optical classification and model of natural waters. II. *Limnol. Oceanogr.* 27:500–509.

Beeton, A. M. 1965. Eutrophication of the St. Lawrence Great Lakes. *Limnol. Oceanogr.* 10:240–254.

Beeton, A. M. 1966. Indices of Great Lakes Eutrophication. Publ. No. 15. University of Michigan Great Lakes Research Division. Lansing. pp. 1–8.

Beeton, A. M. 1969. Changes in the environment and biota of the Great Lakes. in *Eutrophication: Causes, Consequences, Correctives,* National Academy of Sciences, Washington, D.C., pp. 150-187.

Berg, K. 1938. Studies on the bottom animals of Esrom Lake. *Kgl. Danske Vidensk. Selsk. Naturv. Math. Afd.* 8:1–255.

Birge, E. A. 1915. The heat budgets of American and European lakes. *Trans. Wis. Acad. Sci. Arts Lett.* 18:166–213.

Boland, D. H. P. 1976. Trophic Classification of Lakes Using LANDSAT–1 (ERTS–1) Multispectral Scanner Data. EPA-600/3-76-037. U.S. Environmental Protection Agency, Washington, D.C. p. 245.

Boland, D. H. P. and R. J. Blackwell. 1977. Trophic classification of lakes utilizing contact data, LANDSAT and aircraft-acquired multi-spectral scanner data. Proc. 4th Jt. Conf. on Sensing of Environmental Pollutants, New Orleans, LA.

Brooks, J. L. 1969. Eutrophication and changes in the composition of the zooplankton.in *Eutrophication: Causes, Consequences, Correctives,* National Academy of Sciences, Washington, D.C. pp. 236–255.

Brooks, J. L. and S. I. Dodson. 1965. Predation, body size and composition of plankton. *Science* 150:28–35.

Brinkhurst, R. O. 1974. *The Benthos of Lakes.* St. Martin's Press, New York.

Brundin, L. 1942. Sur Limnologie jamtlandischer Seen. *Mitt. Anst. Binnenfisch. Drottningholm, Stockholm.* 20:1–104.

Brundin, L. 1949. Chironomiden und andere Bodentiere der Sudchweden Urgebirsseen. *Rep. Inst. Freshwat. Res. Drottningholm.* 30:1–914.

Brundin, L. 1956. Die bodenfaunistischen Seetypen und ihre Anwendbarkheit auf die Sudhalkugel. Zugleich ein Theorie der produktionbiologischen Bedeutung der glazialen Erosion. *Rep. Inst. Freshwat. Res. Drottningholm.* 37:186–235.

Brundin, L. 1958. The bottom faunistical lake type system and its application to the southern hemisphere. Moreoever a theory of glacial erosion as a factor of productivity in lakes and oceans. *Verh. Int. Ver. Limnol.* 13: 288–297.

Brylinsky, M. and K. H. Mann. 1973. Analysis of factors governing productivity in lakes and reservoirs. *Limnol. Oceanogr.* 18:1–14.

Burns, N. M. 1976. Temperature, oxygen and nutrient distribution patterns in Lake Erie, 1970. *J. Fish. Res. Bd. Can.* 33:485–511.

Canfield, D. E., K. A. Langeland, M. J. Maceina, W. T. Haller, J. V. Shireman, and J. R. Jones. 1983. Trophic state classification of lakes with aquatic macrophytes. *Can. J. Fish. Aquat. Sci.* 40:1713–1718.

Carlander, K. D. 1955. The standing crop of fish in lakes. *J. Fish. Res. Bd. Can.* 12:543–570.

Carlson, R. E. 1977. A trophic state index for lakes. *Limnol. Oceanogr.* 22:361–369.

Chapra, S. C. and H. F. H. Dobson. 1981. Quantification of the lake trophic typologies of Naumann (surface quality) and Thienemann (oxygen) with special reference to the Great Lakes. *J. Great Lakes Res.* 7:182–193.

Chapra, S. C. and A. Robertson. 1977. Great Lakes eutrophication: the effects of point source control of total phosphorus. *Science* 196:1448–1450.

Colby, P. J. and S. J. Nepszy. 1981. Variation among stocks of walleye (*Stizostedion vitreum vitreum*): management implications. *Can. J. Fish. Aquat. Sci.* 38:1814–1831.

Cole, G. A. 1977. Lake classification — good and bad. Classification, Inventory and Analysis of Fish and Wildlife Habitat. FWS/OBS-78/76:67–78. U.S. Fish and Wildlife Service, Washington, D.C.

Cook, D. G. and M. G. Johnson. 1974. Benthic macro-invertebrates of the St. Lawrence Great Lakes. *J. Fish. Res. Bd. Can.* 31:763–782.

Decksbach, N. K. 1929. Uber verschiedene Typenfolge der Seen. *Arch. Hydrobiol.* 20:65–80.

Deevey, E. S. 1941. Limnological studies in Connecticut. VI. The quantity and composition of the bottom fauna of thirty-six Connecticut and New York lakes. *Ecol. Monogr.* 11:413–455.

Dillon, P. J. and F. H. Rigler. 1975. A simple method for predicting the capacity of a lake for development based on lake trophic status. *J. Fish. Res. Bd. Can.* 32:1519–1531.

Dobson, H. F. H., M. Gilbertson, and P. G. Sly. 1974. A summary and comparison of nutrients and related water quality in Lakes Erie, Ontario, Huron, and Superior. *J. Fish. Res. Bd. Can.* 31:731–738.

Eberly, W. R. 1975. The use of oxygen deficit measurements as an index of eutrophication in temperate limnetic lakes. *Verh. Int. Ver. Limnol.* 19:439–441.

Elster, H. J. 1958. Lake classification, production and consumption. *Verh. Int. Ver. Limnol.* 13:101–120.

Elster, H. J. 1974. History of limnology. *Mitt. Int. Ver. Theor. Angew. Limnol.* 20:7–30.

Entz, B. 1977. Environmental conditions of percid waters in central Europe. *J. Fish. Res. Bd. Can.* 34:1586–1591.

Forel, F. A. 1892, 1895, 1904. *Le Léman: Monographie Limnologique.* F. Rouge, Lausanne.

Forsberg, C. and S. Ryding. 1980. Eutrophication parameters and trophic state indices in 30 Swedish waste–receiving lakes. *Arch. Hydrobiol.* 80:189–207.

Håkanson, L. 1984. On the relationship between lake trophic level and lake sediments. *Water Res.* 18:303–314.

Harvey, H. H. 1975. Fish populations in a large group of acid-stressed lakes. *Verh. Int. Ver. Limnol.* 19:2406–2417.

Harvey, H. H. 1978. Fish communities of the Manitoulin Island lakes. *Verh. Int. Ver. Limnol.* 20:2031–2038.

Harvey, H. H. 1981. Fish communities of the lakes of the Bruce Peninsula. *Verh. Int. Ver. Limnol.* 21:1222–1230.

Hayes, F. R. 1957. On the variation in bottom fauna and fish yield in relation to trophic level and lake dimensions. *J. Fish. Res. Bd. Can.* 14:1–32.

Hayes, F. R. and E. H. Anthony. 1964. Productive capacity of North American lakes as related to the quantity and the trophic level of fish, the lake dimensions, and the water chemistry. *Trans. Am. Fish. Soc.* 93(1):53–57.

Herdendorf, C. E. 1982. Large lakes of the world. *J. Great Lakes Res.* 8:379–412.

Hooper, F. F. 1956. Some Chemical and Morphometric Characteristics of Southern Michigan Lakes. Pap. No. 41, Michigan Academy of Science, Ann Arbor, MI. pp. 109–130.

Hornstrom, E. 1981. Trophic characterization of lakes by means of qualitative phytoplankton analysis. *Limnologica* 13(2):249–261.

Hutchinson, G. E. 1938. On the relation between the oxygen deficit and the productivity and typology of lakes. *Int. Rev. Hydrobiol.* 36:336–355.

Hutchinson, G. E. 1957. *Treatise on Limnology,* Vol. 1. John Wiley & Sons, New York.

Hutchinson, G. E. 1973. Eutrophication: the scientific background of a contemporary practical problem. *Am. Sci.* 61:269–279.

Hutchinson, G. E. and H. Loffler. 1956. The thermal classification of lakes. *Proc. Natl. Acad. Sci. U.S.A.*, 42:84–86.

Jarnefelt, H. 1925. Zur Limnologie einiger Gewasser Finnlands. I. *Ann. Soc. Zool-Bot. Fenn. Vanamo.* 2:185–356.

Jarnefelt, H. 1953. Die Seetypen in Bodenfaunistischer Hinsicht. *Ann. Soc. Zool-Bot. Fenn. Vanamo.* 15:1–38.

Jarnefelt, H. 1958. On the typology of the northern lakes. *Verh. Int. Ver. Limnol.* 13:228–235.

Jensen, S. 1979. Classification of lakes in southern Sweden on the basis of their macrophyte composition by means of multivariate methods. *Vegetatio* 39(3):129–146.

Johnson, M. G., J. H. Leach, C. J. Minns, and C. H. Olver. 1977. Limnological characteristics of Ontario lakes in relation to associations of walleye (*Stizostedion vitreum*), northern pike (*Esox lucius*), lake trout (*Salveninus namaycush*), and smallmouth bass (*Micropterus dolomieui*). *J. Fish. Res. Bd. Can.* 34:1592–1601.

Kemp, P. H. 1971. Chemistry of natural waters. VI. Classification of waters. *Water Res.* 5:945–956.

Kratzer, C. R. and P. L. Brezonik. 1981. A Carlson-type trophic state index for nitrogen in Florida lakes. *Water Resour. Bull.* 17(4):713–715.

Larkin, P. A. and T. G. Northcote. 1958. Factors in lake typology in British Columbia, Canada. *Verh. Int. Ver. Limnol.* 13:252–263.

Leach, J. H., L. M. Dickie, B. J. Shuter, U. Borgmann, J. Hyman, and W. Lysack. 1987. A review of methods for prediction of potential fish production with application to the Great Lakes and Lake Winnipeg. *Can. J. Fish. Aquat. Sci.* 44(Suppl. 2):471–485.

Lewis, W. M. 1983. A revised classification of lakes based on mixing. *Can. J. Fish. Aquat. Sci.* 40:1779–1787.

Lueschow, L. A., J. M. Helm, D. R. Winter, and G. W. Karl. 1970. Trophic nature of selected Wisconsin lakes. *Wisc. Acad. Sci. Arts Lett.* 58:237–264.

Lundbeck, J. 1926. Die Bodentierweld Norddeutscher Seen. *Arch. Hydrobiol.* 7(Suppl.):1–473.

Lundbeck, J. 1936. Untersuchungen uber die Bodenbesidlung der Alpenrandseen. *Arch. Hydrobiol.* 10(Suppl.):208–358.

Margalef, R. 1958. "Trophic" typology versus biotic typology, as exemplified in the regional limnology of Northern Spain. *Verh. Int. Ver. Limnol.* 13:339–349.

Margalef, R. 1975. Typology of reservoirs. *Verh. Int. Ver. Limnol.* 19:1841–1848.

Marshall, T. R. and P. H. Ryan. 1987. Abundance patterns and community attributes of fishes relative to environmental gradients. *Can. J. Fish. Aquat. Sci.* 45(Suppl. 2):198–215.

Matuszek, J. E. 1978. Empirical predictions of fish yields of large North American lakes. *Trans. Am. Fish. Soc.* 107:385–394.

Michalski, M. F. P. and N. Conroy. 1972. Water Quality Evaluation for the Lake Alert Study. Water Quality Branch, Biology Section. Ontario Ministry of the Environment, Toronto.

Miyadi, D. 1933. Studies on the bottom fauna of Japanese lakes. X. Regional characteristics and a system of Japanese lakes based on the bottom fauna. *Jpn. J. Zool.* 4:417–437.

Moyle, J. B. 1945a. Classification of lake waters upon the basis of hardness. *Proc. Minn. Acad. Sci.* 13:8–12.

Moyle, J. B. 1945b. Some chemical factors influencing the distribution of aquatic plants in Minnesota. *Am. Midl. Nat.* 34:402–420.

Moyle, J. B. 1946. Some indices of lake productivity. *Trans. Am. Fish. Soc.* 76:322–334.

Moyle, J. B. 1956. Relationships between the chemistry of Minnesota surface waters and wildlife management. *J. Wildl. Manage.* 20:303–320.

Naumann, E. 1932. Grundzuge der regionalen Limnologie. *Binnengewasser.* 11:1–176.

Newton, M. E. and C. M. Fetterolf, Jr. 1966. Limnological Data from Ten Lakes, Genesee and Livingston Counties, Michigan. September 1965. Water Resources Commission, Bureau of Water Management, Michigan Department of Natural Resources, Lansing.

Nicholls, K. H. and P. J. Dillon. 1978. An evaluation of phosphorus-chlorophyll-phytoplankton relationships for lakes. *Int. Rev. Ges. Hydrobiol.* 63:141–154.

Nygaard, G. 1949. Hydrobiological study of some Danish ponds and lakes. II. The quotient hypothesis and some new or little known phytoplankton organisms. *Kgl. Danske. Videnskab. Selskab, Biol. Skr.* 7.

Odum, E. P. 1971. *Fundamentals of Ecology.* W.B. Saunders, Philadelphia.

Patterson, J. C., P. F. Hamblin, and J. Imberger. 1984. Classification and dynamics simulation of the vertical density structure of lakes. *Limnol. Oceanogr.* 29(4):845–861.

Pearsall, W. H. 1923. A theory of diatom periodicity. *J. Ecol.* 11:165–183.

Pennak, R. W. 1958. Regional lake typology in northern Colorado, USA. *Verh. Int. Ver. Limnol.* 13:264–283.

Pitblado, J. R., W. Keller, and N. I. Conroy. 1980. A classification and description of some northeastern Ontario lakes influenced by acid precipitation. *J. Great Lakes Res.* 6(3):247–257.

Prat, N. 1978. Benthos typology of Spanish reservoirs. *Verh. Int. Ver. Limnol.* 20:1647–1651.

Rai, H. and G. Hill. 1980. Classification of Central Amazon lakes on the basis of their microbiological and physico-chemical characteristics. *Hydrobiologia* 72(1–2):85–99.

Rao, S. S. and C. G. M. Berkel. 1978. Bacterial index ratio: a novel water quality assessment technique. *J. Great Lakes Res.* 4:106–109.

Rawson, D. S. 1939. Some physical and chemical factors in the metabolism of lakes. *Am. Assoc. Adv. Sci.* 10:9–26.

Rawson, D. S. 1952. Mean depth and fish production of large lakes. *Ecology* 33:513–521.

Rawson, D. S. 1955. Morphometry as a dominant factor in the productivity of large lakes. *Verh. Int. Ver. Limnol.* 12:164–175.

Rawson, D. S. 1956. Algal indicators of trophic lake types. *Limnol. Oceanogr.* 1:18–25.

Rawson, D. S. 1960. A limnological comparison of twelve large lakes in northern Saskatchewan. *Limnol. Oceanogr.* 5:195–211.

Rawson, D. S. 1961. A critical analysis of the limnological variables used in assessing the productivity of northern Saskatchewan lakes. *Verh. Int. Ver. Limnol.* 14:160–166.

Reynoldson, T. B. 1978. Observation on typology of some Alberta lakes with special reference to their oligochaete fauna. *Verh. Int. Ver. Limnol.* 20:190–191.

Rigler, F. H. 1975. Nutrient kinetics and the new typology. *Verh. Int. Ver. Limnol.* 19:197–210.

Rodhe, W. 1958. Primarproduktion und Seetypen. *Verh. Int. Ver. Limnol.* 13:121–141.

Rodhe, W. 1969. Crystallization of eutrophication concepts in northern Europe, in *Eutrophication: Causes, Consequences, Correctives,* National Academy of Sciences, Washington, D.C. pp. 50–64.

Rodhe, W. 1975. The SIL founders and our fundament. *Verh. Int. Ver. Limnol.* 19:16–25.

Rott, E. 1984. Phytoplankton as biological parameters for the trophic characterization of lakes. *Verh. Int. Ver. Limnol.* 22:1078–1085.

Round, F. E. 1958. Algal aspects of lake typology. *Verh. Int. Ver. Limnol.* 13:306–310.

Ryder, R. A. 1964. Chemical Characteristics of Ontario Lakes with Reference to a Method for Estimating Fish Production. Sect. Rep. (Fish) No. 48. Ontario Department of Lands and Forests, Toronto.

Ryder, R. A. 1965. A method for estimating the potential fish production of north-temperate lakes. *Trans. Am. Fish. Soc.* 94:214–218.

Ryder, R. A. 1982. The morphoedaphic index — use, abuse and fundamental concepts. *Trans. Am. Fish. Soc.* 111:154–164.

Ryder, R. A. and S. R. Kerr. 1990. Harmonic communities in aquatic ecosystems: a management perspective, in *Management of Freshwater Fisheries. Proc. EIFAC Symp. Goteborg, Sweden, May 31–June 3, 1988.* W. L. T. von Densen, B. Steinmetz, and R. H. Hughes, Eds. pp. 594–623.

Ryder, R. A., S. R. Kerr, K. H. Loftus, and H. A. Regier. 1974. The morphoedaphic index, a fish yield estimator—review and evaluation. *J. Fish. Res. Bd. Can.* 31:663–688.

Saether, O. A. 1975. Nearctic chironomids as indicators of lake typology. *Verh. Int. Ver. Limnol.* 19:3127–3133.

Schlesinger, D. A. and A. M. McCombie. 1983. An evaluation of climate, morphoedaphic and fisheries data as predictors of yields from Ontario sport fisheries. *Ont. Fish. Tech. Rep. Ser.* 10.

Schlesinger, D. A. and H. A. Regier. 1982. Climatic and morphoedaphic indices of fish yields from natural lakes. *Trans. Am. Fish. Soc.* 111:141–150.

Schneider, J. C. 1975. Typology and fisheries potential of Michigan lakes. *Mich. Acad.* 8:59–84.

Schneider, J. C. 1981. Fish Communities in Warmwater Lakes. Michigan Department of Natural Resources, Fisheries Res. Rep. 1980, Lansing.

Shannon, E. E. and P. O. Brezonik. 1972. Relationships between lake trophic state and nitrogen and phosphorus loading rates. *Environ. Sci. Technol.* 6:719–725.

Shapiro, J. 1975. The Current Status of Lake Trophic Indices — A Review. Interim Rep. No. 15. Limnological Research Center, University of Minnesota, Minneapolis.

Smith, R. C. and K. S. Baker. 1978. Optical classification of natural waters. *Limnol. Oceanogr.* 23:260–267.

Sprules, W. G. 1980. Zoogeographic patterns in the size structure of zooplankton communities, with possible applications to lake ecosystem modeling and management, in *Evolution and Ecology of Zooplankton Communities.* W. C. Kerfoot, Ed. University Press of New England, Hanover, NH. pp. 642–656.

Steedman, R. J. and H. A. Regier. 1987. Ecosystem science for the Great Lakes: perspectives on degradative and rehabilitative transformations. *Can. J. Fish. Aquat. Sci.* 44(Suppl. 2):95–103.

Stockner, J. G. 1971. Preliminary characterization of lakes in the experimental lakes area, northwestern Ontario, using diatom occurrence in sediments. *J. Fish. Res. Bd. Can.* 28:265–275.

Strahler, A. N. and A. H. Strahler. 1973. *Environmental Geoscience: Interaction between Natural Systems and Man.* John Wiley & Sons, New York.

Strom, K. M. 1931. Feforvatn. A Physiograpical and biological study of a mountain lake. *Arch. Hydrobiol.* 22:491–536.

Thienemann, A. 1909. Vorlaufige Mitteilung über Probleme Ziele der biologischen Erforschung der neun westfalischen Talsperren. *Ber. Versamml. Bot. Zool. Ver. Rheinl.-Westf.* pp. 101–108.

Thienemann, A. 1925. Die Binnengewasser Mitteleuropas: Eine Limnologische Einfuhrung. *Binnengewaesser.* 1.

Thienemann, A. 1928. Der Sauerstaff im Eutrophen und Oligotrophen Sen ein Beitrag zur Seetypenlehre. *Binnengewaesser.* 4.

Tonn, W. M. and J. J. Magnuson. 1982. Patterns in the species composition and richness of fish assemblages in northern Wisconsin lakes. *Ecology* 63:1149–1166.

Tonn, W. M., J. J. Magnuson, and A. M. Forbes. 1983. Community analysis in fishery management: an application with northern Wisconsin lakes. *Trans. Am. Fish. Soc.* 112:368–377.

U.S. Environmental Protection Agency. 1974. An Approach to a Relative Trophic Index System for Classifying Lakes and Reservoirs. Working Pap. No. 24. National Eutrophication Survey, Pacific Northwest Environmental Research Laboratory, Corvallis, OR.

Uttormark, P. D. and J. P. Wall. 1975. Lake Classification — A Trophic Characterization of Wisconsin Lakes. EPA-660/3-75-033. U.S. Environmental Protection Agency, Washington, D.C. p. 165.

Valle, K. J. 1927. Okologische-Limnologische Untersuchungen über die Boden und Tiefenfauna in einigen Seen nordliche vom Lagoda-See. *Acta Zool. Fenn.* 4:1–231.

Vallentyne, J. R., J. Shapiro, and A. M. Beeton. 1969. The process of eutrophication and criteria for trophic state determination. in *Modelling the Eutrophication Process. Proc. Workshop St. Petersburg, Forida, November 19 to 21, 1969.* pp. 57–67.

Vollenweider, R. A. 1968. Scientific Fundamentals of the Eutrophication of Lakes and Flowing Waters, with Particular Reference to Nitrogen and Phosphorus as Factors in Eutrophication. OECD Rep. DAS/SCI/ 68.27. U.N. Organization for Economic and Cultural Development. Paris.

Vollenweider, R. A., M. Munawar, and P. Stadlemann. 1974. A comparative review of phytoplankton and primary production in the Laurentian Great Lakes. *J. Fish. Res. Bd. Can.* 31:739–762.

Walker, W. W. 1979. Use of hypolimnetic oxygen depletion rate as a trophic state index for lakes. *Water Resour. Res.* 15:1463–1470.

Walker, K. F. and G. E. Likens. 1975. Meromixis and a reconsidered typology of lake circulation patterns. *Verh. Int. Ver. Limnol.* 19:442–458.

Warren, C. E. 1971. *Biology and Water Pollution Control.* W. B. Saunders. Philadelphia.

Weber, C. A. 1907. Aufbau und Vegetation der Moore Norddeutschlands. *Bot. Jahrb. Beibl.* 90:19–34.

Welch, D. M. 1978. Land/Water Classification. A Review of Water Classifications and Proposals for Water Integration into Ecological Land Classification. Ecol. Land Classif. Ser., Environment Canada. Ottawa.

Welch, P. S. 1948. *Limnological Methods.* Blakiston Company. Philadelphia.

Welch, E. B. and L. T. Lindell. 1978. Phosphorus loading and response in Lake Vanern nearshore areas. *Environ. Sci. Technol.* 12:321–327.

Wetzel, R. G. 1975. *Limnology.* W. B. Saunders. Philadelphia.

Whipple, G. C. 1898. Classification of lakes according to temperature. *Am. Nat.* 32(373):25–33.

Williams, W. D. 1964. A contribution to lake typology in Victoria, Australia. *Verh. Int. Ver. Limnol.* 15:158–168.

Winberg, G. G. 1961. Modern conditions and problems in the study of primary production of waters. *Ref. Zh. Biol. Minsk.* 22 Zh 329:11–24.

Witzig, A. S. and C. A. Whitehurst. 1981. Current use and technology of Landsat MSS data for lake trophic classification. *Water Resour. Bull.* 17(6):962–970.

Zafar, A. R. 1959. Taxonomy of lakes. *Hydrobiologia* 13:287–299.

Zimmermman, A. P., K. M. Noble, M. A. Gates, and J. E. Paloheimo. 1983. Physicochemical typologies of south-central Ontario lakes. *Can. J. Fish. Aquat. Sci.* 40:1788–1803.

Zumberge, J. H. 1952. *The Lakes of Minnesota: Their Origin and Classification.* University of Minnesota Press, Minneapolis.

Chapter 4

GREAT LAKES AQUATIC HABITAT CLASSIFICATION BASED ON WETLAND CLASSIFICATION SYSTEMS

P. M. McKee, T. R. Batterson, T. E. Dahl, V. Glooschenko, E. Jaworski, J. B. Pearce, C. N. Raphael, T. H. Whillans, and E. T. LaRoe

TABLE OF CONTENTS

ABSTRACT

A review of wetland classification systems was carried out to identify attributes of the systems that may be appropriate in classifying wetlands and other aquatic habitats in the Great Lakes, and to suggest means of developing an aquatic habitat classification system for the Great Lakes with a focus on fisheries. The Ontario Ministry of Natural Resources/ Canadian Wildlife Service system classifies wetlands based on a variety of biological, social, hydrological, and special features components. This system is neither readily disaggregated nor modified, and the specific habitat evaluation system incorporated within this classification is judged to be inappropriate for the diverse habitat management needs that may exist in the various jurisdictions of the Great Lakes basin. A classification system for Great Lakes aquatic habitats should have sufficient structural flexibility in order to permit the addition of a separate habitat evaluation system. Geomorphic classifications applied to Great Lakes coastal wetlands are based on erosional and depositional features of shorelines and river mouths that are characteristic of the Great Lakes, but are generally not taken into consideration in other wetlands classifications. The U.S. Fish and Wildlife (FWS) classification for wetlands and deepwater habitats is organized within an hierarchical framework that may be readily disaggregated and modified for revised application. Although lacking sufficient regional resolution to provide an optimal classification system for Great Lakes aquatic habitats, the FWS system does incorporate a benthic habitat classification that is not included in other wetland classifications. The FWS classification was judged to provide a useful starting point for the development of a Great Lakes aquatic habitat classification system because its hierarchical organization permits the addition of new classification components and because both benthic and wetland classifications are already in place. Development of a Great Lakes aquatic habitat classification, based on the FWS framework, would entail the elimination of strictly marine habitat components, the addition of pelagic habitats, and the incorporation of new habitat classification categories at lower hierarchical levels. These new additions might include categories for Great Lakes ecoregions, water depth, geomorphic features, human modification, temperature-oxygen regimes, nutrient regimes, measures of fish community structure, stream order, and stream gradient.

I. INTRODUCTION

A primary goal of wetland classification "is to impose boundaries on natural ecosystems for the purposes of inventory, evaluation and management" (Cowardin et al., 1979). The broad objectives of most wetland classification systems are to describe ecological units having certain homogeneous natural qualities, to arrange these units into systems useful for wetlands management, to provide useful units for inventory and mapping, and to establish uniformity in wetland habitat taxonomy. For the most part, wetland classifications have been developed primarily for waterfowl and soil management and secondarily for management of other resources such as fish, furbearers, wild rice, forests, and water resources.

Wetlands are typically described as areas that are transitional between terrestrial and aquatic systems, have a water table close to or at the surface, or are covered either seasonally or permanently by shallow water. Wetlands are typically dominated by hydrophytes and the substrate usually consists of hydric soils, although not all definitions of wetlands include a requirement for hydric soils or hydrophytes.

Many systems have been developed for classifying wetland habitats and shallow water ecosystems. These have been developed by wetland scientists and environmental managers, and may be regional or national in scope. Wetlands classification has its roots in classification of peatlands in northern Europe and North America beginning in the early part of the 20th century (Mitsch and Gosselink, 1986). The first national wetlands classification system in the U.S., known as the Circular 39 Classification (Shaw and Fredine, 1956), described 20 types of wetlands under 4 major categories: inland freshwater areas, inland saline areas, coastal freshwater areas, and coastal saline areas. This system relied primarily on vegetation form and depth of flooding to identify wetland type. It was widely applied until the adoption of the present FWS system in 1979 (Cowardin et al., 1979). The Canadian Wetland Registry (Tarnocai, 1979) has developed a national wetland classification system based largely on the work of Jeglum et al. (1974), Tarnocai (1970 and 1974), Zoltai et al. (1975), and Zoltai and Tarnocai (1975). This system is hierarchical in structure and classification is carried out on two levels within five wetland classes: bog, fen, marsh, swamp, and shallow water. Some other wetland classification systems include those dealing with coastal wetlands (Odum et al., 1974), Canadian peatlands (Tarnocai, 1980), hydrodynamic energy gradients (Gosselink and Turner, 1978), as well as those used by state and provincial jurisdictions (e.g., Wharton et al., 1976; Pollett, 1979).

No wetlands classification system has been developed specifically for Great Lakes wetlands, although a great deal of work has been done in the inventory and classification of wetlands in the Great Lakes basin, most notably by FWS and the Ontario Ministry of Natural Resources (OMNR). Difficulties in developing a uniform system can be traced to the varying wetland management objectives peculiar to the different regional and national jurisdictions that border the Great Lakes.

Wetlands of the coastlines and connecting channels of the Great Lakes share certain features that are somewhat unique among freshwater wetlands in North America. Great Lakes coastal wetlands are strongly influenced by natural cyclic and noncyclic changes in water level and by the effects of storm surge and water currents. The connecting channels of the Great Lakes may be more aptly described as straits rather than as true rivers, and the wetlands of these straits are affected by depositional patterns and discharge regimes that are unique. The structure and composition of aquatic plant communities in these wetlands are, in turn, controlled by the dynamic physical processes of current action, water level fluctuation, sediment transport, and deposition.

The authors' objectives are to: (1) review a variety of wetland classification systems for possible application or adaptation to the Great Lakes, (2) summarize the various bases for classification, and (3) identify the attributes of these classification schemes that would be useful in a Great Lakes habitat classification system. The focus is on three established classification systems having a history of application in the Great Lakes region. The systems reviewed include a national system, a system applied within regional political boundaries, and a system developed to describe Great Lakes coastal wetlands.

II. SELECTED WETLAND CLASSIFICATION SYSTEMS

A. AN EVALUATION SYSTEM FOR THE WETLANDS OF ONTARIO

The OMNR developed a system for evaluating wetlands in southern Ontario (OMNR/CWS, 1984). This system is designed to measure wetland values using a scoring system for biological, social, hydrological, and special attributes of wetlands and is intended for use in land use planning and habitat management (Figure 1). The OMNR system is a combined wetland classification and evaluation system. Wetland values assigned in the OMNR system are based on worth in terms of wildlife habitat and their value to naturalist groups including recreational users, educators, scientists, local residents, etc.

The OMNR system defines wetlands as "lands that are seasonally or permanently covered by shallow water as well as lands where the water table is close to or at the surface; in either

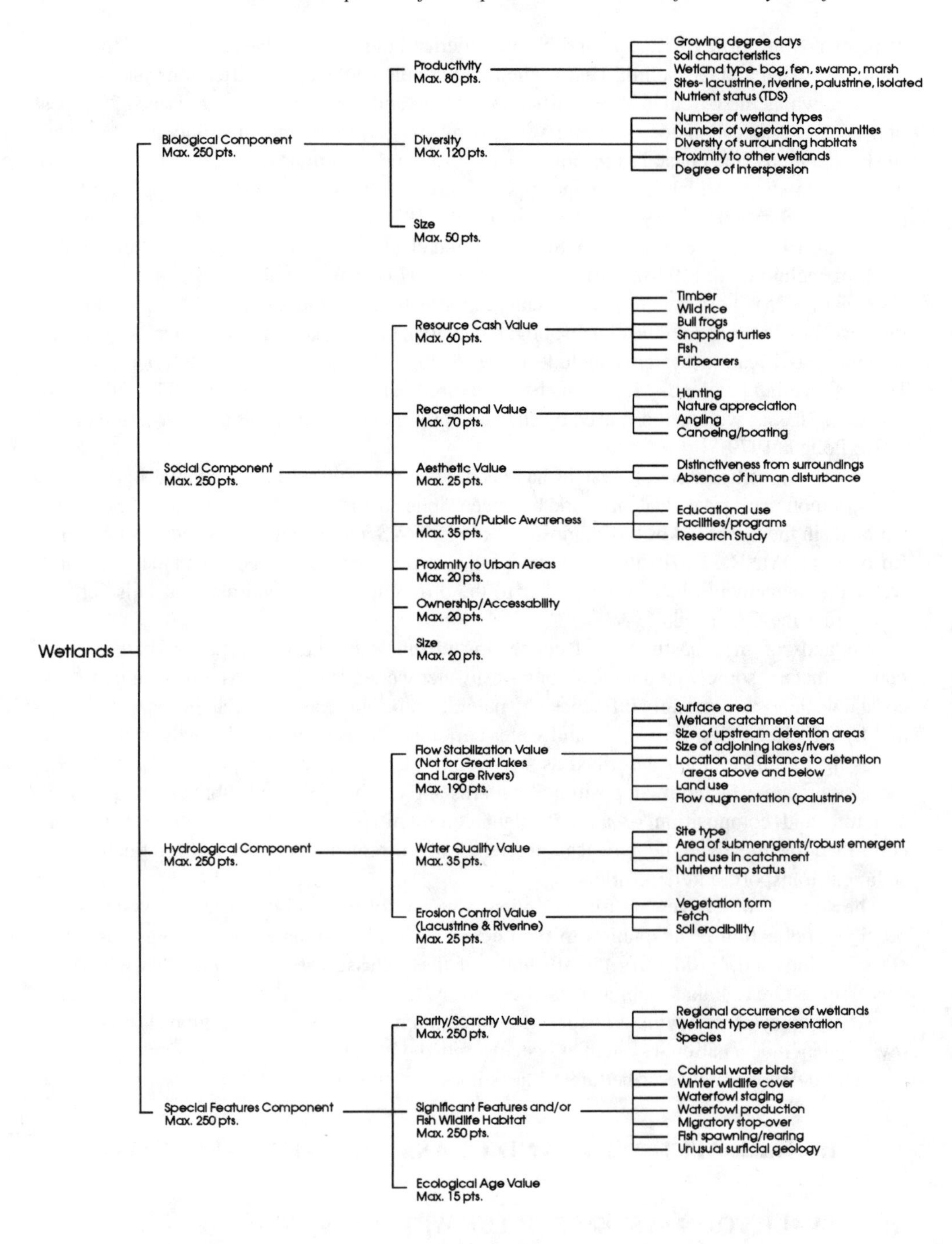

FIGURE 1. Habitat evaluation system for southern Ontario wetlands. (From OMNR/CWS, 1984.)

case, the presence of abundant water has caused the formation of hydric soils and have favored the dominance of either hydrophytic or water tolerant plants." Wetland boundaries are drawn at the deepwater and upland limits of dominance by wetland plants. The minimum wetland area considered is 2 ha. Wetlands may be defined as distinct units or as complexes of units located in close proximity to one another.

Each of the four major components of the OMNR classification system is scored out of a maximum of 250 points. Wetlands are ranked into classes (from 1 to 7) based on the combined score from each of the biological, social, hydrological, and special features components.

The biological component is subdivided into evaluation categories for productivity, diversity, and size. Productivity values are based on climate (degree-days), soil types (organic, inorganic, or undesignated), wetland types (bog, fen, swamp, or marsh), sites (lacustrine, riverine, palustrine, or isolated), and nutrient status (based on total dissolved solids, TDS). The highest score is given to those wetlands exhibiting characteristics of habitats supporting high primary production. Diversity values are based on the number of wetland types (bog, fen, swamp, marsh) within a wetland complex and the number of vegetational structure zones present (e.g., trees living and dead, mosses, narrow and broadleaf emergents, free-floating plants, etc.). Also considered are the diversity of the surrounding plant habitat, proximity to other wetlands, the degree of interspersion between wetland vegetation zones, and the amount and distribution of open water area within the wetland. Size value is determined on the basis of wetland surface area, and is weighed by the wetland diversity score. In this manner, a small but diverse wetland may score higher than a large, low-diversity wetland. In general, however, high scoring wetlands are larger in size (Glooschenko et al., 1987).

The social value component of the OMNR system is subdivided into seven categories. The resource cash value category is scored on the basis of the presence and abundance of harvestable renewable natural resources (timber, wild rice, bull frogs, snapping turtles, fish, furbearers). Scoring in the recreational value category is based on the type and intensity of use for recreation (hunting, fishing, hiking, etc.). Scoring for the remaining five categories within the social component is based on aesthetic values, uses for education or research, proximity to urban areas, ownership/accessibility, and size.

The hydrological component of the evaluation system consists of three categories: flow stabilization, water quality, and erosion control. Most of the total score for the hydrological component is assigned for flow stabilization, which is based on wetland area, wetland catchment, relative position in a watershed, and function in streamflow augmentation. OMNR/CWS (1984) acknowledges that flow stabilization is an insignificant function of wetlands in Great Lakes coastal areas or in very large rivers. The score for control of water quality is based on wetland function in phosphorus and nitrogen accumulation, on the type of wetland site (palustrine, isolated, riverine, or lacustrine), and on land use practices in the wetland catchment. Erosion control value is scored on the basis of wetland vegetation form, fetch, and soil erodibility.

The fourth component of the OMNR system is for special features and consists of categories for rarity or scarcity, significant features and/or fish and wildlife habitat, and ecological age. Rarity/scarcity values are assigned on the basis of regional abundance of similar habitats and on the presence of regionally rare plant and animal species. Significant features and fish and wildlife habitat values are assessed on the basis of the occurrence of regionally unusual or unique geological features, and the quality and quantity of habitat provided to fish and wildlife. Ecological age is scored on the basis of wetland type (bog, fen, swamp, marsh) and the time scale that is required for each habitat type to form.

B. CLASSIFICATION OF WETLANDS AND DEEPWATER HABITATS OF THE U.S.

FWS initiated development of a wetland classification system for nationwide application in 1974. The system was developed, refined, and field tested over a 5-year period and was published in 1979 as a "Classification of Wetlands and Deepwater Habitats of the United States" (Cowardin et al., 1979). Because wetlands are continuous with deepwater habitats, both categories are addressed in the FWS classification; thus, the system is designed for classification of a wide range of continental aquatic and semiaquatic habitats. Between 1979 and late 1991, the FWS classification has been used to inventory 55% of the wetland ecosystems within the lower 48 states, including 95% of the Great Lakes coastal zone located within the U.S.

The purpose of the classification system is to:

1. Describe ecological units having certain common natural attributes
2. Arrange these units in a system that will facilitate resource management decisions
3. Furnish units for inventory and mapping
4. Provide a nationwide uniformity in wetland concepts and terminology

The classification system that was developed defines wetlands according to fundamental ecological characteristics rather than according to the occurrence of specific biotic components of management interest. It does not define the limits of proprietary jurisdiction of any federal, state, or local government, nor does it establish the geographical scope of the regulatory programs of governmental agencies. Unlike the OMNR system, the FWS system does not incorporate a habitat evaluation component, although development of a separate wetland evaluation technique (WET) is underway.

A wetland must have enough water at some time during the growing season to cause physiological problems for plants and/or animals not adapted to live in water or in saturated soils. No minimum mapping unit exists for wetlands under the FWS classification, although it is generally assumed that wetlands 1 ha and larger have been mapped. The most precise wetland definition would be based on hydrology, but that would require detailed measurements over periods of time to determine if an area was indeed a wetland. Plants and soils do provide clues as to the hydrologic history of a site. Both wetland plants and hydric soils are usually present in most wetlands. However, few plants are found in wetlands that are subject to drastic fluctuations in water level, wave action, turbidity, or high concentrations of salts, and hydric soils are not found on rocky shores, rock bottom, streambeds, etc. For these reasons, FWS defines wetlands as

> "lands transitional between terrestrial and aquatic systems where the water table is usually at or near the surface or the land is covered by shallow water. For purposes of this classification a wetland must have one or more of the following three attributes: (1) at least periodically, the land supports predominantly hydrophytes, (2) the substrate is predominantly undrained hydric soil, and (3) the substrate is nonsoil and is saturated with water or covered by shallow water at some time during the growing season of each year" (Cowardin et al., 1979)

The FWS classification system is hierarchical, with wetlands assigned to five major systems at the broadest level: marine, estuarine, riverine, lacustrine, and palustrine. Systems are further subdivided into subsystems that reflect hydrologic conditions (e.g., littoral vs. limnetic in the lacustrine system); the palustrine system is not divided into subsystems. Below the subsystem level is the class level, which describes the appearance of the wetland in terms of vegetation (e.g., emergent, aquatic bed, forested) or substrate, where vegetation is inconspicuous or absent (e.g., unconsolidated shore, rocky shore, streambed). A depiction of the FWS classification system to the class level is given in Figure 2. Each class can be further subdivided into subclasses and subclasses into dominance types. The classification also includes modifiers to describe hydrology (water regime), water chemistry (pH, salinity, and halinity), and special modifiers relating to human activities (e.g., impounded, partly drained, farmed, artificial). The FWS system can also be applied with biogeographic modifiers such as Bailey's ecoregions (Bailey, 1978).

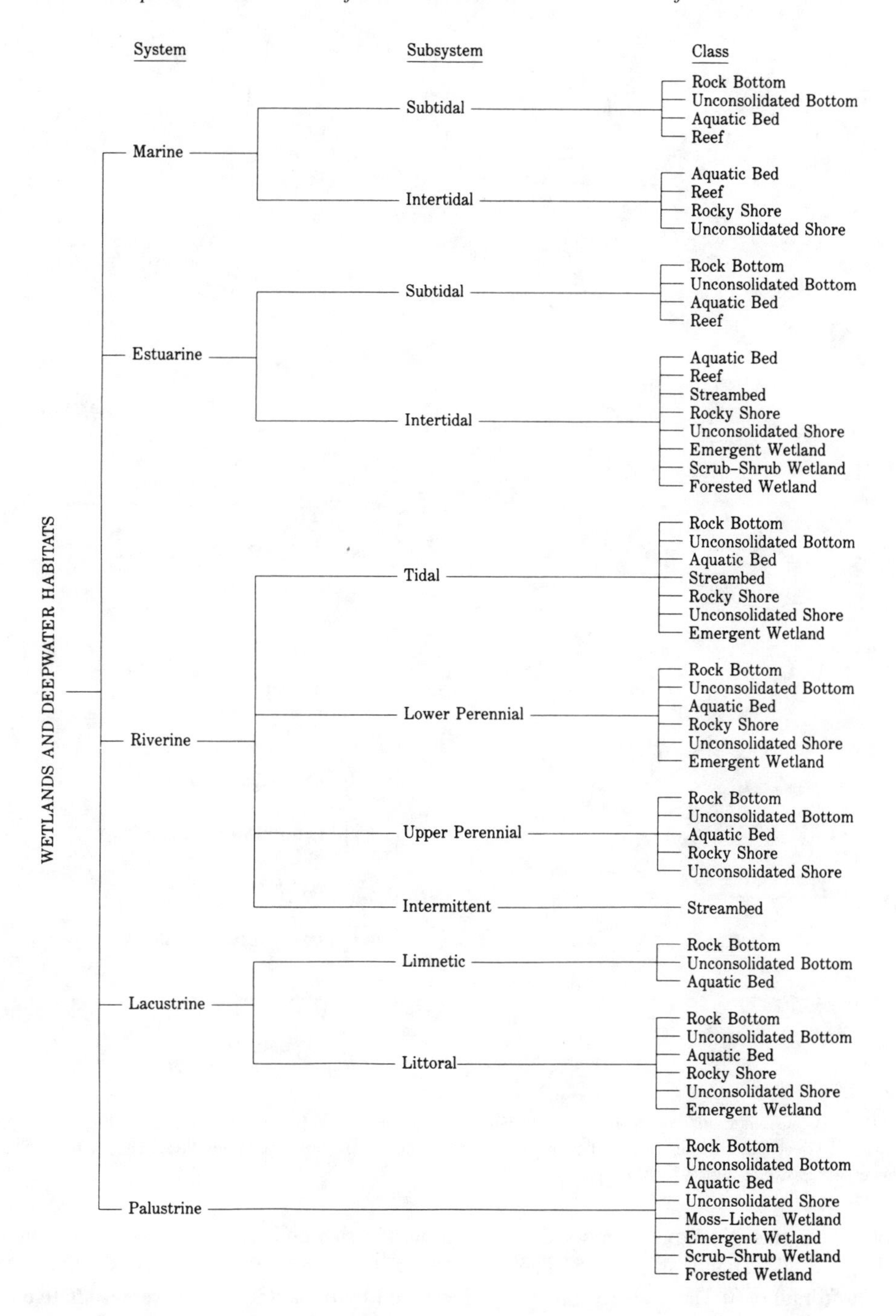

FIGURE 2. Classification hierarchy of wetlands and deepwater habitats, showing systems and classes. The palustrine system does not include deepwater habitats. (From Cowardin, L. M. et al., 1979.)

C. GEOMORPHIC CLASSIFICATION OF GREAT LAKES COASTAL WETLANDS

Jaworski and Raphael (1979) described several coastal Great Lakes wetlands on the basis of geomorphic structure. This system views wetlands as basins occupied by plant communities. Wetland boundaries are to a large extent dictated by landform. These simple "geomorphic

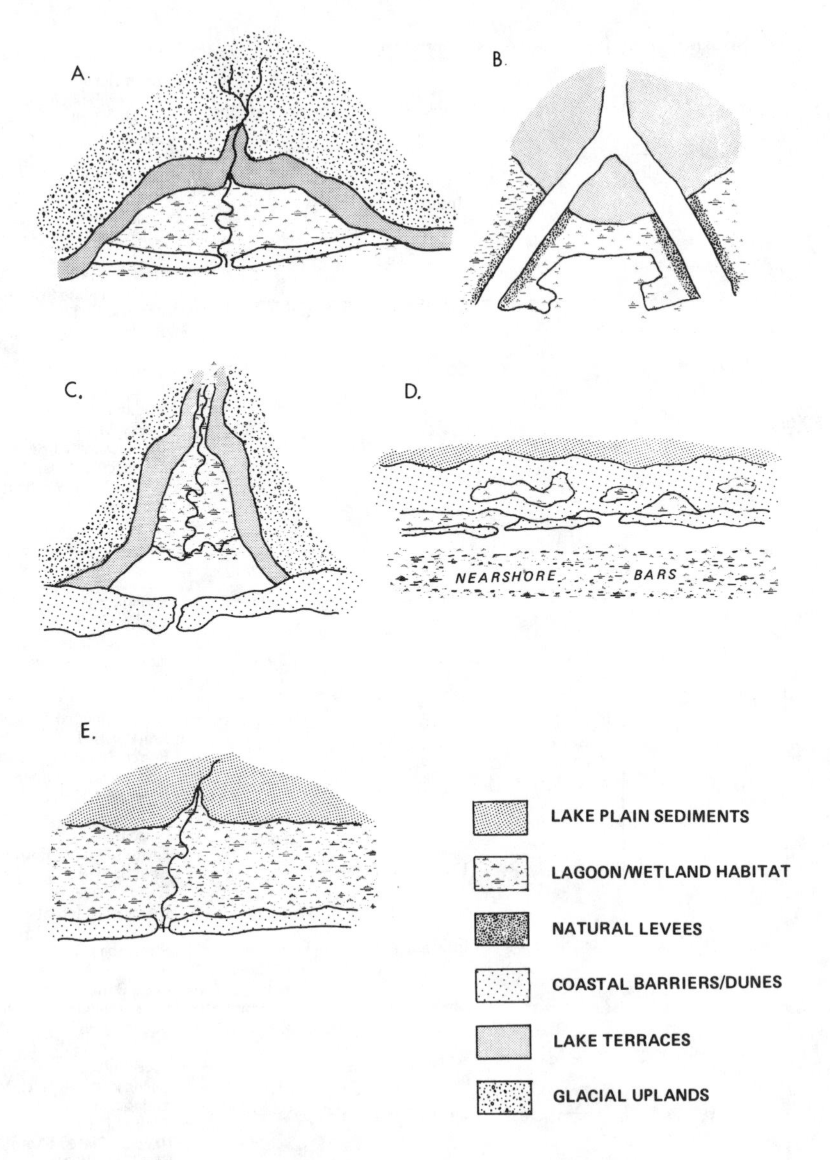

FIGURE 3. Geomorphic models of coastal wetlands in the Great Lakes. (From Jaworski, E. and Raphael, C. N., Impact of Great Lakes Water Level Changes on Coastal Wetlands, Institute of Water Research, Michigan State University, East Lansing, 1979. With permission.)

models" or classifications were used in describing the fish and wildlife resources of Great Lakes coastal wetlands in the U.S. The five fundamental geomorphic settings described are shown in Figure 3. These structural forms are judged by the authors to be representative of the most common coastal wetland morphologies found in the Great Lakes.

Figure 3A illustrates a sheltered lagoon bounded by a sand barrier on the lakeward side and by upland on its flanks. This configuration appears to occur along shorelines having high wave energies, such as those of Lake Michigan. Marsh plants develop in the lagoon most extensively when wave energies are low or when water levels fall. The sand barrier is formed from nearshore currents transporting littoral sediments along the shoreline.

The bird foot delta of Figure 3B forms in places where sediments are deposited by a river along the banks of distributary channels. Marshy areas develop in protected waters around these estuarine structures. In the Great Lakes, the St. Clair delta is the best-known example of this type of structure.

The coastal wetland form shown in Figure 3C is a flooded river estuary. These environments form in deep river valleys excavated during periods of low water level. Sand barriers are typically deposited by nearshore currents to form protective structures across the mouth, favoring the development of wetland vegetation within the estuary.

Wetlands formed along shoreline complexes of the type shown in Figure 3D are typically found along hummocky shorelines where nearshore sandbars provide protective sandbars. Wetland vegetation develops in depressions in the hummocky landscape and in the nearshore zone, as is found in Saginaw Bay.

The wetland form shown in Figure 3E is relatively common in the Great Lakes. It is characterized by an upland boundary, often a glacial lakeshore terrace, and a coastal barrier such as a spit, beach ridge, or transgressive sand barrier. Examples include Long Point, Ontario and wetlands at Erie, PA.

III. DISCUSSION

Each of the three classification systems reviewed has its own strengths and weaknesses in terms of its applicability to the classification of Great Lakes aquatic habitats. The FWS system is broader than the OMNR or the geomorphic systems in that in its present form it can be used to classify deepwater benthic habitats as well as the wetlands associated with the Great Lakes.

The framework of the three classifications vary considerably. The FWS classification system is open ended and flexible. It can be readily adapted and applied without major restructuring of the basic framework. The hierarchical framework of the FWS system permits classification at higher or lower levels of taxonomic precision, as well as disaggregation and restructuring to suit the needs of the habitat manager. The OMNR system is close ended and its application requires the completion of all steps of the evaluation process. Thus, the OMNR framework is less flexible than the FWS framework, although it does incorporate an evaluation scheme for wetlands that is lacking in the other classification systems. The geomorphic classification system is simple and is based on broad measures of the physical features of Great Lakes coastal wetlands. This latter "system" is less elaborate than either of the former two systems, but does provide examples of geomorphic classification, which is lacking within the frameworks of the other two systems. It is particularly important in explaining some peculiar features of Great Lakes wetlands.

The inventory procedures required in applying the three classification systems vary. In the OMNR system, considerable field work and consultation with experts and public interest groups are required for successful application. Wetland boundaries and zones are also delimited with the aid of topographic maps and aerial photographs. In the FWS system, habitat management needs determine the level of detail required in data collection, and the system may be applied at the subsystem or class level solely through the use of topographic maps and aerial photos or satellite images. The simpler geomorphic classifications can also be made using aerial photographs, topographic maps, and satellite images, with the need for only limited field verification. Applications of remote sensing analyses of wetlands are given by Roller (1977), Lyon (1979), and Lyon and Drobney (1984).

The degree of habitat description in terms of taxonomy and forcing functions varies greatly in these three systems. The OMNR classification system incorporates a detailed examination of measures of ecosystem structure and function in its biological and special features component, and classifies both biotic and abiotic habitat parameters. Forcing functions include temperature (degree-days) and nutrient status. The social component of the OMNR system is unique among the three systems and relates to the societal value of wetland resources rather than to habitat structure. Habitat evaluation, in this sense, is a management procedure that should not be directly integrated with the classification framework for Great Lakes wetlands because management objectives and social values will vary strongly from jurisdiction to jurisdiction and with time. However, the ability to link separate habitat evaluation systems with the classification system will be crucial, so that management of habitats for various purposes (e.g., fisheries, waterfowl, water retention, water quality, etc.) can be facilitated. The OMNR system is restricted to wetlands as defined in the traditional sense. It does not accommodate the effects of habitat alteration by humans and is not useful for describing other aquatic habitats in the Great Lakes. The final result of an OMNR classification is a numeric classification (classes 1 to 7) based on the total habitat score and cannot be used to describe biotic or abiotic habitat features.

The FWS classification, like the OMNR system, also incorporates a considerable number of biotic and abiotic habitat parameters within its framework. At the system level, the FWS classification splits wetlands into five wetland systems. Only three of these, riverine, lacustrine, and palustrine are applicable to freshwater habitats in the Great Lakes basin. The degree of habitat splitting by the FWS system increases with progression from system to subsystem and so on down the hierarchy. Because of its hierarchical structure, this system permits a greater range of biotic and abiotic habitat parameters. The FWS system also incorporates important forcing functions at or below the class level in the form of measures of water regime permanence. Modifiers that have been used by the FWS system also address habitat alteration by human intervention in the forms of excavation, diking, impoundment, drainage, farming, artificial wetland development, and spoils disposal. Many Great Lakes wetlands, particularly in the lower lakes, have been altered by human intervention; thus, provision for habitat alteration within a Great Lakes habitat classification system is seen as a necessity. The Great Lakes habitats that may be classified by the existing FWS system include all riverine habitats, coastal wetland habitats, and offshore benthic habitats. However, detailed classifications within the class, subclass, or modifier levels of the system are appropriate only for benthic or wetland habitats. They provide little or no descriptive resolution of open water, limnetic, or riverine habitat parameters such as thermal stratification patterns or stream gradients that are important in controlling fish distribution and abundance. Therefore, the FWS system could not be used in its present form as a classification scheme for all of the aquatic habitats in the Great Lakes. Its hierarchical structure, however, would allow the addition of new or modified habitat classifications without a major change in the existing classification scheme.

Geomorphic habitat classification provides information only on the physical structure of coastal wetland habitats and provides no information on the biotic features. However, physical structure may be important in determining the composition of the fish community and providing useful information, particularly to habitat managers concerned with nearshore currents and beach-building processes. This classification does provide clues as to the overall stability of wetlands. For example, delta wetlands will not be destroyed by lake level changes due to constant nourishment of the system by fresh sediments. On the other hand, wetlands sheltered by transgressive barriers formed by coastal erosion are more subject to erosion or displacement occurring from changes in lake level. Geomorphic features are distinctive features of Great Lakes coastal wetlands and therefore warrant attention in habitat classification.

In the last analysis, several criteria are fundamental in the development of a successful Great Lakes aquatic habitat classification system:

1. The system must be simple enough so that it can be mastered and easily applied by environmental managers and scientists.
2. The system must be able to stand alone, independent of the tools of inventory.

TABLE 1
Potential Classification for Great Lakes Aquatic Habitats

System	Subsystem	Class	Subclass	Modifiers
Lacustrine	Littoral	*	*	Great Lakes ecoregions Water level regimes Fish community structure Geomorphic structure Human modification
	Limnetic-benthic	(As in Limnetic subsystem of Cowardin et al.)		Great Lakes ecoregions Fish community structure Temperature-oxygen regimes Water depth TDS/phosphorus Human modification
	Limnetic-pelagic	Unstratified	—	Great Lakes ecoregions Fish community structure Temperature regimes
		Stratified	Epilimnion	Great Lakes ecoregions Temperature-oxygen regimes Fish community structure Depth TDS/phosphorus
			Metalimnion	Great Lakes ecoregions Temperature-oxygen regimes Fish community structure Depth
			Hypolimnion	Great Lakes ecoregions Temperature-oxygen regimes Depth Fish community structure
Riverine	*	*	*	Great Lakes ecoregions Water level regimes Fish community structure Human modification Temperature regimes Velocity regimes or gradients Stream order Conductivity Turbidity
Palustrine	*	*	*	Great Lakes ecoregions Fish community structure Human modification

Note: Asterisks indicate classification as given by Cowardin et al. (1979).

Data based on or modified from Cowardin et al. (1979).

REFERENCES

Bailey, R. G. 1978. Ecoregions of the United States. U.S. Forest Service, Intermountain Region, Ogden, UT.

Balon, E. K. 1975. Reproductive guilds of fishes: a proposal and definition. *J. Fish. Res. Bd. Can.* 32:821–864.

Cowardin, L. M., V. Carter, F. C. Golet, and E. T. LaRoe. 1979. Classification of Wetlands and Deepwater Habitats of the United States. FWS/OBS-79/31. U.S. Fish and Wildlife Service. Washington, D.C.

Geis, J. W. and J. L. Kee. 1977. *Coastal Wetlands Along Lake Ontario and the St. Lawrence River in Jefferson County, New York.* State University of New York. Syracuse.

Glooschenko, V., J. H. Archbold, and D. Herman. 1987. The Ontario wetland evaluation system: replicability and bird habitat selection, in *Proc. Int. Symp. Ecology and Management of Wetlands.* Croom Helm. Beckenham, Kent, U.K.

Gosselink, J. G. and R. E. Turner. 1978. The role of hydrology in freshwater wetland ecosystems, in *Freshwater Wetlands: Ecological Processes and Management Potential.* R. E. Good, D. F. Wingham, and R. L. Simpson, Eds., Academic Press. New York. pp. 63–78.

Herdendorf, C. E. 1987. The Ecology of the Coastal Marshes of Western Lake Erie: A Community Profile. Biol. Rep. 85(7.9). U.S. Fish and Wildlife Service. Washington, D.C.

Herdendorf, C. E., S. M. Hartley, and M. D. Barnes, Eds. 1981. Fish and Wildlife Resources of the Great Lakes Coastal Wetlands Within the United States. Vol. 1. Overview. FWS/OBS-81/02-VL. U.S. Fish and Wildlife Service. Washington, D.C.

Jaworski, E. and C. N. Raphael. 1979. *Impact of Great Lakes Water Level Changes on Coastal Wetlands.* Institute of Water Research, Michigan State University, East Lansing.

Jeglum, J. K., A. N. Boissoneau, and V. G. Haavisto. 1974. Toward a Wetland Classification System for Ontario. Info. Rep. 0-X-215. Canadian Forestry Service, Canadian Department of the Environment. Ottawa.

Lyon, J. G. 1979. Remote sensing analysis of coastal wetland characteristics: the St. Clair Flats, Michigan. Paper presented at 13th Int. Symp. Remote Sensing of the Environment. Enviromental Research Institute, Ann Arbor, MI. pp. 1117–1129.

Lyon, J. G. and R. D. Drobney. 1984. Lake level effects as measured from aerial photos. *J. Surv. Eng.* 110:103–111.

Mitsch, W. J. and J. G. Gosselink. 1986. *Wetlands.* Van Nostrand Reinhold. New York.

Odum, H. T., B. J. Copeland, and A. E. McMahan, Eds. 1974. *Coastal Ecological Systems of the United States.* Vol. 1 to 4. The Conservation Foundation, Washington, D.C.

Ontario Ministry of Natural Resources/Canadian Wildlife Service (OMNR/CWS). 1984. An Evaluation System for Wetlands of Ontario South of the Precambrian Shield. 2nd ed. Wildlife Branch, OMNR/CWS, Environment Canada. Ottawa.

Patch, S. P. and D. Busch. 1984. The St. Lawrence River — Past and Present. U.S.F.W.S. Spec. Rep. Buffalo District. U.S. Army Corps of Engineers. Buffalo, NY.

Pollett, F. C. 1979. Report on wetland activities across Canada, in *Proc. Workshop on Canadian Wetlands, Saskatoon, Saskatchewan, June 11 to 13, 1979.* C. D. A. Rubec and F. C. Pollett, Eds., Lands Directorate, Ecological Land Classification Ser. No. 12. Environment Canada. Ottawa. pp. 69–77.

Roller, N. G. 1977. *Remote Sensing of Wetlands.* Environmental Research Institute of Michigan, Tech. Rep. No. 193400-14-T, Ann Arbor.

Ryder, R. A. 1965. A method for estimating the potential fish production of north-temperate lakes. *Trans. Am. Fish. Soc.* 94:214–218.

Shaw, S. P. and C. G. Fredine. 1956. Wetlands of the United States, Their Extent and Their Value for Waterfowl and Other Wildlife. Circular 39. U.S. Department of the Interior, Fish and Wildlife Service. Washington, D.C.

Tarnocai, C. 1970. Classification of Peat Landforms in Manitoba. Canadian Department of Agriculture, Research Station, Pedology Unit, Winnipeg.

Tarnocai, C. 1974. Peat landforms and associated vegetation in Manitoba. in *Proc. Can. Soil Survey Committee Organic Soil Mapping Workshop, Winnipeg.* J. H. Day, Ed., Soil Research Institute. Ottawa. pp. 3–20.

Tarnocai, C. 1979. Canadian Wetland Registry. in *Proc. Workshop on Canadian Wetlands, Saskatoon, Saskatchewan, June 11 to 13, 1979.* C. D. A. Rubec and F. C. Pollett, Eds. Lands Directorate, Ecological Land Classification Ser. No. 12. Environment Canada. Ottawa. pp 9–38.

Tarnocai, C. 1980. Development, age and classification of Canadian peatlands. in Proc. Workshop on an Organic Soil and Mapping Interpretation in Newfoundland. Land Resources Research Institute, Research Branch. Agriculture Canada. Ottawa.

Terrell, J. W., T. E. McMahon, P. D. Inskip, R. F. Raleigh, and K. L. Williamson. 1982. Habitat Suitability Index Models. Appendix A. Guidelines for Riverine and Lacustrine Applications of Fish HSI Models with the Habitat Evaluation Procedures. FWS/OBS/10a. U.S. Fish and Wildlife Service. Washington, D.C.

Wharton, C. H., H. T. Odum, K. Ewel, M. Duever, A. Largo, R. Boyt, J. Bartholomew, E. DeBellevue, S. Brown, M. Brown, and L. Duever. 1976. *Forested Wetlands of Florida — Their Management and Use.* Center for Wetlands, University of Florida Press, Gainesville.

Zoltai, S. C. and C. Tarnocai. 1975. Perennially frozen peatlands in the western arctic and subarctic of Canada. *Can. J. Earth Sci.* 12:28–43.

Zoltai, S. C., F. C. Pollett, J. K. Jeglum, and G. D. Adams. 1975. Developing a wetland classification for Canada, in *Forest Soils and Forest Land Management, Proc. 4th North Am. Forest Soils Conference,* B. Bernier and C. H. Winget, Eds. Les Presses de l'Universite Laval, Quebec. pp. 497–511.

Chapter 5

REVIEW OF HABITAT CLASSIFICATION SCHEMES APPROPRIATE TO STREAMS, RIVERS, AND CONNECTING CHANNELS IN THE GREAT LAKES DRAINAGE BASIN

P. L. Hudson, R. W. Griffiths, and T. J. Wheaton

TABLE OF CONTENTS

ABSTRACT

Studies of lotic classification, zonation, and distribution carried out since the turn of the century were reviewed for their use in developing a habitat classification scheme for flowing water in the Great Lakes drainage basin. Seventy papers, dealing mainly with fish but including benthos, were organized into four somewhat distinct groups. The largest group of papers deals with longitudinal replacement or addition of species from headwater to lower river reaches. The second group, dealing with large rivers, emphasizes lateral zonation. Another scheme categorizes habitat regionally on the basis of climatic and terrestrial characteristics of the watershed or region without regard to longitudinal differences within the watershed. In the last and most recent group of studies (1973 to 1988) multivariate statistics were used to develop relations between habitat variables and fish biomass or occurrence. A total of 92 habitat attributes were summarized in these studies, 49 of which were common and fell into four categories: a biological group, a chemical group, and two geomorphological groups — one associated with a longitudinal succession (e.g., stream order, flow, volume) and the other independent of river length (e.g., gradient, velocity, substrate). Overall geomorphological attributes made up 85% of the most important descriptors, among which position-independent and -dependent factors were evenly split. In addition, a hierarchical scale of habitat measurements is suggested, and sources of data and inventory methods, including statistical treatment, are reviewed. Finally, an outline is provided for developing a classification system for riverine habitat in the Great Lakes drainage basin.

I. INTRODUCTION

The distribution and abundance of biota in streams and rivers are controlled by the many chemical and physical properties that change along the courses of the streams. The resulting biotic zonation has led to many river classification schemes in which scientists have visualized rivers as having assemblages of species that respond, by their occurrence and relative abundance, to the physical and chemical gradients present. The reasons for classifying lotic habitat are many, but a major focus should be the systematic ordering of habitat so that streams and rivers may be inventoried quantitatively. From a utilitarian point of view, a classification scheme should provide information that will enable scientists and others to determine the identity, location, and magnitude of resources. In addition, ordering should provide an improved perspective for thinking about streams and rivers and serve as a guide for understanding relations between abiotic and biotic factors. The goal of a classification scheme should be to separate streams and rivers into subjectively defined, isolated reaches without losing sight of interrelations between reaches and obscuring important ecological commonalities. The river continuum concept of Vannote et al. (1980) provides this framework theoretically, but its practical value is limited. A lotic classification scheme should be similar to land-based systems so that it can integrate land management processes, for example, the forest industry has a timber classification system that provides tree species, composition, and biomass for any land unit. Quantification of lotic systems on an areal basis is needed from a fishery perspective.

The physical boundaries of the habitat units should have biological meaning from several perspectives. Thus, a classification scheme should be usable by three groups: (1) state and provincial managers in ascertaining the area and production they have to work with in developing management plans, (2) regulatory agencies in determining losses and gains in rehabilitation and mitigation activities, and (3) researchers in studying the effects that the public and private sector or the managers and regulators have on habitat by their use, manipulation, or regulation of flowing water. A master classification scheme for the Great Lakes basin is certainly not immediately achievable, but through an iterative process based on a hierarchical framework a broadly based classification system could eventually become usable by all three of the groups. The goal of the authors is to provide some generalizations about how to circumscribe or bound habitats and apply them in developing a hierarchical classification for riverine (lotic) systems. This scheme should be tested, and if all or most of the deductions fail to agree with field data collected later, the evidence should be reexamined and an attempt made to develop a new classification by inductive inference.

The authors' primary intent here is to review and summarize classification schemes used for streams and rivers, mainly in North America and Europe, and comment on their applicability to the Great Lakes basin. To introduce this portion we provide a brief description of classification concepts and a definition of lotic habitat. A secondary intent is to discuss the habitat requirements of the fishes of interest, and to provide information about the catchment and the data sets available to the classifier. Gathering additional data and manipulating these data sets are often required. This is dealt with in an inventory section. Finally, an outline is provided for developing a classification scheme for streams and rivers in the Great Lakes basin.

II. LOTIC HABITAT DEFINED

Habitat is simply the place where an organism lives. Expansion of this concept of the term is marked with confusion (Whittaker et al., 1973), and quantification is lacking (Southwood, 1977). Coulombe (1978), who reviewed the history and evolution of the word, discovered 31 terms or phrases containing the word "habitat". In the following statements, we attempt to establish a standard to follow in a discussion of habitat classification schemes. Physical, chemical, and biological variables (the environment) define the place where an organism lives. Niche, a closely related term, defines the way a species adjusts to other related species in this space. If these two concepts are used together, the term ecotope might be used (Whittaker et al., 1973). Habitat can be used for communities or species, i.e., the environment of the community at a given place or the environment of a species (including the range of situations in which it occurs). When distinction is desired, "biotopes" should apply to the community environment and "habitat" to the environment of the species (Udvardy, 1959). Habitat may best be described as the environment of the community at a specific spatial scale. This concept allows the investigator to bound the place of interest. In this way a series of units or names are produced that can be conveniently ranked, accounted for, and inventoried. Variables used in bounding lotic habitat can be grouped into four broad categories (Anderson and Wallace, 1984): (1) physiological constraints (e.g., oxygen, temperature, hardness), (2) trophic considerations (e.g., kinds of food, acquisition of food), (3) physical aspects (e.g., depth, substrate), and (4) biotic interactions (e.g., predation, competition). These categories must be considered in terms of their heterogeneity in time and space, and the fact that organisms can directly influence their own habitat. Habitat and organisms are thus part of a system linked by feedback (Southwood, 1977). The distribution pattern resulting from habitat selection by a given aquatic species or community reflects the optimal overlap of these categories.

On a broad geographic scale, geology and climate are the ultimate determinants of river macrohabitat (Minshall et al., 1985). Climate affects the type and density of vegetation, and vegetation affects runoff and erosion, and hence the sediment yield and organic matter loading. Geology affects the relative erosiveness of the parent material, basin topography, chemical load, and bed composition. On the other end of the scale, local lithology and geomorphology determine unique microhabitat, such as the presence of a spring or the hydraulic stress feature of a particular riffle. On a biological scale, daily feeding forays contrast with annual reproductive runs. This scale of heterogeneity in physical, chemical, and biological variables provides the detail to be used for classifying lotic habitat.

III. THE NATURE AND USE OF CLASSIFICATION

Many needs for classification exist and, consequently, many different classification systems are needed (Davis and Henderson, 1978). It is most improbable that management needs can be met by a single approach to habitat classification. Many aquatic classification schemes that have been advanced provide only a general framework, hence their usefulness for inventorying resources is limited (Pennak, 1978; Warren, 1979; Lotspeich, 1980; Lotspeich and Platts, 1982; Frissell et al., 1986). Problems also arise from differences in philosophy and scale. Two important philosophical stances are classifications such as "hypotheses" and classifications such as "tools" (Warren, 1979). From the hypothetical view, the conceptual classification says something about nature that is either true or false. In the other view, conceptual devices are little more than tools for accomplishing particular tasks, judged primarily on the basis of their utility. Should thoughts be ordered? Is it desirable to show relations between theory and empiricism? Is it desirable to predict, explain, and understand the behavior of the system, or know where to plant fish, locate a power plant, or implement sea lamprey control? The answer to all of these questions should be yes. Also, should a classification scheme have high stability but low resolution, or low stability and high resolution? Most scientists would agree that a hierarchical approach would allow the resolution of some of these philosophical and scaling questions.

On a conceptual basis, classification has several important uses (Davis and Henderson, 1978) as it serves at least four functions: (1) a basis for cataloging the status of current resources, (2) a means of transferring experience and knowledge of a studied area to a similar but unstudied area, (3) a framework for assessing local management opportunities and predicting the outcomes of treatments or actions, and (4) a vocabulary for communication between managers, between managers and researchers, and between managers and the public. Once developed, the system must be validated through peer review, and it should be open ended and flexible. The classification system should be easily adaptable to mapping operations and the logic and structure of the system should be suitable for building and operating a comprehensive, computer-based management and information system.

Some researchers believe that environmental components used in classification schemes should be place-independent (Davis and Henderson, 1978), i.e., geographic location is unimportant to classification of a site. Place independence requires that the classification be determined only by the characteristics that can be perceived, counted, or measured on the site being classified. Rocks or soils are good examples of place-independent environmental attributes, whereas the brown trout (*Salmo trutta*) habitat cannot meaningfully be classified without knowledge of the geographic comparison in time and space of a diverse set of factors relating to food, shelter, and reproduction. Should biological communities be classified and then physical factors to delimit them be elucidated, or should habitats be classified physically and then the community that occurs in this area described? As an approximate test of place independence, one might describe a 1-km stretch of stream somewhere in the Great Lakes, and then determine limits of location.

Attempts have been made to delineate large areas within the Great Lakes that show ecological similarity on the basis of climate, landforms, soils, vegetation, hydrology, and wildlife (Anonymous, 1987). Bailey et al. (1978) termed these areas ecoregions. However, ecoregions are not place independent. This classification does not show whether a section of stream flowing over bedrock with an average slope of 10% in upstate New York (Northeastern Highlands ecoregion) is the same structurally and functionally as one in northeast Minnesota (Northern Lakes and Forest ecoregion).

IV. CHARACTERISTICS OF LOTIC CLASSIFICATION SCHEMES

The authors reviewed more than 70 papers encompassing stream zonation or classification schemes dealing mainly with fish, but including benthos. Many physical, chemical, and biological characteristics were used as measures of habitat in the various studies. Some of these characters are listed in Table 1. To condense this list and identify the most important characteristics, we selected up to four or five habitat attributes that the author of each work believed to be most important in establishing the presence, absence, or abundance of fish or invertebrates. This exercise elucidated 49 attributes in four categories — a biological and a chemical group, and two geomorphological groups. Of the two geomorphological groups, one was associated with a longitudinal succession (e.g., stream order, flow, volume) and the other was independent of length (e.g., gradient, velocity, substrate). Overall geomorphological attributes made up 85% of the most important descriptors, among which position-independent and position-dependent attributes were about evenly divided.

We must next discuss temporal and spatial scale. Many of the site characteristics in Table 1 could also be sampled at the catchment level, raising the question of what level of detail is appropriate when one measures habitat attributes. Six hierarchical levels that a regional classification scheme should include (Table 2) were modified from Frissell et al. (1986), who provided an excellent conceptual framework for a hierarchical habitat classification scheme. A prominent feature to be noted in Table 2 is persistence. In many of the studies reviewed, samples were taken at the relatively unstable pool-riffle level. Yet one can probably assume that the habitat attributes associated with a pool-riffle sequence, and even with a reach, are randomly distributed and representative of the more persistent segment level (100 to 1000 years). The length of sampling sites in 17 studies that provided a spatial scale ranged from 20 to 2500 m (average 340 m). Most of these sites were chosen by reviewing maps of the ecoregion or watershed, and by aerial photos and field reconnaissance at the segment. Most of the catchment attributes in Table 2 can be obtained from standard maps or other existing data sources, and most site characteristics are or can be obtained by site-specific sampling.

The studies listed in Table 3 can be placed into four distinctive groupings. The first group consists of longitudinal zonation studies that deal with length dimensions, either explicitly or implied, although one may question the demarcation of the zones and whether a continuum is or is not involved. A second group deals specifically with lateral zonation in large rivers. The third combines streams into large physiographic areas of an ecoregion. The fourth group enables estimation of the presence or absence of a species in a particular reach and estimates biomass independent of zonation, although some predictor or discriminating variable may be related to longitudinal position. Stream classification systems for West Virginia (Bailey et al., 1975) and South Carolina (Beasley et al., 1988) did not fit any of the categories. They were unique in employing aesthetics, access, and use as classification factors in a systematic and comprehensive treatment of the state's river resources.

Table 3 lists eight of the more common habitat attributes used by the various authors to delineate riverine habitat and thus help categorize the approaches the authors took. These attributes are the more common length independent variables used, plus a few composites to identify less frequent attributes. Length-dependent variables were not listed because they are usually incorporated into most studies. Biological variables could include macrophytes, invertebrates, or fishes.

A. LONGITUDINAL ZONATION SCHEMES

The community structure of the fish fauna typically varies from the headwaters to the lower reaches of streams because of the progressive replacement or addition of species. Scientists have recognized this phenomenon since the turn of the century (Hawkes, 1975) and still debate

TABLE 1
Habitat Attributes Used in Classification, Zonation, or Distribution Studies[a]

Catchment Attributes

Geomorphological	Hydrological	Meteorological	Socioeconomic
Bifurcation ratio	Extreme flow variation	Frost-free period	Access
Catchment area	Groundwater levels	Precipitation	Availability
Drainage density	Mean daily flow	Snowbelt areas	Land use
Elevation	Mean seasonal flow	Summer temp	Stream developments
Forest ratio	Mean annual flow/unit area	Winter temp	Success
Fractional ordering	Runoff	Langleys	Users
Geology	Source of water		
Mean basin length	Stability of flow regime		
Mean basin slope			
No. storms resulting in "x" flow			
Relief ratio			
Stream frequency			
Stream-length/watershed area			
Stream link			
Stream order			
Total channel lengths			
Valley form (shape)			

Site Attributes

Physical	Chemical	Biological
Bank coefficient	Alkalinity	Abundance (fish, benthos)
Bank stability	Chlorides	Bank vegetation
Bankside cover	CO_2	Benthos composition
Canopy cover	Conductivity	Biomass (fish, benthos)
Degree of embeddedness	Dissolved oxygen	BOD
Depth (mean, max, min)	Dissolved solids	Diversity (fish, benthos)
Distance from source	Hardness	Fish community
Flow (mean, max, min, max/mean)	Magnesium	Floodplain vegetation
Growing season	NO_2/NO_3	Functional groups (benthos)
Gradient	NH_4	Indicator species (fish, benthos)
Hydraulic radius	Phosphorus	Macrophytes
Instream cover	pH	Periphyton
Reach in pools, riffles, runs (%)	Sulfate	Particulate organic matter
Permanence	Suspended solids	Productivity
Riffle: pool ratio		Respiration
Sinuosity		
Soils		
Substrate composition		
Substrate roughness		
Temp (annual, etc.)		
Turbidity		
Water surface area		
Width (mean, max, min)		
Width depth ratio		
Width (pool, riffle)		
Velocity		

[a] Modified from Milner et al. (1985).

TABLE 2
Hierarchical Organization of River Systems within the Great Lakes Basin and Its Habitat Subsystems with Potential Variables for Classifying Each Level

System level	Linear spatial scale (m)	Persistence (years)	Variables used	Data source	Example from Great Lakes basin
Ecoregion	10^6	10^6–10^5	Growing season Landform Soils Vegetation Hydrology	National data maps, LANDSAT 1:1,000,000	South Michigan/Northern Indiana Clay Plains (Anonymous, 1987)
Stream system (watershed)	10^5	10^5–10^4	Mean basin slope Valley form Runoff	Regional data map, USGS maps, 1:500,000	Grand River (Great Lakes Basin Commission, 1975)
Segment system (stream order)	10^4	10^3–10^2	Channel floor lithology Stream order Lakes Reservoirs Soils Gradient	USGS maps, aerial photos 1:50,000	Red Cedar River, river segment B (Mattingly et al., 1981)
Reach system	10^3	10^2–10^1	Bank side cover Bank material gradient Riffle:pool:run Temperature Flow Hardness	Aerial photos, field survey 1:24,000	Red Cedar River, study section C-L (Mattingly et al., 1981)

TABLE 2 (Continued)
Hierarchical Organization of River Systems within the Great Lakes Basin and Its Habitat Subsystems with Potential Variables for Classifying Each Level

System level	Linear spatial scale (m)	Persistence (years)	Variables used	Data source	Example from Great Lakes basin
Pool/riffle system	10^2–10^1	10^1–10^0	Bank erosion-deposition Degree of embeddedness Depth Instream cover Substrate Macrophytes	Field surveys	AugustaCreek, Smith site (Cummins et al., 1981)
Microhabitat	10^0	10^0–10^{-1}	Instream structure Hydraulic radius Periphyton Benthos	Field surveys	Augusta Creek, Smith site, snag (Cummins et al., 1981)

Note: Also included in this table is an approximate linear spatial scale and a time scale of continuous potential persistence.

Modified from Frissell et al. (1986).

ways to describe and classify it (Matthews, 1986). Studies dealing with this subject can be divided into two parts: (1) those that emphasize the continuous sequential zonation, and (2) those that mostly illustrate zonation, but choose not to emphasize an orderly progression from the headwaters to the sea. Both kinds of studies are listed in Table 3 under the longitudinal category.

The longitudinal zonation approach was used by early workers who classified river zones on the basis of the dominant fish species present, such as trout, grayling, barbel, and bream. Most of the early work, such as that of Shelford (1911), Carpenter (1928), Berg (1948), and Illies and Botosaneanu (1963), was descriptive and no empirical relations were developed between zones and the physical, chemical, and biological characteristics of the rivers. Some of the early workers did graph or provide ranges of habitat variables in relation to abundance and diversity (Thompson and Hunt, 1930; Burton and Odum, 1945; Huet, 1959). Kuehne (1962) was the first to introduce stream order as a surrogate variable for several physical characteristics, and Sheldon (1968) was the first to use multivariate techniques to document observed relations. Hughes and Gammon (1987) used indices of fish assemblages (index of well-being and index of biotic integrity) to document longitudinal changes. Zonation studies of the type just described have been done for other groups such as algae, macrophytes, and benthos (Hynes, 1970). To facilitate comparisons, we included several examples of stream benthos distribution studies in Table 3. Those by Wright et al. (1984) and Omerod and Edwards (1987) seem to have included thorough analyses and covered large geographic areas.

Longitudinal zonation studies have been criticized because the transition areas between zones can be as long as the zone; consequently, zonation has sometimes been arbitrarily defined (Hawkes, 1975). Biological zonation concepts are usually applicable to limited geographical areas and application to broader areas must deal with historical differences in faunal distribution. Where no convenient natural discontinuities exist, certain arbitrary limits can be used if they are useful and readily definable. The problems of earlier zonation attempts can be corrected and applied to modern classification schemes. One of the more serious problems with zonation is the concept of a sequential continuum. Stream order has been used in North America more than any other scheme in recent decades as a framework to evaluate this continuum (Matthews, 1986), and has been shown by many workers to be correlated with successional changes in fishes (e.g., Harrel and Dorris, 1968; Lotrich, 1973; Platts, 1979; Beecher et al., 1988), whereas others have found no correlation (e.g., Oliff, 1960; Balon and Stewart, 1983; Matthews, 1986; Hughes and Gammon, 1987). Correlation probably occurs in some geographical areas where streams arise in mountainous regions, flow through foothill areas onto the plains and into the ocean, without crossing areas of rejuvenation. On the other hand, no strong correlation has been evident where the landform is fairly homogenous or modified by other events such as glaciation. For example, a first-order stream on the Fox River in Wisconsin is very different from one on the Cuyahoga River in Ohio. In addition, each stream system has its own rate of downstream change that results in segments of different sizes, which may or may not coincide with a defined stream order, within the system.

Other workers (e.g., Ricker, 1934; Van Deusen, 1954; Hallam, 1959; Maciolek, 1978) established attributes, some of which were related to stream size, but stream sections were subdivided into various categories irrespective of a continuum. Ricker and Hallam emphasized specific indicator species, whereas the other investigators focused on general community structure. Van Deusen and Maciolek addressed large geographic areas, whereas Ricker and Hallam dealt with a single basin.

Fish sampling in these studies typically dealt with the entire community, and gear types ranged from explosives to trap nets, though seines were used most often. The sampling methods and designs used by Balon and Stewart (1983) and Gorman and Karr (1978) are good examples of diversified approaches to a variety of circumstances. Analysis was usually simple (tabulating or graphing). Multivariate techniques were only used sparingly. Van Deusen

TABLE 3
Comparisons of Some Lotic Classification and Zonation Studies

		Habitat attributes							
Author	**Location**	**Grad-ient**	**Sub-strate**	**Vel-ocity**	**Temp**	**Cover**	**Bio-logical**	**Chem-ical**	**Land form**
				Longitudinal zonation					
Shelford (1911)	Illinois						X		
Carpenter (1928)	Wales	X	X	X			X		
Thompson and Hunt (1930)	Illinois		X	X			X		
Ricker (1934)	Ontario		X	X	X		X	X	
Burton and Odum (1945)	Virginia	X			X		X		
Trautman (1942)	Ohio	X							
Berg (1948)	Denmark		X	X			X		
Van Deusen (1954)	Maryland		X	X	X	X	X		
Harrison and Elsworth (1958)	S. Africa		X			X	X		
Hallam (1959)	Ontario				X	X	X		
Huet (1959)	W. Europe	X					X		X
Oliff (1960)	S. Africa	X	X	X			X		X
Illies (1961)	Worldwide		X	X	X		X		
Browzin (1962)	Great Lakes								X
Larimore and Smith (1963)	Illinois	X	X	X			X		
Minckley (1963)	Kentucky		X	X			X		
Usinger (1963)	California								X
Harrison (1965)	S. Africa		X	X	X		X		
Maitland (1966)	U.K.	X	X				X	X	
Anonymous (1967)	Michigan				X		X		
Harrel et al. (1967)	Oklahoma						X		
Harrel and Dorris (1968)	Oklahoma						X		
Sheldon (1968)	New York						X		
Carter and Jones (1969)	Kentucky						X		
Whiteside and McNatt (1972)	Texas						X		
Lotrich (1973)	Kentucky						X		
Cummins (1975)	N. America						X		
Maciolek (1978)	Hawaii	X	X		X		X		
Evans and Noble (1979)	Texas						X		
Platts (1979)	Idaho	X	X			X	X		
Mundy and Boschung (1981)	Alabama						X		
Culp and Davies (1982)	Alberta				X		X	X	X
Cushing et al. (1983)	U.S.	X					X		
Balon and Stewart (1983)	Zambia	X					X		
Fausch et al. (1984)	Midwest						X		
Zelewski and Naiman (1985)	Worldwide	X			X				

TABLE 3 (Continued)
Comparisons of Some Lotic Classification and Zonation Studies

		Habitat attributes							
Author	Location	Gradient	Substrate	Velocity	Temp	Cover	Biological	Chemical	Land form
				Longitudinal zonation					
Hughes and Gammon (1987)	Oregon		X				X	X	
Naiman et al. (1987)	Quebec						X		
Perry and Schaeffer (1987)	Colorado						X		
				Lateral zonation					
Shelford and Boesel (1942)	Lake Erie		X				X		
Geis and Hyduke (1978)	St. Lawrence						X		
Wells and Demas (1979)	Mississippi R.		X	X			X		
Beckett et al. (1983)	Mississippi R.		X	X			X		
Anderson and Day (1986)	Mississippi R.		X				X		
Edwards et al. (1989)	Great Lakes						X		
				Physiographic classification					
Pflieger (1971)	Missouri						X		
Platts (1974)	Idaho						X		
Menzel and Fierstine (1976)	Iowa	X					X		
Hughes and Omernik (1982)	Midwestern U.S.					X	X		
Pflieger et al. (1982a)	Missouri						X		
Matthews (1985)	Midwestern U.S.		X	X			X		
Hawkes et al. (1986)	Kansas	X	X				X		
Larsen et al. (1986)	Ohio						X		
Rohm et al. (1987)	Arkansas	X				X	X	X	
				Predictive classification					
Ziemer (1973)	Alaska								X
Swanston et al. (1977)	Alaska								X
Binns (1978)	Wyoming				X				
Binns and Eiserman (1979)	Wyoming		X	X	X	X		X	
Paragamian (1981)	Iowa		X						
Barber et al. (1982)	Alaska	X	X			X			

TABLE 3 (Continued)
Comparisons of Some Lotic Classification and Zonation Studies

		Habitat attributes							
Author	Location	Grad-ient	Sub-strate	Vel-ocity	Temp	Cover	Bio-logical	Chem-ical	Land form
			Predictive classification						
Graham et al. (1982)	Montana	X	X			X			
Fraley and Graham (1982)	Montana		X			X			
Wright et al. (1984)	U.K.		X					X	
Milner et al. (1985)	Wales					X		X	
Layher and Maughan (1985)	Kansas	X			X			X	
Lanka et al. (1987)	Wyoming	X							X
Layher et al. (1987)	Kansas				X			X	
McClendon and Rabeni (1987)	Missouri		X			X		X	
Ormerod and Edwards (1987)	Wales	X						X	
Scarnecchia and Bergersen (1987)	Colorado		X					X	
Wesche et al. (1987)	Wyoming					X			

(1954) was the only member of the group who mapped his classification. Papers concerning longitudinal fish zonation studies have declined in the literature; in contrast, detailed benthos zonation studies continue to be reported (Griffiths, 1987, 1988).

B. LARGE-RIVER SCHEMES

Most river zonation and classification schemes deal with streams of order 7 or less. Large rivers such as the connecting channels of the Great Lakes may appear monotypic for tens of kilometers. Manmade structures and impacts are the only discontinuities in the longitudinal direction. Several authors have noted that habitat changes in large rivers change from longitudinal to lateral differences. The analyses of smaller systems normally ignore lateral differences (microhabitat) and focus on longitudinal characters (macrohabitat). On the other hand, microhabitats in large rivers are measurable. Geis and Hyduke (1978) described three lotic habitats in the St. Lawrence River: type 1 was freely flowing open water deeper than 5.5 m; type 2 was deep littoral, 1.8 to 5.5 m deep; and type 3 consisted of shallow littoral areas <1.8 m deep. Edwards et al. (1989) used slightly different depth contours (0 to 1.8, 1.9 to 3.7, and 3.8 to 5.5 m) to estimate standing crop and production of benthos, macrophytes, and periphyton. These are 6-ft depth contours from charts of the National Oceanic and Atmospheric Administration (NOAA), and are easily measured. The depth strata used vary, depending on the cross-sectional character of the river, navigational modifications, and photic depth. In a more complex system such as the Mississippi River, a mosaic of lateral and longitudinal stratification produced five different habitats (Anderson and Day, 1986). Shelford and Boesel (1942) divided a portion of western Lake Erie, which mimics a large river, into seven types of habitat on the basis of depth and substrate.

C. LARGE-SCALE PHYSIOGRAPHIC SCHEMES

In this scheme, rivers are grouped regionally — usually in entire basins — without regard to differences that may occur along their lengths. The regional differences are generally based on climatic and terrestrial characteristics. For example, Rohm et al. (1987) used Omernik's (1987) ecoregions of the conterminous U.S. to divide Arkansas, and then used fish assemblages and physicochemical variables to characterize the six regions in that state. Similarly, Larsen et al. (1986) derived ecoregions for Ohio based on land use, landform, potential natural vegetation, and soil. They then used fish assemblages to characterize the resulting areas and called them aquatic ecoregions. In contrast fish distribution was used independently to regionalize streams in Missouri (Pflieger, 1971), Kansas (Hawkes et al., 1986), and Oregon (Hughes et al., 1987), and physiographic terms were then used to identify the regions. Hawkes et al. (1986) also used physical and chemical variables to characterize each region.

Existing data were used in some of these studies, whereas in others, sampling was conducted specifically for the classification. If sampling was to be carried out, candidate catchments were identified from 1:500,000 scale maps and disturbed catchments were identified (based on land use, population density, etc.) and eliminated. Ground reconnaissance and aerial photos were then used to further refine the process. Sampling involved the collection of data on 10 to 24 physicochemical variables, and fish from as few as 22 sites to as many as 1608. In some studies only smaller streams were used to characterize the region, whereas in others, all sizes of permanent streams were included. Statistical procedures varied, but basically sites were clustered on similarity, the classification scheme was tested for accuracy, and chemical or physical variables were correlated with the various clusters of sites. Some of the problems encountered in classifying sites (correct classification 50 to 75% of the time) were related to longitudinal differences, but all authors nevertheless felt that this concept was useful for integrating land-water systems. Most authors used the models to predict the type of fish community and ranges of abiotic variables that could be expected in a moderately undisturbed stream based on its location, and further used it as a measure of what might be attainable in rehabilitating a damaged stream in that region. Contiguous fish ecoregions are often more useful than interspersed sites in managing and planning exercises.

D. PREDICTIVE SCHEMES

Predictive classification methods rely on the relation that can be developed between habitat variables and fish occurrence or biomass. The basic assumption is that habitat conditions control the occurrence or absence of a species at a stream site and that these habitat conditions are also indicative of standing crop or biomass of fish in a stream. Biologists usually want to use habitat models to predict what will happen at another place or in the future, when conditions change. Many diverse models have been developed since 1970, though fishery biologists have searched for variables closely linked to fish abundance for at least 35 years (Fausch et al., 1988). One of the earliest models developed for a lotic context was that of Ziemer (1973). Model development today has evolved into a sophisticated procedure.

As an example of this evolution, Layher et al. (1987) used the following analysis sequence in developing a single species classification scheme and biomass prediction equation. They divided locations into two categories according to the presence or absence of the species, tested distributions of variables for homogeneity, and used Student *t*-tests to statistically determine significant differences. Variables that showed statistically significant differences between groups were then used in discriminant function analyses to develop predictive models for occurrence. Data were plotted to determine if a relation existed between standing crop and physical or chemical variables. To ensure linearity, the investigator divided the range of each variable into increments and calculated the mean standing crop within each increment. The habitat variable increments were scaled from 0 to 1, and these index values were used in a stepwise multiple regression to identify variables that explained variations in standing crop.

Corroboration of the models with other data sets were based on Pearson and Spearmen correlation coefficients calculated between observed and predicted values.

The number of sites analyzed in these predictive schemes ranged from 30 to 209, and the number of habitat variables from 19 to 38. Fish sampling involved a mark-and-recapture or depletion method of estimation, in which electrofishing, seining, or rotenone (or a combination) was used. Accuracy in classifying streams ranged from 80 to 90%, but the geographical range of the predictions was limited. Collection methods and techniques also affected accuracy. In biomass equations, the R^2 usually exceeded 50%. The number of variables in the models ranged from 1 to 10. Synergistic effects and independence of variables were mentioned as problems but little was done (except for combining similar variables into a single term) in most studies to address these concerns. The matter of placing and estimating abundance of "broad niche" vs. "narrow-niche" species was also discussed.

Fausch et al. (1988) reviewed 99 models that predict a standing crop of fish from habitat variables. To summarize the models, they grouped the habitat variables according to measurement scale, such as at the drainage basin (coarse), channel morphometry (medium), or chemical measurements (fine). Most (25) of the models were based on fine measurements of habitat structure — biological, physical, and chemical variables. Measures of instream, overhead cover were most frequently significant (22 models), followed by depth (20); alkalinity, hardness, and conductivity (15); mean streamflow (14); width (14); surface area (12); and dissolved oxygen (11 models). A major assumption common to all models reviewed was that the fish population was limited by the set of habitat variables in the model, and most authors chose to ignore other limiting factors such as fishing mortality, interspecific competition, or predation. In fact, Rakocinski (1988) noted that fish populations that are subjected to frequent catastrophic mortality should only rarely reach levels high enough to allow population and habitat variables to be correlated. Finally, none of the models was developed through manipulative experiments, and each of the uses of the models involved a risky inductive leap, often beyond the defined population of streams for which inferences are valid. This induction may lead to inaccurate prediction and faulty management decisions.

V. DATA REQUIREMENTS AND SOURCES

In many of the studies reviewed (longitudinal, large rivers, physiographic schemes) physical and chemical variables or even the fish community have been used to define the particular habitat a stream reach represents. In contrast, the predictive schemes defined the requirements of a particular species and these were used to determine their presence or absence in a particular reach. The species or species group requirements would form the basis of a hierarchical system that would classify aquatic areas at different levels of resolution, e.g., coldwater fishery habitat, trout habitat, brook trout (*Salvelinus fontinalis*) habitat, brook trout rearing site, etc.

This hierarchical approach would require determining the major habitat attributes of important fish species in the Great Lakes region. The species of fish should include residents of the Great Lakes proper that use the rivers as spawning areas or whose young use the river as a nursery area. Species should include trouts, salmons, sea lamprey (*Petromyzon marinus*), walleye (*Stizostedium v. vitreum*), and forage species. Important species endemic to the rivers, such as smallmouth bass (*Micropterus dolomieui*) and others, should also be considered. The habitat requirements of most of these species are well documented in the literature and many are extensively summarized in the habitat suitability index (HSI) model series of the U.S. Fish and Wildlife Service (Terrell et al., 1982). These reports include at least 1 dozen species of interest in the Great Lakes basin: pink salmon (*Oncorhynchus gorbuscha*), chinook salmon (*Oncorhynchus tshawytscha*), lake trout (*Salvelinus fontinalis*), brown trout, rainbow trout

TABLE 4
Habitat Variables Used in Determining Suitability of Stream Reaches to Different Life Stages of Brown Trout

Life stage and habitat variables
Adult
Instream cover (%)
Pools (%)
Pool class rating
Juvenile
Instream cover (%)
Pools (%)
Pool class rating
Fry
Substrate size class (%)
Pools (%)
Fines (%)
Embryo
Average maximum temperature
Average minimum dissolved oxygen (DO)
Gravel in spawning areas (%)
Fines (%)
Other[a]
Maximum temperature
Average minimum DO
pH
Average base flow
Dominant substrate type[b]
Streamside vegetation[b] (%)
Fines (%)
Percent streamside stability[b]
Midday shade[b] (%)
Nitrate-nitrogen (mg/l)[b]
Peak flow

[a] Variables that affect all life stages.
[b] Optional variables.

Adapted from Raleigh et al. (1986).

(*Oncorhynchus mykiss*), brook trout, smallmouth bass, largemouth bass (*Micropterus salmoides*), white sucker (*Catostomous catostomus*), yellow perch (*Perca flavescens*), walleye, and gizzard shad (*Dorosoma cepedianum*).

The HSI models are mechanistic, and thus involve aggregation of variables that affect life requisites of a species (Layher and Maughan, 1984). A suitability index curve is developed for each variable through a literature review, expert opinion, or direct measurements. The curves describe the response of a species over a known or measurable range of a variable. An example of the variable used for various life stages of the brown trout is illustrated in Table 4. A site is visited, the variables are measured, and an evaluation is made by assessing the lowest score or by weighing each variable to reflect its relative importance, and then summing the scores to arrive at a single index. This index theoretically reflects the carrying capacity of the habitat for the population under study. Presence or absence is not addressed, since many suitability curves exceed zero over the measurement range of the variable. The application of the HSI models to field conditions or the transfer of field data for application to other geographic areas requires the use of considerable caution (Layher and Maughan, 1985; Layher et al., 1987; Pajak and Neves, 1987).

Empirically based habitat requirements of Great Lakes species can be derived from information available in the Great Lakes by using a procedure such as that described by Lahyer et al. (1987). Because field collections are time consuming and expensive, available data should first be consulted. Many states have data sets based on surveys that, with proper interpretation, yield much information on presence, absence, or abundance (Fausch et al., 1984). Some of the habitat attributes can be determined from maps and routine monitoring data. Initial relations and tests should be developed with available data sets to guide future sampling. Thus, when variables are chosen to be measured, the geographic area covered and the number of sites visited are reduced, but the area of application is increased.

Another source of information and technique is a computerized retrieval and analysis of habitat information on various species of algae, aquatic insects, and fish developed by Dawson and Hellenthal (1986) for the north-central U.S. The database contains about 58,000 values representing environmental characterizations of 402 genera and 1694 species of aquatic organisms. The system can be used interactively to predict and evaluate aquatic habitats and organism assemblages. Development of a Great Lakes river classification scheme should include this habitat-requirement database and an exploration of its usefulness for analysis.

VI. INVENTORY METHODS

A. GROUND TRUTH MEASUREMENTS

Stream inventory methods range from direct collections in microhabitats to remote sensing techniques. The sampling scale depends on the precision and accuracy of spatial patterns that must be detected. Trade offs between costs in time and money are usually involved. A computerized system with the capacity for storage, retrieval, and analysis of these data is an important component. Data collection methods, analysis, and information management systems are briefly dealt with here.

Sampling in most of the studies reviewed was carried out on a site-specific basis (pool-riffle or segment) within designated reaches selected by using maps (at scales up to 1:500,000) and aerial photos. Many workers have tried to avoid areas with moderate to heavy human impacts. The authors recommend that if researchers need to identify such areas, they are to use methods such as those developed by Fajen and Wehnes (1982). Stream partitioning or the locations of reach breaks may be based on a variety of factors (such as discharge, stream topography, and land use), but stream gradients can easily be determined from large-scale topographical maps (1:24,000 to 1:50,000) and usually reflect distinguishable reaches or habitat types in the field (de Leeuw, 1982). Working from topographic maps, workers identify reach breaks where gradients change from one interval (i.e., 0%) to another (0.0 to 0.5%). By using these measurements, one can partition an entire stream system into its component reaches before entering the field. Before data collection, a visual assessment is made in the field to verify the identification of reaches on maps, as well as to assess the uniformity of the particular reaches under study.

Sampling techniques, physical and chemical variables that can be measured, and methods for the inventory of fish were described by Demarchi and Chamberlain (1978) and de Leeuw (1982) for British Columbia, by Imhof and Biette (1982) for Ontario, and by Collotzi and Dunham (1978) and Platts et al. (1983) for the western intermountain region of the U.S. Terrell et al. (1982) provided guidance in the field application of HSI models, especially techniques for collecting field data, including the selection of sites and the recording of field data. Novel approaches such as the use of helicopters in British Columbia to sample areas with limited access (Shera and Harding, 1982) may be useful elsewhere. Eight-member crews gathered extensive data on the channel characteristics and fish populations of an area at the rate of 200 km^2 per air hour and a cost of \$56 km^{-1} per channel surveyed. In contrast, the use of scuba to measure microhabitat components was described by Gosse and Helm (1982), and Tyus (1987) used radiotelemetry to describe the habitat of the rare razorback sucker (*Xyrauchen texanus*). Whatever the mode of collection, the problem of developing reproducible results remains. Platts (1982) addressed this facet in an article entitled "Stream inventory, garbage in — reliable analysis out: only in fairy tales".

B. REMOTE METHODS

Remote sensing methods, many of which were developed and refined in the 1980s, are at the other end of the spatial scale from ground-based surveys. Studies in which this technique has been used range from the analysis of continents with a resolution of 1 km, to small subbasins with the more recently developed airborne imagery spectrometers that have a resolution of 8 m. Hardisky et al. (1986) reviewed the performance and availability of various remote sensors carried by aircraft and satellites. The coarser systems, especially color infrared (IR) photography, have been used routinely to delimit boundaries of wetlands (Gammon et al., 1977). LANDSAT and high-altitude aircraft have been used to evaluate drainage patterns, land use, etc., in ecoregions or catchments (Table 2). Intermediate resolution techniques have been used to discriminate between species of plants and animals and estimate biomass and productivity of wetlands (Hardisky et al., 1986). Low-altitude aerial photographs have been used to identify submergent and emergent vegetation to species (Minor et al., 1977; Harwood, 1975). Color IR photography from both the dormant (leaves-off) and the growing seasons (leaves-on) may be used advantageously in the evaluation of riverine habitat, as it has been in wetland studies (Gammon et al., 1977; Jensen et al., 1986). Typically, hand-held radiometers have been used to measure at ground level and evaluate spectral output, and these instruments have been used in the Great Lakes basin to evaluate suspended sediments in rivers (Haugen et al., 1976). Remote sensing technology will continue to improve. Several habitat attributes once measurable only by ground-based surveys may now be distinguished by their particular reflective signatures. Vande Castle et al. (1988) have used the Thematic Mapper of LANDSAT-5, the High Resolution Visible sensor of SPOT-1, and the Advanced Very High Resolution Radiometer of the NOAA meteorological satellites in the Great Lakes to provide image maps of water quality characteristics, including chlorophyll *a*, total suspended solids, turbidity, Secchi disk depth, and water temperature with high accuracy ($0.80 < r^2 < 0.99$). They used a geographic information system (GIS) procedure to integrate this spatial information attained from remote sensing technology with the more conventional forms of data.

LANDSAT material and analytical services are available from the Earth's Resource Observation System Data Center, Sioux Falls, SD. Seasonal high- and low-altitude color IR photographs at scales of 1:100,000 and 1:24,000 are available from National Aeronautics and Space Administration (NASA). Various state organizations (Geological Survey, Highway Commission, Soil Conservation Service) and private firms (aerial surveying) are potential sources of photos. The U.S. Environmental Protection Agency (EPA) Environmental Photographic Interpretation Center, Warrenton, VA, facilitates the application of aerial photography and environmental remote sensing to environmental problems. This group operates on a contractual basis and has access to the aerial photography resources of other federal agencies (Getz et al., 1983).

Side scan sonar and hydroacoustic techniques are also available for mapping surficial features and estimating the density, distribution, and movement of fish in larger rivers (Edsall et al., this volume). Side scan sonar is capable of producing geological mosaics resembling high-quality aerial photographs of land masses (Klein, 1985) and have been used to survey and map freshwater habitat (Sly and Widmer, 1984). Fixed-location hydroacoustic systems are well suited for studying fish in rivers. Vertical soundings from surface or bottom-mounted transducers at multiple sampling stations across a river give estimates of distribution and abundance of fish (Raemhild, 1985). Another promising technology is the doppler river counter, which effectively monitors passage rates through restricted areas such as wiers (Johnston and Hopelain, 1990).

The use of meteorological data from weather stations can also be thought of as remote sensing. These data can be used both for ecoregions (regional winter and summer temperatures, ice conditions, growing season) and at the reach level. Simple models are available in which daily mean temperatures at a particular reach are described as a function of daily mean air temperature (Jeppesen and Iversen, 1987). These models are accurate and precise. Predicted temperatures deviate <1°C from actual measurements 90% of the time. Thermal imagery based on low or high altitude flights may also be a useful tool in defining habitat reaches based on temperature.

C. STATISTICAL METHODS

Most of the early fish distribution studies and even recent longitudinal schemes are descriptive. Fish distribution and habitat features are described subjectively, either verbally or graphically. With the increased speed and capacity of computers and the availability of prepared multivariate statistical programs, scientists can now manipulate and analyze large data sets. Multivariate programs allow a somewhat objective method of examining patterns in the distribution of species against an environmental background. The importance of multivariate techniques lies not in the statistical significance of the components, but in the capacity for data exploration or classification of objectives. The sequence of the analysis and the addressing of assumptions associated with it vary in the papers we reviewed, and this aspect requires further evaluation. Green and Vascotto (1978) provided a comparative example of two approaches, and Fausch et al. (1988) addressed several important aspects of model building, including proper statistical methods, error in habitat measurements, and major biological assumptions of models. More evaluations of this kind are needed if this field is to develop on a firm statistical footing. Each statistical package uses slightly different terms to describe a procedure that may include several techniques, test statistics, cutoff options, etc. The general procedure and a comment on certain discrepancies is provided below.

In the first stage of any analysis, each variable under consideration is evaluated by calculating simple statistics, examining histograms and scatter plots of dependent vs. independent variables, and examining simple correlation coefficients between all variables. Also sought are normality, homogeneity of variances, independence (multicollinearity), and linearity between variables. In descriptive studies such as those used in classification schemes, large departures from standard assumptions are tolerable. However, if significant testing is to be done, these assumptions must be met. Environmental variables are typically transformed (log), or nonparametric techniques (Rakocinski, 1988) are used. An alternative is to use principal component analysis (PCA) scores as independent variables. The dependent variables are sometimes transformed or linearized by indexing to a scale of 0 to 1. Fish variables used in the analysis include species lists of abundance (numbers or weights), presence or absence, trophic guilds, Karr's index of biotic integrity, diversity, and other biotic indices.

The next step is typically cluster analysis to reduce data complexity, since some large data sets include >1000 sites or >100 species. Sites are usually clustered on the basis of fish composition, but may also be characterized by physical or chemical variables or species clustered by sites. Determining the number of clusters is the key in the analysis. In a classification scheme, a compromise is reached between reducing localities and forming locality groups with more than one habitat type. A general discussion of the problem was given by the SAS Institute (1985), and Pflieger et al. (1982b) provided a method for dealing specifically with fish communities. Some authors have subjectively designated a level based on ecological information. A compromise between objective rules and subjective sense is necessary. Another use of the cluster procedure is in conjunction with PCA. The PCA scores are plotted on two-dimensional PCA axes and the species or location factor scores are grouped by cluster analysis (Carpenter et al., 1982). The operation of cluster analysis, in this sense, is similar to that of a multirange test used with an analysis of variance procedure in providing an objective method of grouping factors.

Several techniques relate the homogenous species assemblages derived from cluster analysis to habitat variables. The clusters can be plotted, sample by sample, on a map illustrating the grouping in relation to different physiographic areas (Pflieger et al., 1982b). A univariate procedure can be used on each environmental variable to determine whether the groups or clusters of samples differ significantly from others (Layher et al., 1987). In most studies we reviewed, multiple discriminant analysis (DA) was performed in the form of a canonical analysis (CA) or DA. Typically, CA isolates the important variables determining the abundance of a species or occurrence of a species cluster. DA is used to establish a classification system and to calculate the percent of observations correctly classified. Dolman (1986) used a leave-out-one cross-validation (Efron and Gong, 1983) to calculate unbiased error rates for classifying new reservoirs when no independent data sets were available to check error rate. Another option would be to use PCA and simple correlation coefficients to relate factor scores to habitat variables (Sprules, 1977). Hawkes et al. (1986) used rotated axes to evaluate their factors, but Green and Vascotto (1978) advise against this method.

Several studies restrict the number of species to be used in the analysis (Pflieger et al., 1982b; Wright et al., 1984; Hawkes et al., 1986). Pflieger et al. (1982b) tried to develop an objective criterion. Reducing numbers of species restricts the possibility that those species could be used to predict unique habitats. Some researchers test as many species or habitats as possible, using a stepwise method, until no further improvement in discrimination results. However, stepwise regression techniques can be misleading because the order in which additional variables enter or leave the model has no functional significance (Johnson, 1981). It is better to calculate all possible models by using "all subsets" regression and to choose the best one for a given number of independent variables (for example, the best regression with three independent variables) by using an objective criterion such as Mallow's Cp statistic (Fausch et al., 1988).

The final step is to use multiple regression (MR) to estimate fish standing crop on the basis of physical, chemical, or biological habitat variables established in the CA or DA procedure. Typically these values are screened before they are used in either the DA or CA. Variables are eliminated if correlation is less than a certain value. This restriction of variables with low correlation coefficients is not recommended because two or more values may have high coefficients but may be strongly correlated with each other, and thus the inclusion of both does not improve the model. The inclusion of other variables with lower but uncorrelated values may produce a much better discrimination (Graham et al., 1982). High multicollinearity is usually not a problem when the purpose of the regression analysis is to infer the response function or prediction of new observations, provided these inferences are made within the range of observation. The effect of an impact on the different independent variables cannot be evaluated, since estimated regression coefficients associated with these variables become imprecise and tend to have no real meaning in relation to other coefficients if the independent variables are strongly correlated (Neter and Wasserman, 1974). This property applies to stepwise DA. A stepwise regression procedure is normally used in the analysis in the form of a forward, backward, or combination method. These procedures should be used with caution (Johnson, 1981). Barber et al. (1982) found no difference between a backward elimination and a forward selection method. Milner et al. (1985) proposed a slightly different stopping rule — the percentage of explained variance, rather than a set R^2. In multiple regression, R^2 increases as variables are added, so that the largest model is always chosen if this is the sole criterion.

Accuracy is usually qualified by checking predicted values against observed values with an independent data set, and precision by regressing predicted against actual values and examining the correlation coefficient (McClendon and Rabeni, 1987). Fraley and Graham (1982) increased precision by fitting the regression model with a zero intercept. Few modelers calculated standard errors for regression coefficients, calculated confidence or prediction intervals around the regression line, or tested their models with other data (Fausch et al., 1988). The percent variance accounted for by any of the above analyses does not necessarily equate to ecological utility. Results still must be associated in terms of ecological insight (Matthews, 1985).

Wright et al. (1984) provided an excellent description of the development of a biological classification of unpolluted running water sites in the U.K. based on the macroinvertebrate fauna. They grouped the sites by using detrended correspondence analysis, a nonparametric analog to PCA. Sites were classified by using two-way indicator species analysis, which classifies both sites and species and constructs ordered two-way tables to show the relation between them. The program also constructs a key to the sample classification by identifying one or more differential species that are particularly diagnostic in each division in the classification. The key can therefore be used to classify new sites without reclassifying all sites. The CA was used to relate the site grouping to the environmental data. The number of discriminating functions used were determined by transforming Wilk's lambda into a chi-square statistic. The percentages of correct placement when the habitat variables were used ranged from 90 at the 2-group level to 73 at the 16-group level.

D. INFORMATION STORAGE METHODS

A computerized information storage, retrieval, and analysis system is needed for a Great Lakes classification scheme to consolidate information already in state, provincial, and federal files, and to handle new or revised information as it becomes available. The system must be customized to meet the needs of the Great Lakes system, but could be developed by incorporating systems already in place. The State of Missouri has implemented a hierarchical series of stream codes that permits the retrieval of data from any combination of 5139 drainage areas and stream segments (Pflieger et al., 1982a). The stream codes interface with files being developed on stream resource data. Shera (1982) described the British Columbia coding system and Chamberlin (1982) described the mapping and aquatic database development that can be sorted and retrieved to serve a number of planning and management functions. A national system exists in the U.S. in the form of a River Reach Fisheries Information System (RRFIS). The reach file is a computerized catalog based on the hydrological features of the U.S. and is designed to organize and analyze water resource data (Olson et al., 1981; Whitworth et al., 1985). This tool for organizing data uses standard U.S. Geological Survey and Water Resources Council hydrologic maps, the EPAs River Reach File numbering system, and existing natural resource information. The software, maintained and operated by the U.S. Fish and Wildlife Service (FWS), is operational, and is supported on a mainframe computer. The RRFIS provides capabilities for interactive database management and a GIS that can be customized to meet the special needs of each user. User manuals for each of the computer technologies used in RRFIS are available. A pilot test program has been run on several streams in the Great Lakes basin for the Sea Lamprey Control Program of the Great Lakes Fishery Commission (Whitworth et al., 1985; Klar, personal communication).

Mapping is an integral part of data information systems, and computer techniques such as those described by Ruediger and Engels (1982) could be used for mapping stream channel topography and habitat. This information or remote sensing data must be transferred to a planimetrically correct orthophotomap base. A map may then be reproduced at any convenient size to illustrate the distribution and relations developed from a classification (Gammon et al.,

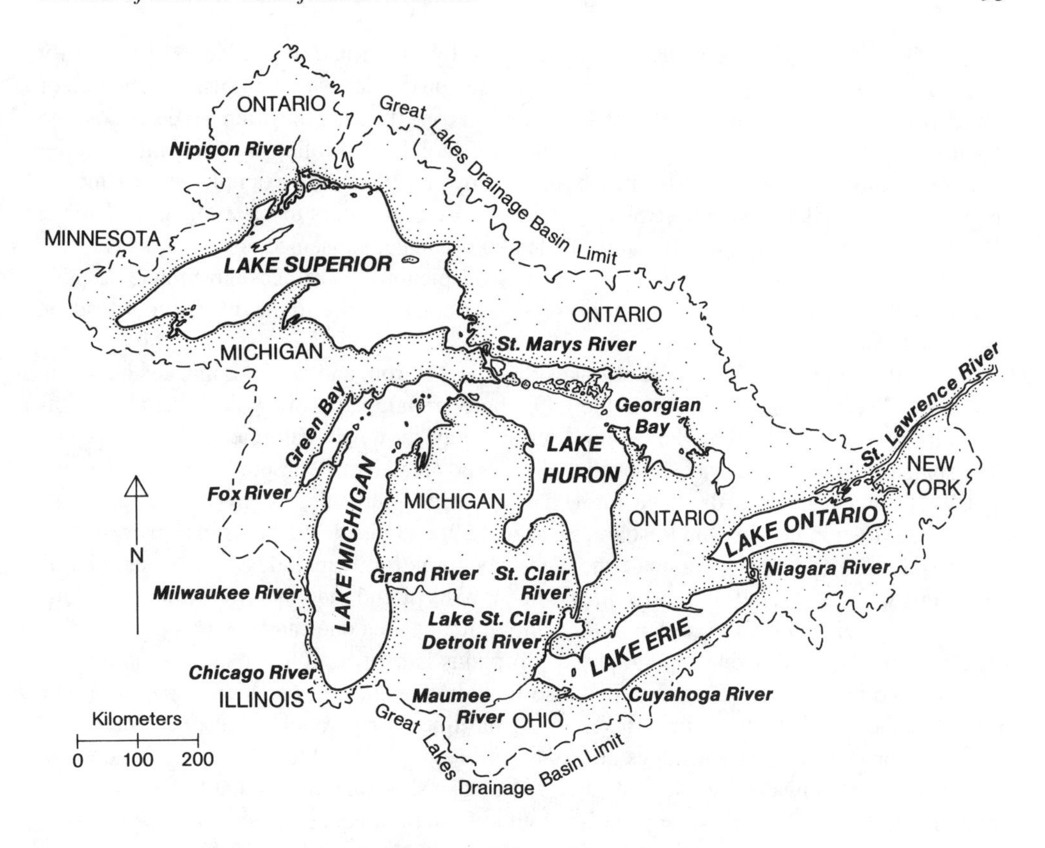

FIGURE 1. Great Lakes watershed.

1977). Thus, a geographic information system should answer such questions as: How many miles of second-order streams are in the Grand River basin? What species of fish are present and what is the standing stock of species of special interest?

VII. GREAT LAKES BASIN CHARACTERISTICS

The total Great Lakes basin land drainage area, including the St. Lawrence River, is 546,000 km^2, of which about 56% is located in Canada (Figure 1). This drainage area has been organized into river basin groups and hydrologic areas (Great Lakes Basin Commission, 1975). For example, in the U.S. each lake basin has been divided into two or more river basin groups. The 15 river basin groups contain anywhere from 1 to 8 hydrologic areas. The 72 hydrologic areas consist of single major catchments or a complex of small catchments draining individual tributaries. The largest catchments are the Maumee River (17,100 km^2) in the U.S. and the Nipigon River (25,258 km^2) in Canada. The many minor streams flowing directly into the major lakes drain the smallest catchments, often only a few square kilometers. In Michigan, there are 513 of these coastal streams, compared with 100 to 150 larger streams (average flow >1 $m^3\ s^{-1}$). The total length of these streams in Michigan exceeds 58,500 km (Brown, 1944). More than 100 streams enter Lake Michigan, of which about 60 are called "rivers" and the rest "creeks". Only eight of the rivers have an average flow of >28 $m^3\ s^{-1}$. The largest is the Fox River in Wisconsin, which has an average flow of 127 $m^3\ s^{-1}$. The largest mean single annual flow into the Great Lakes basin, apart from the connecting rivers, is 358 $m^3\ s^{-1}$ from the Nipigon River (Powers, 1962).

The Great Lakes drainage basin is characterized by ill-defined boundaries, the proximity of the inland boundary of the watershed to the lakes proper at several points, the absence of large tributary rivers, and the relatively large portion (one third) of the drainage basin covered by the lakes themselves. The drainage pattern is largely controlled by such minor relief features as moraines and lacustrine beach ridges (Powers, 1962). For example, in the Chicago region, the original watershed between the Chicago River tributary to Lake Michigan and the Des Plaines River tributary to the Gulf of Mexico drainage, a distance of 18 km, traversed a marsh about 3 m above Lake Michigan. Such inconspicuous watersheds have made possible the diversion of Great Lakes water out of the system at Chicago, and the diversion of Hudson Bay water into Lake Superior.

In the Precambrian shield north of Lakes Superior, Huron, and Ontario, in Canada, and in northern Wisconsin and Michigan in the U.S., there are large areas in which glacial deposits are thin or absent, and where the bedrock surface may show vertical relief as great as 90 m. The sparse glacial deposits have blocked drainage and created many minor lakes, swamps, and muskegs. Of the Ontario drainage into Lake Superior, wetlands occupy 6 to 25% of the watershed (Energy, Mines, and Resources Canada, 1986). Outside the Precambrian crystalline areas, glacial drift deposits dominate the landscape. Landforms include extensive undulating till plains with local relief of 3 to 15 m, and linear belts of end moraines, generally rough and with local relief of 9 to 30 m. Even in the Ohio drainage area one can find gradients as steep as 25 m/km. The undulating topography of till plains is expressed in the aimless wandering back and forth of many streams, which also include stretches of wetlands in their courses. Many of the streams have a pattern reflecting the linear or parallel arrangement of glacial moraines or lacustrine beach ridges and spits. For example, the Milwaukee River flows for nearly 48 km southward, parallel to the Lake Michigan shore and behind the Valders terminal moraine; at Milwaukee, the river rounds the end of the moraine and enters the lake (Powers, 1962). The classic view of a river system progressively changing from high-gradient, turbulent, rocky headwaters, to low-gradient, sluggish, mudbottomed rivers, and with a channel pattern progressing from straight, through braided, to meandering is the exception rather than the rule in the Great Lakes basin.

The many lakes, wetlands, closed basins, and flat areas in the catchment tend to retard runoff, trap transported sediment, and to smooth out stream flow. The melting of winter snow and ground ice, together with the general prevalence of a warm season precipitation maximum, tends to give most streams a peak stage in March to June. The maximum flow of the larger streams, where uncontrolled, is generally three to seven times their average flow. In contrast, the connecting channels of the Great Lakes are large rivers (1200 to 9900 $m^3 s^{-1}$), with limited intrinsic tributary drainage systems and relatively stable hydrology (about a 2:1 ratio of maximum to minimum flow). A summary of flow, catchment, and water quality characteristics of the connecting channels is given in Table 5.

Twenty ecoregions (areas that show broad ecological unity based on such characteristics as climate, topography, soils, natural vegetation, and land use) have been described within the Great Lakes basin (Anonymous, 1987). The development of the Canadian and U.S. ecoregions was done independently and thus could be reduced to a more manageable number of four to eight. There seem to be three or four groups of major summer or frost-free zones, four land uses, five elevations, and four soil zones that could be integrated and used in a stream classification scheme. Northern spruce fir forests cover 91% of the Lake Superior basin, whereas only 21% of the Lake Erie basin is covered by forests (northern hardwoods). Catchments with surface soils that contain considerable amounts of clay-sized particles tend to contribute higher loads of suspended solids and phosphorus per unit area than do those that have more coarse grained (sandy) soils. Streams draining clay soils also are more vulnerable to flash floods. Lake Erie streams, at least those of the western basin, are good examples of streams draining surface soils dominated by clay. Parts of the Lake Michigan basin (Wiscon-

TABLE 5
Summary of Flow, Basin, and Mean Water Quality Characteristics for the Connecting Rivers of the Great Lakes[a]

Characteristic	River				
	St. Marys	**St. Clair**	**Detroit**	**Niagara**	**St. Lawrence**
River length (km)	121	63	41	58	869
Drop in elevation (m)	6.7	1.5	1.0	99.3	74.0
Mean discharge (m^3 s^{-1})					
Annual	2100	5097	5210	5692	6739
Maximum	3738	6570	6654	7505	9911
Minimum	1161	3002	3171	3285	4361
Max/min ratio	3.2	2.2	2.1	2.3	2.3
Basin area (10^3 km^{-2})					
Lake(s) above	82	117	1	26	64
Drainage basin	128	252	18	59	19
Cumulative total	210	579	598	683	766
Water quality					
Conductance (μmhos cm^{-1})	105	208	256	274	329
Total phosphorus (ppb)	3.6	5.5	9.0	9.4	13.2
Total kjeldahl nitrogen (ppb)	99	188	180[b]	170	232
$NO_3 + NO_2$ (ppb)	302	310	298	265	256
Si (ppm)	2.3	1.06	0.67	0.46	0.24
pH (range)	(7.9–8.3)	(7.7–8.7)	(7.0–8.6)	(7.3–8.5)	(8.2–8.7)
Chloride (ppm)	1.28	5.94	6.90	15.30	24.20
Dissolved oxygen (ppm)	11.2	11.2	9.3	—	12.2
Alkalinity (ppm)	41.6	80.0	81.2	97.5	92.4
Chlorophyll *a* (ppb)	0.83	2.10	3.40	1.10	3.19

[a] Data collected by R. Stevens, Canada Centre for Inland Waters (CCIW), Burlington, Ontario for the St. Marys River (1983), the St. Clair River (1980), and the St. Lawrence River (1985). Data for the Detroit River are mean values for 1967 to 1982 from the STORET data system (Herdendorf, 1987) or those collected by M. Charlton, CCIW, Burlington, Ontario in 1985. Data for the Niagara River collected by D. Rathke, Center for Lake Erie Area Research, Columbus, OH, in 1985.

[b] Average value from Vaughan and Harlow (1965).

sin side), Lake Superior basin, Lake Huron basin (thumb area), and Lake Ontario basin also have a high clay content. Sandy soils are prevalent in the Lake Michigan basin, particularly in the northeast of the basin. Water quality in streams from sandy areas is generally good (Sonzogni et al., 1978).

The native fishes of the Great Lakes basin consist of 153 species, 64 genera, and 25 families (Bailey and Smith, 1981). Many of the streams and rivers in the Great Lakes basin contain excellent populations of nonmigratory fish species such as brook trout and smallmouth bass. In addition, many of the river and stream tributaries to the Great Lakes historically supported migratory spawning runs of coldwater fishes from the lake, such as whitefishes and chars (Loftus, 1958; Lawrie and Rahrer, 1972), and cool- or warmwater runs of species such as the walleye and lake sturgeon (*Acipenser fulvescens*) (Ryder, 1968). Degradation of the outlets of many of the larger rivers continues to prevent some of these runs. The rainbow trout, an introduced species, now has one of the largest migratory spawning runs in the lakes, with the possible exception of two species of suckers and the ubiquitous rainbow smelt. Other more recently introduced migratory exotics are the brown trout and three Pacific salmons (pink, chinook, coho). The most troublesome is the migratory run of sea lampreys in the northern streams that provide particularly good spawning habitat for this species (Manion and Hanson, 1980).

TABLE 6
Examples of References by Basin that Contain Data Usable for Classifying Lotic Habitat in the Great Lakes Basin

Basin	Reference
Lake Superior	McLain et al. (1965) B
	Brown (1944) P
	Manion and McLain (1971) B
	Moore and Braem (1965) B
	Schneberger and Hasler (1946) B, C, P
	Smith (1962) P
	Thomson (1944) B
	Zimmerman (1968) C
	Ryder (1964) P
Lake Michigan	Applegate (1961) B
	Brown (1944) P
	Cummins et al. (1981) B, P
	Mattingly et al. (1981) B, C, P
	MI Dept. of Conservation (1957) P
	Minshall et al. (1983) B, C, P
	Schelske et al. (1980) B, P, C
	Stauffer (1972) B
	Velz and Gannon (1960) P
	Zimmerman (1968) C
Lake Erie	Baker et al. (1975) C, P
	Brown (1953) B
	Gordon and MacCrimmon (1982) P
	Horwitz (1978) B, P
	Langlois (1965) B, P
	Mahon et al. (1979) B, P
	NY Conservation Dept. (1929) B, C, P
	Ohio DNR (1953) C, P
	Smith et al. (1981) B
	Tramer and Rodgers (1973) B
	Van Meter and Trautman (1970) B
Lake Huron	Adams (1937) P
	Applegate (1950) B
	Applegate and Smith (1950) B
	Brown (1944) P
	Brown and King (1987) B
	Hartig and Stifler (1980) C
	Newell and Buda (1976) P
	Ricker (1934) B, C, P
	Water Resources Commission (1973) P
Lake Ontario	Barton et al. (1985) B
	Hallam (1959) B
	Hill (1983) C
	NY Conservation Dept. (1927, 1928, 1929, 1931, 1932, 1940) B, C, P
	Steedman (1988) B, P
Connecting channels	Beak Consultants Limited (1988) B
	Creese (1987) B
	Duffy et al. (1987) B, C, P
	Edsall et al. (1988) B, C, P
	Edwards et al. (1989) B, C, P
	Griffiths (1987, 1988) B, C, P
	Haas et al. (1985) B
	Hudson et al. (1986) B,C,P
	Liston and McNabb (1986) B, C, P
	Manny et al. (1988) B, C, P
	Patch and Busch (1986) B, C, P
	Thornley and Hamdy (1983) B, C, P

Note: Data include P = physical, C = chemical, and B = biological.

Reasonable amounts of physical, chemical, and biological data exist on the streams and rivers in the Great Lakes basin. However, they are widely dispersed among journals, government reports, and file cabinets. An example of the diversity of material is given in Table 6. Many data were collected by workers of the U.S. Bureau of Commercial Fisheries through the Great Lakes Fishery Commission while research was being conducted on the sea lamprey and on the operation of lamprey control devices. Some of these data have been published, but most remain on file at the FWS, Marquette Biological Station. Much information can be found in publications of the U.S. Geological Survey, NOAA (e.g., Great Lakes Water Supply Forecasting), U.S. Soil Conservation Service, Federal Insurance Administration (floodplain mapping), Environment Canada, Ontario Ministry of the Environment, International Joint Commission (International Reference Group on Great Lakes Pollution From Land Use Activities; Reports of the Upper Lakes Reference Group), state or provincial Departments of Natural Resources (e.g., Ontario Fisheries Information System), and water resources commissions or divisions.

For example, a review of reports prepared by Michigan's Institute for Fisheries Research revealed 303 reports on streams and rivers in that state alone (Zurek, 1987). The EPA has a computerized data management system, known as BIOS, that contains data on the distribution, abundance, and physical condition of aquatic organisms and their habitats. The system also supports a broad range of analytical capabilities. It would seem prudent to thoroughly review existing data sets before additional data are collected on Great Lakes rivers and streams.

VIII. OUTLINE FOR DEVELOPING A CLASSIFICATION SYSTEM FOR AQUATIC SYSTEMS

Although several classification schemes for aquatic systems have been proposed (e.g., Illies, 1961; Pennak, 1971; Persoone, 1979; Frissell et al., 1986), no single scheme has met with wide acceptance or use. This nonacceptance might imply that the present schemes do not meet the practical needs of fisheries biologists or environmental regulators. Lack of acceptance might be due to one or more of the following reasons: (1) the schemes were conceptually developed, but no practical guidance was provided, (2) classifications that go beyond a single watershed or species are too difficult to use, (3) the classification was not developed broadly enough to satisfy the needs of a diverse set of users, and (4) no institutional arrangements were made for long-term commitment to the project and maintenance of the data set. There is no question that the development of a Great Lakes stream classification system will be difficult. It will require a major commitment of money and manpower, and the establishment of an administrative unit to provide the direction and leadership required to secure broad interjurisdictional acceptance. Christie et al. (1986) suggested initiatives for improving the acquisition and use of scientific data. These problems are beyond the scope of the present paper but a general set of guidelines for developing a Great Lakes classification scheme are suggested.

Many of the techniques used in the studies reviewed here can be incorporated into a Great Lakes classification scheme. The basic sampling unit for all studies reviewed is a field survey of 100 to 200 m sections of river. The connecting channels would need to be scaled upward. Sampling is the expensive part of a classification and the number and distribution of these basic sampling units would establish the range and accuracy of prediction or discrimination. The number and placement of basic sampling units will depend on collating easily accessible physical and chemical variables needed to stratify the Great Lakes rivers and streams at several spatial scales. Defining these relatively homogeneous tracts is fundamental to any classification scheme. This definition will allow the prediction from the basic sampling areas to other areas. After sampling, it is primarily a matter of blocking the spatial patterns and determining what to classify. Do we cluster or discriminate at the level of the ecoregion or the reach, or somewhere in between? This blocking technique should allow both place-dependent and place-independent classification or stream reach or brown trout habitat to be classified with the same data set. In fact, the process reduces to iterative runs using place-dependent physiochemical data vs. species or community habitat requirements.

Suggested steps to follow in developing a hierarchical classification scheme are given in Table 7. Representative hydraulic units could be chosen by clustering the 20 Great Lakes ecoregions into 3 to 5 groups and randomly picking a hydraulic unit within each group. This random selection would provide a broad range of conditions for evaluating the first classification attempt. Representative watersheds are chosen within each hydraulic unit by studying topographic maps, aerial photos, and historical data sets. These representative watersheds are then divided into segments and reaches based on classifying variables taken from existing maps and photos (Table 2). The segments and reaches are clustered and sampling sites randomly picked from these homogeneous groups. Selected sites are field checked for the

TABLE 7
Steps in Developing a Classification System

Sampling steps

1. Select representative hydraulic units
2. Study topography maps, aerial photos, etc.
3. Search and collate historical data
4. Stratify watershed at segment and reach level
5. Select sampling sites at pool-riffle level
6. Field reconnaissance check
7. Select physical, chemical, and biological variables
8. Select sampling design and gear
9. Establish statistical procedures
10. Collect data

Analysis steps

1. Select a species or community
2. Stratify sampling sites into presence/absence
3. Test for statistically different habitat variables
4. Develop discriminant predictive equation
5. Develop biomass predictive equation
6. Develop models by watershed and then interchange for collaboration

existence and adequacy of past studies. HSI models and other literature are reviewed to determine the best physical, chemical, and biological variables to be measured, taking into consideration variables unique to Great Lakes species and life stages. Finally, select sampling design and gear, establish statistical procedures, and collect data.

In analyzing the data to develop a key for a species, community, or group, the sites are divided into two groups, based on the presence or absence of a taxon or taxa. The two groups of sites are tested for statistically different habitat variables and a discriminant and biomass predictive equation is developed. These equations are used to predict presence or absence and biomass on unsampled areas of the watershed. The areas are then sampled for corroboration. The different watershed models are then compared for compatability among ecoregions. Model variability and efficiency of collecting predictive variables are evaluated. Model refinement continues, and the database is maintained and made available to all researchers in the Great Lakes basin.

The ease of constructing a classification system would depend on the availability of the necessary data and their application. The fisheries profession must be careful not to dismiss a system simply because the data are currently unavailable. The sooner a functional system is established, the sooner we will know what data to collect in the inventory process. Application of the classification system will depend on the distribution of three model attributes: realism, precision, and generality. The authors hope that the classification system would incorporate all the important functional attributes of fish population response to habitat. A balance needs to be struck between generality and precision. Some models, although relatively imprecise, may be useful for basinwide planning and analysis in fisheries management. In contrast, when fishery biologists are faced with managing a highly valued fishery resource in a limited area, they can afford to gather detailed information about factors that limit the population and will probably be successful in developing a relatively precise model. Let researchers begin by collating the easily measured and available data at the ecoregion to segment level, and let future users determine the more refined needs of a Great Lakes basin stream classification system.

ACKNOWLEDGMENTS

We thank R. A. Ryder and R. Biette for contributions to the manuscript, W. D. Busch for guiding its completion, and M. Murphy for typing the manuscript.

This is Contribution 776 of the National Fisheries Research Center-Great Lakes, U.S. Fish and Wildlife Service, 1451 Green Road, Ann Arbor, MI 48105.

REFERENCES

Adams, M. P. 1937. Saginaw Valley Report. Michigan Stream Control Commission, Lansing.

Anderson, N. H. and J. B. Wallace. 1984. Habitat, life history, and behavioral adaptations of aquatic insects. in *An Introduction to the Aquatic Insects of North America*. R. W. Merritt and K. W. Cummins, Eds. Kendall/Hunt, Dubuque, IA. pp. 38–58.

Anderson, R. V. and D. M. Day. 1986. Predictive quality of macroinvertebrate-habitat associations in lower navigation pools of the Mississippi River. *Hydrobiologia* 136:101–112.

Anonymous. 1967. Michigan Stream Classification; 1967 System. Michigan Department of Natural Resources. Lansing.

Anonymous. 1987. The Great Lakes: An Environmental Atlas and Resource Book. U.S. Environmental Protection Agency, Chicago/Environment Canada, Toronto.

Applegate, V. C. 1950. Natural History of the Sea Lamprey, *Petromyzon marinus,* in Michigan. Spec. Sci. Rep. Fish. 55. U.S. Fish and Wildlife Service. Washington, D.C.

Applegate, V. C. 1961. Downstream Movement of Lampreys and Fishes in the Carp Lake River, Michigan. Spec. Sci. Rep. Fish. 387. U.S. Fish and Wildlife Service. Washington, D.C.

Applegate, V. C. and B. R. Smith. 1950. Sea Lamprey Spawning Runs in the Great Lakes in 1950. Spec. Sci. Rep. Fish. 61. U.S. Fish and Wildlife Service. Washington, D.C.

Bailey, R. E., O. E. Maughan, and R. A. Whaley. 1975. Stream classification system for West Virginia. in *Symposium on Trout Habitat Research and Management*. Appalachian Consortium Press, Boone, NC. pp. 78–86.

Bailey, R. G., R. D. Pfister, and J. A. Henderson. 1978. Nature of land resource classification — a review. *J. For.* 76:650–655.

Bailey, R. M. and G. R. Smith. 1981. Origin and geography of the fish fauna of the Laurentian Great Lakes basin. *Can. J. Fish. Aquat. Sci.* 38:1539–1561.

Baker, D. B., W. B. Jackson, and B. L. Prater, Eds. 1975. Proc. Sandusky River Basin Symp. International Reference Group on Great Lakes Pollution from Land Use Activities. International Joint Commission. Windsor, Canada.

Balon, E. K. and D. J. Stewart. 1983. Fish assemblage in a river with unusual gradient (Luongo, Africa-Zaire system), reflections on river zonation, and description of another new species. *Environ. Biol. Fish.* 9:225–252.

Barber, W. E., M. W. Oswood, and S. J. Deschermeier. 1982. Validation of two fish habitat survey methods in southeast Alaska, in *Proc. Symp. Acquisition and Utilization of Aquatic Habitat Inventory Information.* Western Division. American Fisheries Society. Bethesda, MD. pp. 225–231.

Barton, D. R., W. D. Taylor, and R. M. Biette. 1985. Dimensions of riparian buffer strips required to maintain trout habitat in southern Ontario streams. *North Am. J. Fish. Manage.* 5:364–378.

Beak Consultants Limited. 1988. Benthic Invertebrate Survey of the St. Marys River, 1985. Ontario Ministry of the Environment. Water Resources Branch. Great Lakes Section. Toronto, Ontario, Canada.

Beasley, B. R., D. A. Lange, K. T. Newland, and W. C. Brittain. 1988. South Carolina Rivers Assessment. Rep. No. 164. South Carolina Water Resources Commission. Columbia.

Beckett, D. C., C. R. Bingham, and L. G. Sanders. 1983. Benthic macroinvertebrates of selected habitats of the lower Mississippi River. *J. Freshwater Ecol.* 2:247–261.

Beecher, H. A., E. R. Dott, and R. F. Fernau. 1988. Fish species richness and stream order in Washington State streams. *Environ. Biol. Fish.* 22:193–209.

Berg, K. 1948. Biological studies on the River Susaa. *Folia Limnol. Scand.* 4:1–318.

Binns, N. A. 1978. Evaluation of habitat quality in Wyoming trout streams, in Classification, Inventory, and Analysis of Fish and Wildlife Habitat. FWS/OBS-78/76. U.S. Fish and Wildlife Service. Washington, D.C. pp. 221–242.

Binns, N. A. and F. M. Eiserman. 1979. Quantification of fluvial trout habitat in Wyoming. *Trans. Am. Fish. Soc.*108:215–228.

Brown, C. J. D. 1944. Michigan Streams—Their Lengths, Distribution, and Drainage Areas. Inst. Fish. Res. Misc. Publ. 1. Michigan Department of Conservation. Ann Arbor.

Brown, E. H. 1953. Survey of the bottom fauna of the mouths of ten Lake Erie south shore rivers: its abundance, composition, and use as index of stream pollution, in Lake Erie Pollution Survey — Final Report. Ohio Department of Natural Resources. Columbus. pp. 156–170.

Brown, S. S. and D. K. King. 1987. Community metabolism in natural and agriculturally disturbed riffle sections of the Chippewa River, Isabelle County, Michigan. *J. Freshwater Ecol.* 4:39–51.

Browzin, B. S. 1962. On classification of rivers in the Great Lakes-St. Lawrence Basin, in Proc. 5th Conf. Great Lakes Res. Great Lakes Research Division. University of Michigan. Ann Arbor. pp. 86–92.

Burton, G. W. and E. P. Odum. 1945. The distribution of stream fish in the vicinity of Mountain Lake, Virginia. *Ecology* 26:182–194.

Carpenter, A. L., W. N. Jessee, and D. A. Rundstrom. 1982. Evaluating community similarity: an exploratory multivariate analysis, in *Proc. Symp. Acquisition and Utilization of Aquatic Habitat Inventory Information.* Western Division. American Fisheries Society. Bethesda, MD. pp. 369–375.

Carpenter, K. E. 1928. *Life in Inland Waters.* Sidgwick & Jackson, London.

Carter, J. P. and A. R. Jones. 1969. Inventory and Classification of Streams in the Upper Cumberland River Drainage of Kentucky. Fish. Bull. 52. Kentucky Department of Fish and Wildlife Resources. Frankfort.

Chamberlin, T. W. 1982. Systematic aquatic biophysical inventory in British Columbia, Canada, in *Proc. Symp. Acquisition and Utilization of Aquatic Habitat Inventory Information.* Western Division. American Fisheries Society. Bethesda, MD. pp. 17–25.

Christie, W. J., M. Becker, J. W. Cowden, and J. R. Vallentyne. 1986. Managing the Great Lakes basin as a home. *J. Great Lakes Res.* 12:2–17.

Creese, E. E. 1987. Report on the 1983 Benthic Invertebrate Survey of the Niagara River and Nearby Lake Ontario. Ontario Ministry of the Environment. Water Resources Branch. Great Lakes Section. Toronto, Ontario, Canada.

Collotzi, A. W. and D. K. Dunham. 1978. Inventory and display of aquatic habitat, in Classification, Inventory, and Analysis of Fish and Wildlife Habitat. FWS/OBS-78/76. U.S. Fish and Wildlife Service. Washington, D.C. pp. 533–542.

Coulombe, H. N. 1978. Summary remarks, in Classification, Inventory, and Analysis of Fish and Wildlife Habitat. FWS/OBS-78/76. U.S. Fish and Wildlife Service. Washington, D.C. pp. 593–604.

Culp, J. M. and R. W. Davies. 1982. Analysis of longitudinal zonation and the river continuum concept in the Oldman-South Saskatchewan River system. *Can. J. Fish. Aquat. Sci.* 39:1258–1266.

Cummins, K. W. 1975. The ecology of running waters, theory and practice, in Proc. Sandusky River Basin Symp. D. B. Baker, W. B. Jackson, and B. L. Prater, Eds. International Reference Group on Great Lakes Pollution from Land Use Activities, International Joint Commission. Windsor, Canada. pp. 278–293.

Cummins, K. W., M. J. Klug, G. M. Ward, G. L. Spengler, R. W. Speaker, R. W. Ovink, D. C. Mahan, and R. C. Petersen. 1981. Trends in particulate organic matter fluxes, community processes and macroinvertebrate functional groups along a Great Lakes Drainage Basin river continuum. *Int. Ver. Theor. Angew. Limnol. Verh.* 21:841–849.

Cushing, C. E., C. D. McIntire, K. W. Cummins, G. W. Minshall, R. C. Petersen, J. R. Sedell, and R. L. Vannote. 1983. Relationships among chemical, physical, and biological indices along river continua based on multivariate analyses. *Arch. Hydrobiol.* 98:317–326.

Davis, L. S. and J. A. Henderson. 1978. Many uses and many users; some desirable characteristics of a common land and water classification system. in Classification, Inventory and Analysis of Fish and Wildlife Habitat. FWS/OBS-78/76. U.S. Fish and Wildlife Service. Washington, D.C. pp. 13–34.

Dawson, C. L. and R. A. Hellenthal. 1986. A Computerized System for the Evaluation of Aquatic Habitats Based on Environmental Requirements and Pollution Tolerance Associations of Resident Organisms. EPA/600/53-86/019. U.S. Environmental Protection Agency. Corvallis, OR.

de Leeuw, A. D. 1982. A British Columbia stream habitat and fish population inventory system, in *Proc. Symp. Acquisition and Utilization of Aquatic Habitat Inventory Information.* Western Division. American Fisheries Society. Bethesda, MD. pp. 32–40.

Demarchi, D. A. and T. W. Chamberlain. 1978. The Canadian experience: an approach toward biophysical interpretation, in Classification, Inventory, and Analysis of Fish and Wildlife Habitat. FWS/OBS-78/76. U.S. Fish and Wildlife Service. Washington, D.C. pp. 145–164.

Dolman, W. B. 1986. Classification of Texas Reservoirs in Relation to Limnology and Fish Community Associations. Federal Aid Project F-31–R-12. Texas Parks and Wildlife Department. Austin.

Duffy, W. G., T. R. Batterson, and C. D. McNabb. 1987. The St. Marys River, Michigan: An Ecological Profile. Biol. Rep. 85(7.10). U.S. Fish and Wildlife Service. Washington, D.C.

Edsall, T. A., B. A. Manny, and C. N. Raphael. 1988. The St. Clair River and Lake St. Clair, Michigan: An Ecological Profile. Biol. Rep. 85(7.3). U.S. Fish and Wildlife Service. Washington, D.C.

Edsall, T. A., R. H. Brock, F. J. Horvath, R. P. Bukata, and W. J. Christie. This volume.

Edwards, C. J., P. L. Hudson, W. G. Duffy, S. J. Nepszy, C. D. McNabb, R. C. Haas, C. R. Liston, B. Manny, and W. N. Busch. 1989. Hydrological, morphometrical, and biological characteristics of the connecting rivers of the international Great Lakes: a review, in Proc. Int. Large River Symposium. Can. Spec. Publ. Fish. Aquat. Sci. 106. Department of Fisheries and Oceans, Ottawa. D. P. Dodge, Ed. pp. 240–264.

Efron, B. and G. Gong. 1983. A leisurely look at the bootstrap, the jacknife and cross-validation. *Am. Stat.* 34:36–48.

Energy, Mines, and Resources Canada. 1986. Canada Wetland Regions. Map produced by Geographic Services Division. Ottawa.

Evans, J. W. and R. L. Noble. 1979. The longitudinal distribution of fishes in an east Texas stream. *Am. Midl. Nat.* 101:333–343.

Fajen, O. F. and R. E. Wehnes. 1982. Missouri's method of evaluating stream habitat, in *Proc. Symp. Acquisition and Utilization of Aquatic Habitat Inventory Information.* Western Division. American Fisheries Society. Bethesda MD. pp. 117–123.

Fausch, K. D., J. R. Karr, and P. R. Yant. 1984. Regional application of an index of biotic integrity based on stream fish communities. *Trans. Am. Fish. Soc.* 113:39–55.

Fausch, K. D., C. L. Hawkes, and M. G. Parsons. 1988. Models that Predict Standing Crop of Stream Fish from Habitat Variables: 1950–85. Gen. Tech. Rep. PNW-GTR-213. Forest Service. Pacific Northwest Research Station. Portland, OR.

Fraley, J. J. and P. J. Graham. 1982. Physical habitat, geologic bedrock types and trout densities in tributaries of the Flathead River drainage, Montana, in *Proc. Symp. Acquisition and Utilization of Aquatic Habitat Inventory Information.* Western Division. American Fisheries Society. Bethesda MD. pp. 178–185.

Frissell, C. A., W. J. Liss, C. E. Warren, and M. D. Hurley. 1986. A hierarchical framework for stream habitat classification: viewing streams in a watershed context. *Environ. Manage.* 10:199–214.

Gammon, P. T., D. Malone, P. D. Brooks, and V. Carter. 1977. Three approaches to the classification and mapping of inland wetlands, in *Proc. 11th Int. Symp. Remote Sensing of Environment.* Environmental Research Institute of Michigan. Ann Arbor. pp. 1545–1555.

Geis, J. W. and N. P. Hyduke. 1978. Habitat mapping and critical habitat studies. Technical Report N. Environmental Assessment of the FY 1979. Winter Navigation Demonstration on the St. Lawrence River. Volume 2. State University of New York, Institute of Environmental Program Affairs, Syracuse.

Getz, T. J., J. C. Randolph, and W. F. Echelberger. 1983. Environmental application of aerial reconnaissance to search for open dumps. *Environ. Manage.* 7:553–562.

Gordon, D. J. and H. R. Mac Crimmon. 1982. Juvenile salmonid production in a Lake Erie nursery stream. *J. Fish Biol.* 21:455–473.

Gorman, O. T. and J. R. Karr. 1978. Habitat structure and stream fish communities. *Ecology* 59:507–515.

Gosse, J. C. and W. T. Helm. 1982. A method for measuring microhabitat components for lotic fishes and its application with regard to brown trout, in *Proc. Symp. Acquisition and Utilization of Aquatic Habitat Information.* Western Division. American Fisheries Society. Bethesda MD. pp. 138–149.

Graham, P. J., B. B. Shepard, and J. J. Fraley. 1982. Use of stream habitat classification to identify bull trout spawning areas in streams, in *Proc. Symp. Acquisition and Utilization of Aquatic Habitat Information.* Western Division. American Fisheries Society. Bethesda MD. pp. 186–190.

Great Lakes Basin Commission. 1975. Great Lakes Basin Framework Study. Appendix 14: Flood Plains. Great Lakes Basin Commission. Ann Arbor, MI.

Green, R. H. and G. L. Vascotto. 1978. A method for the analysis of environmental factors controlling patterns of species composition in aquatic communities. *Water Res.* 12:583–590.

Griffiths, R. W. 1987. Environmental Quality Assessment of Lake St. Clair in 1983 as Reflected by the Distribution of Benthic Invertebrate Communities. Ontario Ministry of Environment, Water Resources Branch, Southwestern Region. London, Ontario, Canada.

Griffiths, R. W. 1988. Environmental Quality Assessment of the St. Lawrence River in 1985 as Reflected by the Distribution of Benthic Invertebrate Communities. Ontario Ministry of the Environment, Water Resources Branch, Great Lakes Section. Toronto.

Haas, R. C., W. C. Bryant, K. D. Smith, and A. J. Nuhfer. 1985. Movement and Harvest of Fish in Lake St. Clair, St. Clair River, and Detroit River. Final Report Winter Navigation Study. Michigan Department of Natural Resources, Fisheries Division. Mt. Clemens.

Hallam, J. C. 1959. Habitat and associated fauna of four species of fish in Ontario streams. *J. Fish. Res. Bd. Can.* 16:147–173.

Hardisky, M. H., M. F. Gross, and V. Klemas. 1986. Remote sensing of coastal wetlands. *Bioscience* 36:453–460.

Harrel, R. C. and T. C. Dorris. 1968. Stream order, morphometry, physio-chemical conditions, and community structure of benthic macroinvertebrates in an intermittent stream system. *Am. Midl. Nat.* 80:220–251.

Harrel, R. C., B. J. Davis, and T. C. Dorris. 1967. Stream order and species diversity of fishes in an intermittent stream. *Am. Midl. Nat.* 78:429–436.

Harrison, A. D. 1965. River zonation in southern Africa. *Arch. Hydrobiol.* 61:380–386.

Harrison, A. D. and J. F. Elsworth. 1958. Hydrobiological studies on the Great Berg River, Western Cape Province. I. General description, chemical studies and main features of the flora and fauna. *Trans. R. Soc. S. Afr.* 35:125–226.

Hartig, J. H. and M. E. Stifler. 1980. Water Quality and Pollution Control in Michigan. Michigan Department of Natural Resources. Lansing.

Harwood, J. E. 1975. The Use of Remote Sensing in a Study of Submerged Aquatic Macrophytes of the Pamlico River Estuary, North Carolina. M.S. dissertation, East Carolina University, Greenville, NC.

Haugen, R. K., H. L. McKim, and T. L. Marlar. 1976. Remote Sensing of Land Use and Water Quality Relationships — Wisconsin Shore, Lake Michigan. Cold Regions Research and Engineering Lab. Rep. 76–30. Corps of Engineers, Hanover, New Hampshire.

Hawkes, C. L., D. L. Miller, and W. G. Layher. 1986. Fish ecoregions of Kansas: stream fish assemblage patterns and associated environmental correlates. *Environ. Biol. Fish.* 17:267–279.

Hawkes, H. A. 1975. River zonation and classification. in *River Ecology.* B. A. Whitton, Ed. Blackwell Scientific. Oxford. pp. 312–374.

Herdendorf, C. E. 1987. The Ecology of the Coastal Marshes of Western Lake Erie: A Community Profile. Biol. Rep. 85(7.9). U.S. Fish and Wildlife Service. Washington, D.C.

Hill, A. R. 1983. Nitrate-nitrogen mass balances for two Ontario rivers. in *Dynamics of Lotic Systems.* T. P. Fontaine and J. M. Bartell, Eds. Ann Arbor Science. Ann Arbor, MI. pp. 457–477.

Horwitz, R. J. 1978. Temporal variability patterns and the distributional patterns of stream fishes. *Ecol. Monogr.* 48:307–321.

Hudson, P. L., B. M. Davis, S. J. Nichols, and C. M. Tomcko. 1986. Environmental Studies of Macrobenthos, Aquatic Macrophytes, and Juvenile Fish in the St. Clair-Detroit River System. Admin. Rep. 86–7, National Fisheries Center-Great Lakes. U.S. Fish and Wildlife Service. Ann Arbor, MI.

Huet, M. 1959. Profiles and biology of western European streams as related to fish management. *Trans. Am. Fish. Soc.* 88:155–163.

Hughes, R. M. and J. M. Omernik. 1982. A proposed approach to determine regional patterns in aquatic ecosystems, in *Proc. Symp. Acquisition and Utilization of Aquatic Habitat Information.* Western Division. American Fisheries Society. Bethesda, MD. pp. 92–102.

Hughes, R. M. and J. R. Gammon. 1987. Longitudinal changes in fish assemblages and water quality in the Willamette River, Oregon. *Trans. Am. Fish. Soc.* 116:196–209.

Hughes, R. M., E. Rexstad, and C. E. Bond. 1987. The relationship of aquatic ecoregions, river basins and physiographic provinces to the ichthyogeographic regions of Oregon. *Copeia* 423–432.

Hynes, H. B. N. 1970. *The Ecology of Running Waters.* University of Toronto Press, Toronto.

Illies, J. 1961. Versuch einer allgemein biozoentischen Gliederung der Fliessgewaesser. *Int. Rev. Ges. Hydrobiol.* 46:205–213.

Illies, J. and L. Botosaneanu. 1963. Problemes et methodes de la classification et de la zonation ecologique des eaux courantes, considerees surtout du point de vue faunistique. *Mitt. Int. Ver. Theor. Angew. Limnol.* 12:1–57.

Imhof, J. and R. M. Biette. 1982. Assessing fluvial trout habitat in Ontario, in *Proc. Symp. Acquisition and Utilization of Aquatic Habitat Information.* Western Division. American Fisheries Society. Bethesda, MD. pp. 197–201.

Jensen, J. R., M. E. Hodgson, E. Christensen, H. E. Mackey, L. R. Tinney, and R. Sharitz. 1986. Remote sensing inland wetlands: a multispectral approach. *Photogram. Eng. Remote Sens.* 52:87–100.

Jeppesen, E. and T. M. Iversen. 1987. Two simple models for estimating daily mean water temperature and diel variations in a Danish low gradient stream. *Oikos* 49:149–155.

Johnson, D. H. 1981. The use and misuse of statistics in wildlife habitat studies. The Use of Multivariate Statistics in Studies of Wildlife Habitat. D. E. Capen, Ed. Gen. Tech. Rep. RM-87, Forest Service, Rocky Mountain Forest and Range Experiment Station, Fort Collins, CO. pp. 11–19.

Johnston, S. V. and J. S. Hopelain. 1990. The application of dual-beam target tracking and Doppler shifted echo processing to assess upstream salmonid migration in the Klamath River, California. *Rapp. P.-V. Reun. Cons. Int. Explor. Mer.*

Klein, M. 1985. High-resolution seabed mapping: new developments. *Instrum. Soc. Am.* 24:65–73.

Klar, G. Sea Lamprey Control, Marquette Biological Station, MI, U.S. Fish and Wildlife Service, personal communication.

Kuehne, R. A. 1962. A classification of streams illustrated by fish distribution in an eastern Kentucky creek. *Ecology* 43:608–614.

Langlois, T. H. 1965. Portage River Watershed and Fishery. Publ. W-130. Ohio Department of Natural Resources. Columbus.

Lanka, R. P., W. A. Hubert, and T. A. Wesche. 1987. Relations of geomorphology to stream habitat and trout standing stock in small Rocky Mountain streams. *Trans. Am. Fish. Soc.* 116:21–28.

Larimore, R. W. and P. W. Smith. 1963. The fishes of Champaign County, Illinois, as affected by 60 years of stream change. *Ill. Nat. Hist. Surv. Bull.* 28:299–382.

Larsen, D. P., J. M. Omernik, R. M. Hughes, C. M. Rohm, T. R. Whittier, A. J. Kinney, A. L. Gallant, and D. R. Dudley. 1986. Correspondence between spatial patterns in fish assemblages in Ohio streams and aquatic ecoregions. *Environ. Manage.* 10:815–828.

Lawrie, A. H. and J. E. Rahrer. 1972. Lake Superior: effects of exploitation and introductions on the salmonid community. *J. Fish. Res. Bd. Can.* 29:765–776.

Layher, W. G. and O. E. Maughan. 1984. Analysis and refinement of habitat suitability index models for eight warm-water fish species, in Proc. Workshop on Fish Habitat Suitability Index Models. Biol. Rep. 85(6). U.S. Fish and Wildlife Service. Washington, D.C. pp. 182–250.

Layher, W. G. and O. E. Maughan. 1985. Relations between habitat variables and channel catfish populations in prairie streams. *Trans. Am. Fish. Soc.* 114:771–781.

Layher, W. G., O. E. Maughan, and W. P. Wurde. 1987. Spotted bass habitat suitability related to fish occurrence and biomass and measurements of physiochemical variables. *North Am. J. Fish. Manage.* 7:238–251.

Liston, C. R. and C. D. McNabb. 1986. Limnological and Fisheries Studies of the St. Marys River, Michigan in Relation to Proposed Extension of the Navigation Season. FWS/OBS-80(62.3). U.S. Fish and Wildlife Service. Washington, D.C.

Loftus, K. H. 1958. Studies on river-spawning populations of lake trout in eastern Lake Superior. *Trans. Am. Fish. Soc.* 87:259–277.

Lotrich, V. A. 1973. Growth, production and community composition of fishes inhabiting a first-, second-, and third-order stream of eastern Kentucky. *Ecol. Monogr.* 43:377–397.

Lotspeich, F. B. 1980. Watersheds as the basic ecosystem: this conceptual framework provides a basis for a natural classification system. *Water Resour. Bull.* 16:581–586.

Lotspeich, F. B. and W. S. Platts. 1982. An integrated land-aquatic classification system. *North Am. J. Fish. Manage.* 2:138–149.

Maciolek, J. A. 1978. Insular aquatic ecosystems: Hawaii. in Classification, Inventory, and Analysis of Fish and Wildlife Habitat. FWS/OBS-78/76. U.S. Fish and Wildlife Service. Washington, D.C. pp. 103–120.

Mahon, R., E. K. Balon, and D. L. G. Noakes. 1979. Distribution, community structure and production of fishes in the upper Speed River, Ontario: a pre-impoundment study. *Environ. Biol. Fish.* 4:219–244.

Maitland, P. S. 1966. *Studies on Loch Lomond.* Vol 2. *The Fauna of the River Endrick.* Blackie & Son. Glasgow.

Manion, P. J. and L. H. Hanson. 1980. Spawning behavior and fecundity of lampreys from the upper three Great Lakes. *Can. J. Fish. Aquat. Sci.* 37:1635–1640.

Manion, P. J. and A. L. McLain. 1971. Biology of Larval Sea Lampreys (*Petromyzon marinus*) of the 1960 Year Class, Isolated in the Big Garlic River, Michigan, 1960–65. Tech. Rep. No. 16. Great Lakes Fishery Commission. Ann Arbor, MI.

Manny, B. A., T. A. Edsall, and E. Jaworski. 1988. The Detroit River, Michigan: An Ecological Profile. Biol. Rep. 85(7.17). U.S. Fish and Wildlife Service. Washington, D.C.

Matthews, W. J. 1985. Distribution of midwestern fishes on multivariate environmental gradients, with emphasis on *Notropis lutrensis. Am. Midl. Nat.* 113:225–237.

Matthews, W. J. 1986. Fish faunal "breaks" and stream order in the eastern and central United States. *Environ. Biol. Fish.* 17:81–92.

Mattingly, R. L., N. R. Kevern, and R. A. Cole. 1981. Re-examination of a culturally-affected Michigan river. *Int. Ver. Theor. Angew. Limnol. Verh.* 21:830–840.

McClendon, D. M. and C. F. Rabeni. 1987. Physical and biological variables useful for predicting population characteristics of smallmouth bass and rock bass in an ozark stream. *North Am. J. Fish. Manage.* 7:46–56.

McLain, A. L., B. R. Smith, and H. H. Moore. 1965. Experimental Control of Sea Lampreys with Electricity on the South Shore of Lake Superior, 1953–60. Tech. Rep. 10. Great Lakes Fishery Commission. Ann Arbor, MI.

Menzel, B. W. and H. L. Fierstine. 1976. A Study of the Effects of Stream Channelization and Bank Stabilization on Warm Water Sport Fish in Iowa: Subproject No. 5. Effects of Long Reach Stream Channelization on Distribution and Abundance of Fishes. FWS/OBS-76(15). U.S. Fish and Wildlife Service. Washington, D.C.

Michigan Department of Conservation. 1957. Watershed Survey Report: Little Manistee River Watershed. Dingell-Johnson Project F4R7. Michigan Department of Natural Resources, Lansing.

Milner, N. J., R. J. Hemsworth, and B. E. Jones. 1985. Habitat evaluation as a fisheries management tool. *J. Fish Biol.* 27(Suppl. A):85–108.

Minckley, W. L. 1963. The ecology of a spring stream Doe Run, Meade County, Kentucky. *Wildl. Soc. Wildl. Monogr.* 11.

Minor, J. M., L. M. Caron, and M. P. Meyer. 1977. Upper Mississippi River Habitat Inventory. Remote Sensing Applications in Agriculture and Forestry, Res. Rep. 77–7, University of Minnesota. St. Paul.

Minshall, G. W., R. C. Petersen, K. W. Cummins, T. L. Bott, J. R. Sedell, C. E. Cushing, and R. L. Vannote. 1983. Interbiome comparison of stream ecosystem dynamics. *Ecol. Monogr.* 53:1–25.

Minshall, G. W., K. W. Cummins, R. C. Petersen, C. E. Cushing, D. A. Bruns, J. R. Sedell, and R. L. Vannote. 1985. Developments in stream ecosystem theory. *Can. J. Fish. Aquat. Sci.* 42:1045–1055.

Moore, H. H. and R. A. Braem. 1965. Distribution of Fishes in U.S. Streams Tributary to Lake Superior. Spec. Sci. Rep. Fish. 516. U.S. Fish and Wildlife Service. Washington, D.C.

Mundy, P. R. and H. T. Boschung. 1981. An analysis of the distribution of lotic fishes with application to fisheries management, in *Proc. Warmwater Stream Symposium.* Southern Division. American Fisheries Society. Bethesda MD. pp. 266–275.

Naiman, R. J., J. M. Melillo, M. A. Lock, T. E. Ford, and S. R. Reice. 1987. Longitudinal patterns of ecosystem processes and community structure in a subartic river continuum. *Ecology* 68:1139–1156.

Neter, J. and W. Wasserman. 1974. *Applied Linear Statistical Models, Regression, Analysis of Variance, and Experimental Designs.* Richard D. Irwin. Homewood, IL.

New York Conservation Department. 1927. A Biological Survey of the Genessee River System. Suppl. 16th Annu. Rep. 1926. Albany.

New York Conservation Department. 1928. A Biological Survey of the Oswego River System. Suppl. 17th Annu. Rep. 1927. Albany.

New York Conservation Department. 1929. A Biological Survey of the Erie-Niagara System. Suppl. 18th Annu. Rep. 1928. Albany.

New York Conservation Department. 1931. A Biological Survey of the St. Lawrence Watershed (Including the Grass, St. Regis, Salmon, Chateaugay Systems and the St. Lawrence between Ogdensburg and the International Boundary). Suppl. 20th Annu. Rep. 1930. Albany.

New York Conservation Department. 1932. A Biological Survey of the Oswegatchie and Black River System (Including Also the Lesser Tributary Streams of the Upper St. Lawrence River and of Northeastern Lake Ontario). Suppl. 21st Annu. Rep. 1931. Albany.

New York Conservation Department. 1940. A Biological Survey of the Lake Ontario Watershed (Including All Waters from Little Sandy Creek Westward Except the Genessee and Oswego River Systems). Suppl. 29th Annu. Rep. 1939. Albany.

Newell, T. A. and S. G. Buda. 1976. Michigan Tributary Loading to the Upper Great Lakes. EPA 905/4-75-001. U.S. Environmental Protection Agency. Chicago.

Ohio Department of Natural Resources. 1953. Lake Erie Pollution Survey. Chemical and Physical Quality of Lake Erie and its Tributaries in Ohio. Ohio Department of Natural Resources. Columbus.

Olson, J. E., E. A. Whippo, and G. C. Horak. 1981. Reach File Phase II Report: A Standardized Method for Classifying Status and Type of Fisheries. FWS/OBS-81/31. U.S. Fish and Wildlife Service. Washington, D.C.

Oliff, W. D. 1960. Hydrobiological studies on the Tugela River system. I. The main Tugela River. *Hydrobiologia* 14:281–385.

Omernik, J. M. 1987. Ecoregions of the conterminous United States. *Ann. Assoc. Am. Geogr.* 77:118–125.

Ormerod, S. J. and R. W. Edwards. 1987. The ordination and classification of macroinvertebrate assemblages in the catchment of the River Wye in relation to environmental factors. *Freshwater Biol.* 17:533–546.

Pajak, P. and R. J. Neves. 1987. Habitat suitability and fish production: a model evaluation for rock bass in two Virginia streams. *Trans. Am. Fish. Soc.* 116:839–850.

Paragamian, V. L. 1981. Some habitat characteristics that affect abundance and winter survival of smallmouth bass in the Maquoketa River, Iowa, in *Proc. Warmwater Stream Symp.*, Southern Division. American Fisheries Society. Bethesda, MD. pp. 45–53.

Patch, S. P. and W. N. Busch. 1986. Biological Survey of the International Section of the St. Lawrence River. Vol. 1. USFWS Spec. Rep. Buffalo District, Final Rep. U.S. Army Corps of Engineers. Buffalo, NY.

Pennak, R. W. 1971. Toward a classification of lotic habitats. *Hydrobiologia* 38:321–334.

Pennak, R. W. 1978. The dilemma of stream classification. in Classification, Inventory, and Analysis of Fish and Wildlife Habitat. FWS/OBS-78/76. U.S. Fish and Wildlife Service. Washington, D.C. pp. 59–66.

Perry, J. A. and D. J. Schaeffer. 1987. The longitudinal distribution of riverine benthos: A river dis-continuum. *Hydrobiologia* 148:257–268.

Persoone, G. 1979. Proposal for a biotypical classification of water courses in European communities. in *Biological Indicators of Water Quality.* A. James and L. Evison, Eds. John Wiley & Sons. Toronto. pp. 7.1–7.32.

Pflieger, W. L. 1971. A distributional study of Missouri fishes. *Univ. Kans. Publ. Mus. Nat. Hist.* 20:225–570.

Pflieger, W. L., P. S. Haverland, and M. A. Schene. 1982a. Missouri's system for storage retrieval and analysis of stream resource data, in *Proc. Symp. Acquisition and Utilization of Aquatic Habitat Inventory Information.* Western Division. American Fisheries Society. Bethesda, MD. pp. 284–290.

Pflieger, W. L., M. A. Schene, Jr., and P. S. Haverland. 1982b. Techniques for the classification of stream habitats, with examples of their application in defining the stream habitats of Missouri. in *Proc. Symp. Acquisition and Utilization of Aquatic Habitat Inventory Information.* Western Division. American Fisheries Society. Lawrence, KS. pp. 362–368.

Platts, W. S. 1974. Geomorphic and Aquatic Conditions Influencing Salmonids and Stream Classification —with Application to Ecosystem Classification. Surface Environment and Mining Program. U.S. Forest Service. Billings, MT.

Platts, W. S. 1979. Relationships among stream order, fish populations, and aquatic geomorphology in an Idaho river drainage. *Fisheries (Bethesda)* 4(2):5–9.

Platts, W. S. 1982. Stream inventory garbage in—reliable analysis out: only in fairy tales, in *Proc. Symp. Acquisition and Utilization of Aquatic Habitat Inventory Information.* Western Division. American Fisheries Society. Bethesda, MD. pp. 75–84.

Platts, W. S., W. F. Megahan, and G. W. Minshall. 1983. Methods for Evaluating Stream, Riparian, and Biotic Conditions. Gen. Tech. Rep. INT-138. U.S. Forest Service. Ogden, UT.

Powers, W. E. 1962. Drainage and climate. in *Great Lakes Basin.* Publ. No. 71. H. J. Pincus, Ed. American Association for the Advancement of Science. Washington, D.C. pp. 29–40.

Raemhild, G. 1985. *The Application of Hydroacoustic Technique in Solving Fishery Problems Related to Hydroelectric Dam.* BioSonics, Inc., Seattle, WA.

Rakocinski, C. 1988. Population structure of stream-dwelling darters: correspondence with habitat structure. *Environ. Biol. Fish.* 23:215–224.

Raleigh, R. F., L. D. Zuckerman, and P. C. Nelson. 1986. Habitat Suitability Index Models and Instream Flow Suitability Curves: Brown Trout Revisited. Biol. Rep. 82(10.124). U.S. Fish and Wildlife Service. Washington, D.C.

Ricker, W. E. 1934. An ecological classification of certain Ontario streams. *Publ. Ontario Fish. Res. Lab.* 49:1–114.

Rohm, C. M., J. W. Giese, and C. C. Bennett. 1987. Evaluation of an aquatic ecoregion classification of streams in Arkansas. *J. Freshwater Ecol.* 4:127–140.

Ruediger, R. A. and J. D. Engels. 1982. A technique for mapping stream channel topography and habitat using the RDS/PAL computer program, in *Proc. Symp. Acquisition and Utilization of Aquatic Habitat Inventory Information.* Western Division. American Fisheries Society. Bethesda, MD. pp. 162–169.

Ryder, R. A. 1964. Chemical characteristics of Ontario lakes as related to glacial history. *Trans. Am. Fish. Soc.* 93:260–268.

Ryder, R. A. 1968. Dynamics and exploitation of mature walleyes, (*Stizostedion vitreum vitreum*), in the Nipigon Bay region of Lake Superior. *J. Fish. Res. Bd. Can.* 25:1347–1376.

SAS Institute, Inc. 1985. *SAS User's Guide: Statistics.* 5th ed. Cary, NC.

Scarnecchia, D. L. and E. P. Bergersen. 1987. Trout production and standing crop in Colorado's small streams as related to environmental features. *North Am. J. Fish. Manage.* 7:315–330.

Schelske, C. L., L. E. Feldt, and M. S. Simmons. 1980. Phytoplankton and Physical-Chemical Conditions in Selected Rivers and the Coastal Zone of Lake Michigan, 1972. Great Lakes Res. Div. Publ. 19. University of Michigan, Ann Arbor.

Schneberger, E. and A. D. Hasler. 1946. Brule River survey: Introduction. *Trans. Wis. Acad. Sci. Arts Lett.* 36: 1–6.

Sheldon, A. L. 1968. Species diversity and longitudinal succession in stream fishes. *Ecology* 49:193–198.

Shelford, V. E. 1911. Ecological succession. I. Stream fishes and the methods of physiographic analysis. *Biol. Bull.* 21:9–35.

Shelford, V. E. and M. W. Boesel. 1942. Bottom animal communities of the islands area of western Lake Erie in the summer of 1937. *Ohio J. Sci.* 42:179–190.

Shera, W. P. 1982. Using the British Columbia hierarchial watershed coding system to organize basin data, in *Proc. Symp. Acquisition and Utilization of Aquatic Habitat Inventory Information.* Western Division. American Fisheries Society. Bethesda, MD. pp. 26–31.

Shera, W. P. and E. A. Harding. 1982. Cost-efficient biophysical stream surveys: a proven approach, in *Proc. Symp. Acquisition and Utilization of Aquatic Habitat Inventory Information.* Western Division. American Fisheries Society. Bethesda, MD. pp. 319–324.

Sly, P. G. and C. C. Widmer. 1984. Lake trout (*Salvelinus namaycush*) spawning habitat in Seneca Lake, New York. *J. Great Lakes Res.* 10:168–189.

Smith, B. R. 1962. Spring and Summer Temperatures of Streams Tributary to the South Shore of Lake Superior, 1950–60. Spec. Sci. Rep. Fish. No. 410. U.S. Fish and Wildlife Service. Washington, D.C.

Smith, G. R., J. N. Taylor, and T. W. Grimshaw. 1981. Ecological survey of fishes in the Raisin River drainage, Michigan. *Mich. Acad.* 13:275–305.

Sonzogni, W. C., T. J. Monteith, W. N. Bach, and V. G. Hughes. 1978. United States Great Lakes Tributary Loading. Prepared by the Great Lakes Basin Commission Staff for the Pollution From Land Use Activities Reference Group, International Joint Commission. Windsor, Ontario.

Southwood, T. R. E. 1977. Habitat, the templet for ecological strategies?. *J. Anim. Ecol.* 46:337–365.

Sprules, W. G. 1977. Crustacean zooplankton communities as indicators of limnological conditions: an approach using principal component analysis. *J. Fish. Res. Bd. Can.* 34:962–975.

Stauffer, T. M. 1972. Age, growth, and downstream migration of juvenile rainbow trout in a Lake Michigan tributary. *Trans. Am. Fish. Soc.* 101:18–29.

Steedman, R. J. 1988. Modification and assessment of an index of biotic integrity to quantify stream quality in southern Ontario. *Can. J. Fish. Aquat. Sci.* 45:492–501.

Swanston, D. N., W. L. Meehan, and J. A. McNutt. 1977. A Quantitative Geomorphic Approach to Predicting Productivity of Pink and Chum Salmon in Southeast Alaska. Res. Pap. PNW-227. Pacific Northwest Forest and Range Experiment Station, U.S. Forest Service. Portland, OR.

Terrell, J. W., J. E. McMahon, P. D. Inskip, R. F. Raleigh, and K. L. Williamson. 1982. Habitat Suitability Index Models: Appendix A. Guidelines for Riverine and Lacustrine Applications of Fish HSI Models with the Habitat Evaluation Procedures. FWS/OBS-82(10.A). U.S. Fish and Wildlife Service. Washington, D.C.

Thompson, D. H. and F. D. Hunt. 1930. The fishes of Champaign County: a study of the distribution and abundance of fishes in small streams. *Ill. Nat. Hist. Surv. Bull.* 19:1–101.

Thomson, J. W. 1944. A survey of the larger aquatic plants and bank flora of the Brule River. *Trans. Wis. Acad. Sci. Arts Lett.* 36:57–76.

Thornley, S. and Y. Hamdy. 1983. An Assessment of the Bottom Fauna and Sediments of the Detroit River. Ontario Ministry of the Environment, Water Resources Branch, Great Lakes Section. Toronto, Ontario.

Tramer, E. J. and P. M. Rogers. 1973. Diversity and longitudinal zonation in fish populations of two streams entering a metropolitan area. *Am. Midl. Nat.* 90:366–374.

Trautman, M. B. 1942. Fish distribution and abundance correlated with stream gradients as a consideration in stocking programs. *Trans. North Am. Wildl. Conf.* 7:211–223.

Tyus, H. M. 1987. Distribution, reproduction, and habitat use of the razorback sucker in the Green River, Utah, 1979–1986. *Trans. Am. Fish. Soc.* 116:111–116.

Udvardy, M. F. D. 1959. Notes on the ecological concepts of habitat, biotope and niche. *Ecology* 40:725–728.

Usinger, R. L. 1963. *Aquatic Insects of California.* University of California Press, Berkeley.

Van Deusen, R. D. 1954. Maryland freshwater stream classification by watersheds. *Contr. Chesapeake Biol. Lab.* 106:1–30.

Van Meter, H. D. and M. B. Trautman. 1970. An annotated list of the fishes of Lake Erie and its tributary waters exclusive of the Detroit River. *Ohio J. Sci.* 72:65–78.

Vande Castle, J. R., R. G. Lathrop, Jr., and T. M. Lillesand. 1988. The significance of GIS and remote sensing technology in Great Lakes Monitoring and Resource Management. in *The Great Lakes: Living with North America's Inland Waters.* D. H. Hickcox, Ed. American Water Resources Association. Bethesda, MD. pp. 155–161.

Vannote, R. L., G. W. Minshall, K. W. Cummins, J. R. Sedell, and C. E. Cushing. 1980. The river continuum concept. *Can. J. Fish. Aquat. Sci.* 37:130–137.

Vaughan, R. D. and G. L. Harlow. 1965. Pollution of the Detroit River, Michigan Waters of Lake Erie, and their Tributaries. U.S. Department of Health, Education, and Welfare. Grosse Ile, MI.

Velz, C. J. and J. J. Gannon. 1960. Drought Flow Characteristics of Michigan Streams. Michigan Water Resource Commission, Ann Arbor.

Warren, C. E. 1979. Toward Classification and Rationale for Watershed Management and Stream Protection. EPA 600/3-79-059. U.S. Environmental Protection Agency. Corvallis, OR.

Water Resource Commission. 1973. Trout Water Quality. Michigan Department of Natural Resources, Lansing.

Wells, F. C. and C. R. Demas. 1979. Benthic invertebrates of the lower Mississippi River. *Water Resour. Bull.* 15:1565–1577.

Wesche, T. A., C. M. Goertler, and W. A. Hubert. 1987. Modified habitat suitability index model for brown trout in southeastern Wyoming. *North Am. J. Fish. Manage.* 7:232–237.

Whiteside, B. G. and R. M. McNatt. 1972. Fish species in relation to stream order and physiochemical conditions in the Plum Creek drainage basin. *Am. Midl. Nat.* 88:90–100.

Whittaker, R. H., S. A. Levin, and R. E. Root. 1973. Niche, habitat, and ecotope. *Am. Nat.* 107:321–338.

Whitworth, M. R., L. S. Ischinger, and G. C. Horak. 1985. Guidelines for Implementing Natural Resources Information Systems: The River Reach Fisheries Information System. Biol. Rep. 85(8). U.S. Fish and Wildlife Service. Washington, D.C.

Wright, J. F., D. Moss, P. D. Armitage, and M. T. Furse. 1984. A preliminary classification of running water site in Great Britain based on macroinvertebrate species and the prediction of community type using environmental data. *Freshwater Biol.* 14:221–256.

Zelewski, M. and R. J. Naiman. 1985. The regulation of riverine fish communities by a continuum of abiotic-biotic factors. in *Habitat Modifications and Freshwater Fisheries*. J. S. Alabaster, Ed. Butterworths. London.

Ziemer, G. L. 1973. Quantitative Geomorphology of Drainage Basins Related to Fish Production. Info. Leafl. No. 162. State of Alaska Department of Fish and Game. Juneau.

Zimmerman, J. W. 1968. Water Quality of Streams Tributary to Lakes Superior and Michigan. Spec. Sci. Rep. Fish. No. 559. U.S. Fish and Wildlife Service. Washington, D.C.

Zurek, G. M. 1987. Index to Reports on Fisheries Research 1930–Present. Michigan DNR, Institute of Fishery Research, Ann Arbor, MI.

Chapter 6

A REVIEW OF THE PHYSICAL AND CHEMICAL COMPONENTS OF THE GREAT LAKES: A BASIS FOR CLASSIFICATION AND INVENTORY OF AQUATIC HABITATS

C. E. Herdendorf, L. Håkanson, D. J. Jude, and P. G. Sly

TABLE OF CONTENTS

ABSTRACT

The following components have been identified as the physical and chemical basis for a classification and inventory of Great Lakes habitat:

Morphometry
- Mean and maximum depth
- Area and volume
- Orientation and fetch characteristics
- Nearshore slope and shoreline configuration
- Morphoedaphic and exposure indices

Sediment
- Edaphic character of watershed
- Texture and composition
- Erosion, transport, and deposition
- Substrate habitat (natural and artificial)

Water Movements
- Waves and currents
- Internal circulation
- Alongshore transport
- Water level fluctuations
- Water retention rates
- Tributary inflows and outflows

Climate and Thermal Regime
- Solar radiation and weather
- Light penetration
- Thermal stratification and mixing
- Ice formation, cover, and breakup
- Seasonal characteristics

Water Constituents
- Suspended solids and turbidity
- Dissolved oxygen and other gases
- Dissolved solids
- Alkalinity and pH
- Nutrients
- Toxic substances
- Tributary inputs

Relationship of Physicochemical Components to Lake Biota
- Vertical zonation of water column
- Horizontal zonation of water column and substrate
- Temporal (seasonal) variation patterns

These components are discussed and models are presented for assessing their relationship to lake biota.

I. INTRODUCTION

The physical forces of the environment that impinge on the lives of fishes are many, complex, and interrelated in their effects. Rawson (1939) was among the first limnologists to articulate the interactions of the major physical and chemical factors that control the productivity of a lake. More recently, Boyce (1974) provided an extensive review of the physical factors that influence chemical and biological processes in the Great Lakes. Figure 1 shows Rawson's concept of the organization of the physical and chemical factors that influence lake metabolism. This organization has stood the test of time and provides us with a useful model for developing a classification system for aquatic habitats. The following sections of this

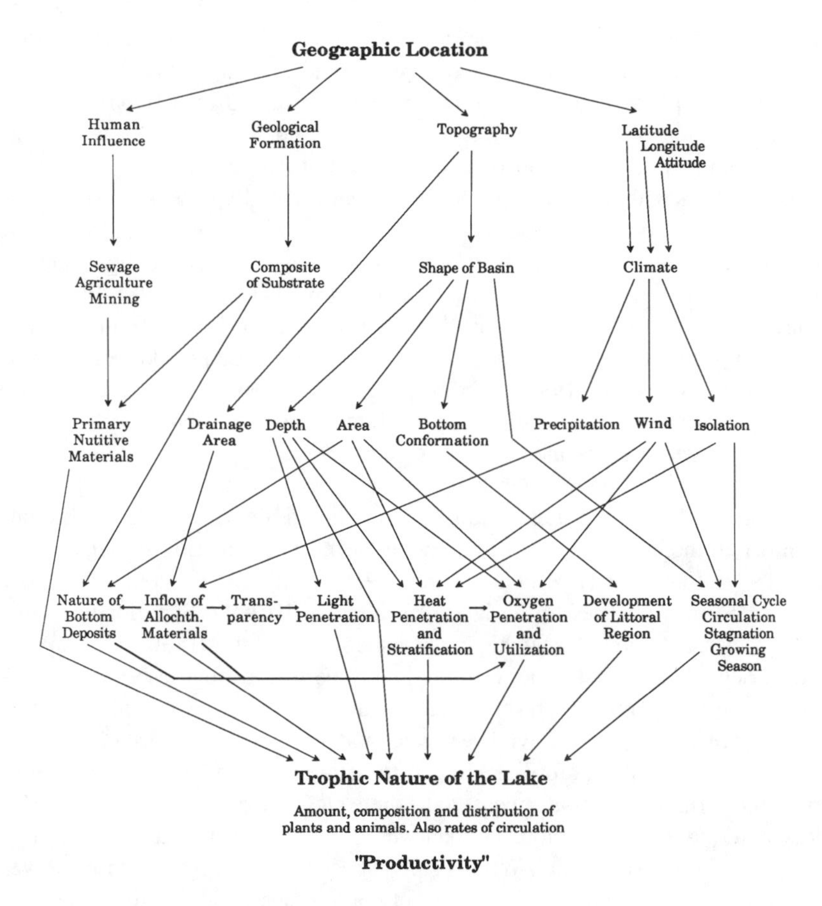

FIGURE 1. Interrelation of physical and chemical factors affecting the metabolism of a lake. (Data source: Rawson, 1939.)

chapter explore the physical and chemical components of the aquatic habitat in terms of their role as confining factors, forcing functions, and response components within the Great Lakes environment.

Physical components of the environment have caused fish to develop many different mechanisms and strategies to overcome the "negative" aspects of habitat. They are particularly important in relation to spawning behavior and the development of early life stages. Many species of fish such as the alewife (*Alosa pseudoharengus*), common carp (*Cyprinus carpio*), and gizzard shad (*Dorosoma cepedianum*) produce large numbers of eggs that hatch rapidly in optimum temperatures. These species usually have protracted spawning seasons to take advantage of optimum environmental conditions when they occur over an extended time period. Other species such as rainbow smelt (*Osmerus mordax*), walleyes (*Stizostedion vitreum*), suckers (Catostomidae), and burbot (*Lota lota*) deposit eggs in tributary streams, beach areas, or rocky shoals. Their larvae are washed into the open water where food availability is considerably enhanced. Deepwater sculpin (*Myoxocephalus thopmsoni*), ninespine stickleback (*Pungitius pungitius*), johnny darters (*Etheostoma nigrum*), and slimy sculpins (*Cottus cognatus*) produce fewer large eggs and fan their nests to ensure adequate oxygenation. Other species such as many salmonids spawn in tributary streams and the young remain there until an optimum size is reached. Lake trout (*Salvelinus namaycush*) use rocky reefs to deposit their eggs deep in the interstices. The eggs are thus protected from being washed out. Bloaters (*Coregonus hoyi*) have developed a strong resistance to starvation (Rice et al., 1985) which enables them to withstand temporary food shortages. Yellow perch (*Perca flavescens*) produce a unique egg matrix, which is accordion-like and expands vertically and horizontally, keeping the eggs suspended far above the substrate. Freshwater drum (*Aplodinotus grunniens*) produce eggs that float near the surface isolating them from any potential adverse impact that might result from resting on the substrate. All these strategies are strongly related to habitat, and their interactions with the physical environment substantially influence the abundance and extent of fish species in a water body.

Recent work has shown that both littoral morphometry and macrophytes are important components of habitat in Michigan and Florida lakes (Werner et al., 1978). Winter-dissolved oxygen levels and macrophytes were important in Wisconsin lakes (Tonn and Magnuson, 1982). Macrophyte diversity was important in Ontario lakes (Eadie and Keast, 1984) and in the St. Clair River (Poe et al., 1986). In addressing a lake ecosystem its tributaries need to be included, especially the lower portions of the larger streams (Herdendorf, 1990b). Many studies of channelized streams and rivers have documented the negative impact of destruction of the habitat diversity in these water bodies.

The fish fauna of the Great Lakes drainage is rich and diverse (Table 1). The number of species known in the lakes ranges from a low of 44 in Lake Superior to highs of 95 in Lake Ontario and 99 in Lake Erie (Hocutt and Wiley, 1986). Similarly, the comparison of numbers of species found in tributaries to the lakes reveals that the Lake Superior watershed is the lowest, 70 species, and the Lake Michigan watershed ranks highest, 130 species. Also, for the three lakes oriented east–west (Erie, Ontario, and Superior), tributaries to the south shore of each lake support larger numbers of species than those entering on the northshore. The southern tributaries to Lake Erie have 109 species and northshore tributaries have 81; for Lake Ontario the ratio is 102:88 and for Lake Superior the ratio is 65:59. Colonization of the Great Lakes from the periglacial region appears to be quite clearly from south to north, and southern tributaries were more readily colonized than were the northern tributaries, assuming a similarity in the tributaries. Hocutt and Wiley (1986) speculate that large expanses of water were barriers to certain stream-inhabiting species but not to others, and that the inhabitants of the southern tributaries are late migrants that for various reasons have not been able to further expand their range.

During their development, fishes are subjected to a wide range of environmental stressors, and the integration of individual effects is an important factor causing fluctuations in population densities. The stressors contribute to the degree of success of spawning, rates of survival of progeny, availability of food, and to the conditions encountered in migration. Temperature, wind, precipitation, turbidity, light, currents, and substances dissolved in the water (Doan, 1942) are among the most important characteristics of the environment.

Langlois (1954) found that the kinds of aquatic animals upon which fishes feed vary with the types of living conditions present in particular areas. In protected bays four positions are used: (1) the upper surfaces of stones and rocks, (2) beneath stones, but not embedded in the substratum, (3) within the substratum, and (4) on or among different kinds of aquatic plants. The struggle for existence among the inhabitants of a shoal area includes: (1) physical conditions such as ice, waves, currents, desiccation, and changes in temperature and (2) competition with other species. The diversity of rock surface and sufficiency of shelter, which modify wave action and reduce erosive power, are favorable to shoal organisms of most types (Langlois, 1954). Hunt (1885) considered the question of wave impact on shore animals and found that some animals seek shelter in rock niches, others have powers of adhesion, while still others have developed a form that offers the least possible resistance to waves and currents. Those animals that live on unstable bottoms come to depend upon the speed with which they can penetrate into such bottoms.

In Lake Erie, Krecker and Lancaster (1933) found that the density of the bottom population and the variety of its forms within the 2 m contour depended upon the type of substratum, the

TABLE 1
Fifty Common Fish Species of the Great Lakes and Their Thermal Preference and Origin in the Lakes (listed in phylogenetic order)

Fish	Adult size (in.)	Thermal preference[a]	Origin[b]
Family Petromyzontidae (lampreys)			
1. Sea lamprey	34	CD	MV
Family Aciperseridae (sturgeons)			
2. Lake sturgeon	48	CL	N
Family Lepisostedae (gars)			
3. Spotted gar	20	W	N
4. Longnose gar	30	CL	N
Family Amiidae (bowfins)			
5. Bowfin	21	W	N
Family Clupeidae (herrings)			
6. Alewife	6	CL	MV
7. Gizzard shad	10	W	FV
Family Hiodontidae (mooneyes)			
8. Mooneye	11	W	N
Family Salmonidae (trouts)			
9. Longjaw cisco	11	CD	N
10. Cisco or lake herring	10	CD	N
11. Lake whitefish	15	CD	N
12. Bloater	9	CD	N
13. Deepwater cisco	11	CD	N
14. Kiyi	10	CD	N
15. Blackfin cisco	13	CD	N
16. Shortnose cisco	10	CD	N
17. Shortjaw cisco	11	CD	N
18. Coho salmon	25	CD	MT
19. Chinook salmon	34	CD	MT
20. Rainbow or steelhead trout	20	CD	FT
21. Atlantic salmon	18	CD	N
22. Brown trout	16	CD	FT
23. Lake trout	18	CD	N
Family Osmeridae (smelts)			
24. Rainbow smelt	7	CD	MT
Family Esocidae (pikes)			
25. Northern pike	25	CL	N
26. Muskellunge	38	CL	N
Family Cyprinidae carps and minnows)			
27. Common carp	17	W	FT
28. Goldfish	10	W	FT
29. Emerald shiner	3	CL	N
30. Spottail shiner	4	CL	N
Family Catostomidae (suckers)			
31. Quillback	15	W	N
Family Ictaluridae (catfishes)			
32. Brown bullhead	12	W	N
33. Channel catfish	20	W	N
Family Gadidae (codfishes)			
34. Burbot	15	CL	N
Family Percichthyidae (temperate basses)			
35. White perch	7	CL	MV
36. White bass	12	W	N
Family Centrarchidae (sunfishes)			
37. Rock bass	9	W	N
38. Pumpkinseed	8	W	N

TABLE 1 (Continued)
Fifty Common Fish Species of the Great Lakes and Their Thermal Preference and Origin in the Lakes (listed in phylogenetic order)

Fish	Adult size (in.)	Thermal preference[a]	Origin[b]
39. Bluegill	8	W	N
40. Smallmouth bass	12	W	N
41. Largemouth bass	12	W	N
42. White crappie	10	W	N
43. Black crappie	9	CL	N
Family Percidae (perches)			
44. Yellow perch	10	CL	N
45. Sauger	14	CL	N
46. Walleye	17	CL	N
47. Blue pike	14	CL	N
Family Sciaenidae (drums)			
48. Freshwater drum	20	W	N
Family Cottidae (sculpins)			
49. Mottled sculpin	3	CL	N
50. Deepwater sculpin	3	CD	N

[a] Thermal preference: CD = coldwater, CL = coolwater, W = warmwater.
[b] Origin: N = native, FT = freshwater introduced, MT = marine introduced, FV = freshwater invader, MV = marine invader.

Note: The scientific names of the fish species listed above can be found in A List of Common and Scientific Names of Fishes from the United States and Canada. American Fisheries Society Spec. Publ. No. 12, 4th ed. (1980).

Data source: Herdendorf (1983).

character of the vegetation, and the depth of the water. Such physical factors as temperature, oxygen, carbon dioxide, and pH were generally so uniform over the area studied that they did not appear to be critical factors.

II. LAKES, CHANNELS, AND TRIBUTARIES

Habitat in some of the Great Lakes is strongly influenced by the contributions of nutrients, sediments, toxic materials, and biological materials that enter via the tributaries. Excessive nutrients have accelerated eutrophication, which has had widespread effects on the biota and physical characteristics (increased sedimentation, more algal blooms, increased macrophyte production, etc.) of the lower lakes, especially Lake Erie (Herdendorf, 1987). Dissolved oxygen depletion in the hypolimnion or degradation of organic matter (Sly, 1988) associated with increased eutrophication could be one of the factors affecting lake trout reproduction. Although Leach et al. (1977) found that yellow perch responded favorably to increased eutrophication, above a certain point, a decline took place in their production, which suggests that a general response curve may exist for all species for increased nutrient levels and other major variables, such as turbidity, climate (temperature profiles), dissolved oxygen, etc. The recent curtailment of nutrient inputs to the Great Lakes and amelioration of toxic contaminant loading, sometimes in conjunction with temporary changes in catch regulation, has allowed many rehabilitation efforts (walleye rebounds in Saginaw Bay and Lake Erie) to be successful.

Sedimentation over spawning areas can be detrimental, as noted by Sly et al. (1983) and Dorr et al. (1981a,b) for lake trout spawning habitat. Biogenic particles formed within the water column or derived from attached vegetation are all important sources of sediment. Some of the sediments in the Great Lakes are derived from riverine inputs, but most materials are derived from resuspension of lake deposits and shore erosion.

Toxic substances are present in the water, sediments, and biota of many rivers that enter the Great Lakes and eventually become part of the chemical medium that fishes must inhabit, along with inputs from the atmosphere (Eisenreich and Hollod, 1979). The impact of these substances is increased through bioaccumulation in various levels of the food chain, including fish (Camanzo et al., 1987). Direct and usually sublethal effects on fish and larval fish should be expected in areas of high concentrations (harbors, certain polluted rivers). These substances may also figure prominently in the lack of reproductive success in fish. Contaminant analyses of lake trout eggs from the various Great Lakes strongly indicts contaminant uptake by Lake Michigan fish as being a major detriment to hatching success (Edsall and Mac, 1982).

Rivers also act as a conduit for biological materials with two-way exchange. Wells (1973) and Perrone et al. (1983) have documented the entry into Lake Michigan of yellow perch produced in connecting water bodies and Mansfield (1984) showed spottail shiners (*Notropis hudsonius*) entering from a small creek in the Lake Michigan watershed. Several species of fish spawn in connecting water bodies, marshes, and rivers (Chubb and Liston, 1986; Jude et al., 1979; Perrone et al., 1983) and spend most of their first year of life in rivers, before migrating into the adjoining Great Lakes (Tesar and Jude, 1985).

III. MORPHOMETRY AND THE LINK TO PRODUCTIVITY THROUGH MODEL CONSTRUCTS

Lake morphometry has a profound influence on quality and quantity of aquatic habitat. The nearshore habitat is often most important for fish, but its subjectivity to human perturbation is greater. Each of the Great Lakes can be divided into: (1) an offshore pelagic zone, characterized by fine sediment, great depths, and generally a low diversity of habitat types and (2) a highly diverse nearshore zone that contains important spawning and nursery habitat for fish. The nearshore zone is also the receptacle for nutrients, contaminants, and sediments from the surrounding watershed. It is prone to development, with its accompanying water intakes, riprapped harbors and shorelines, power plants, sewage treatment plants, industrial outfalls, and loss of wetlands.

Basin morphometry acts as a confining factor that sets the limits of habitat within which aquatic species can reproduce, feed, and compete. Almost every month of the year some species of fish in the Great Lakes spawn, and each of these species has specific spawning requirements. For example, the open water pelagic zone is used by such species as bloaters and deepwater sculpin (Mansfield et al., 1983), the nearshore pelagic zone is used by alewife, rainbow smelt, yellow perch, white bass, and gizzard shad (*Dorosoma cepedianum*), while the nearshore benthic environment is used by slimy sculpins, trout-perch, and johnny darters (Jude et al., 1979). Connecting rivers, lakes, marshes, and embayments are important spawning and nursery areas for such species as white and longnose suckers, salmonids, emerald shiners (*Notropis atherinoides*), yellow perch, walleye, and northern pike (*Esox lucius*) (Chubb and Liston, 1986; Perrone et al., 1983; Jude et al., 1979). Thus, the quantity and quality of these habitat types are critical elements affecting the final species composition in each of the Great Lakes.

The development of simple models and/or indices useful in estimating potential fish productivity of a particular freshwater system is an important goal of resource managers. Boonyuen and Herdendorf (1986) reviewed studies that have attempted to identify the variables describing the morphological, chemical, and physical characteristics of lakes with the objective of using them to estimate lake productivity (Table 2). Rawson (1955) showed

TABLE 2
Physical and Chemical Variables Used in Construction of Productivity Models

Variable	Ref. code	Investigators
Air temperature	1, 10	1 – Brylinsky and Mann (1973)
Altitude	1, 5	2 – Carlander (1955)
Alkalinity		3 – Findenegg (1965)
Total phosphate	1, 4	4 – Moyle (1956)
Total nitrogen	1, 4	5 – Northcote and Larkin (1956)
TDS	5, 8	6 – Rawson (1952, 1955)
Limiting nutrient	10	7 – Rounsefell (1946)
Latitude	1	8 – Ryder et al. (1974)
Lake area	5,7	9 – Schindler (1971)
Mean depth	1, 5, 6, 8, 12	10 – Schlesinger & Regier (1982)
Maximum depth	2	11 – Talling (1969)
Mixing type	1, 3, 9	12 – Thienemann (1927)
		13 – Turner (1960)

Data obtained from Boonyuen and Herdendorf (1986).

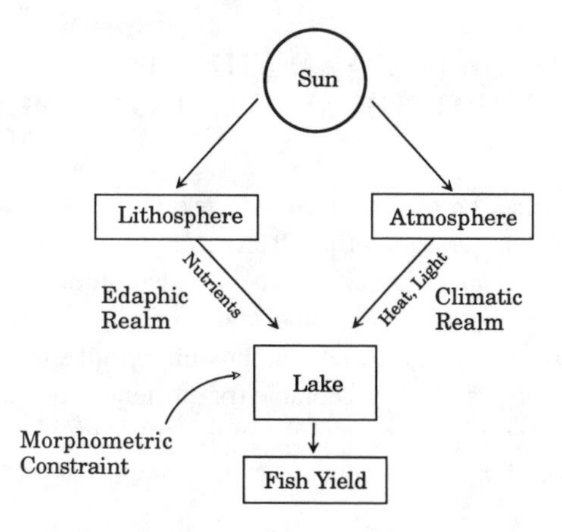

FIGURE 2. Schematic representation of an aquatic ecosystem showing the major energetic and materials pathways resulting in fish yield. (Data source: Håkanson, 1981.)

that in large lakes, morphometry is a dominant factor in fish productivity. He found that the standing crop of fish per unit area decreased with an increase in maximum depth. Rawson also stated that the factors influencing lake productivity could be grouped into three categories: (1) edaphic, e.g., total dissolved solids (TDS), and nutrients, (2) morphometric, e.g., mean depth and lake surface area, and (3) climatic, e.g., air temperature and latitude. He has proposed that given adequate edaphic, morphometric, and climatic information, a multiple correlation analysis would demonstrate a relationship between fish production and physical parameters.

Ryder (1982) presented a schematic of a lake showing the major energetic and material pathways (Figure 2). In the climatic realm he identified solar energy as the most important factor due to its effect on autotrophic production. For the edaphic realm, nutrients and oxygen are the prominent representatives of the material input. The morphometric factors play an important role not only in channeling energy and matter in predictable patterns, but also in acting as a sink for both nutrients and energy. From a conceptual model of this type, a practical numerical model can be constructed.

TABLE 3
Great Lakes Drainage Areas (km²)[a]

Component	Canada		U.S.		Total	
	Land	Water	Land	Water	Land	Water
Lake Superior	83,900	28,700	43,800	53,400	127,700	82,100
St. Mary's River	2,150	106	448	124	2,598	230
Lake Michigan	—	—	118,000	57,800	118,000	57,800
Lake Huron	89,900	36,000	41,400	23,600	131,300	59,600
St. Clair River	228	21	3,060	34	3,288	55
Lake St. Clair	9,790	694	2,640	420	12,430	1,114
Detroit River	552	41	1,680	60	2,232	101
Lake Erie	12,200	12,800	46,600	12,900	58,800	25,700
Niagara River	1,320	26	2,050	34	3,370	60
Lake Ontario	28,200	10,000	32,400	8,960	60,600	18,960
St. Lawrence River (above Iroquois)	1,700	269	4,820	223	6,520	492

[a] Water areas are those only of named body of water, smaller lakes, etc., within the basin being included with the land portion.

Data source: Herdendorf (1990a,b).

Wetzel (1983) pointed out that the morphology of a lake basin is best described by a detailed bathymetric map and that such information is essential for the evaluation of the morphological effects on nearly all of the physical, chemical, and biological processes of a lake. Useful morphometric parameters for "whole lake" assessment include (1) area, (2) mean and maximum depth, (3) volume, (4) fetch length, (5) shoreline length, (6) basin slope, and (7) orientation (Hutchinson, 1957). Examples of morphometric measurements for each of the Great Lakes are given in Tables 3 through 6. Håkanson (1981) provides a useful manual for calculating these and other parameters, and he notes that concerns associated with individual bays, sub-basins, or coastal areas are not dealt with because they require site-specific evaluation. Unfortunately, it is this type of site-specific detail which may be required to construct a useful habitat classification system.

Hypsographic (or depth-area) curves for each of the Great Lakes are shown in Figure 3. These hypsographic curves plot lake depth against cumulative area and provide a characterization of the large elements of basin form. They also provide a means whereby the area of any depth interval or lake volume may be determined.

Mean depth is the most frequently used morphometric variable in the construction of productivity models. Mean depth was first suggested by Thienemann (1927) as the most important morphometric factor in determining lake productivity, but he was unable to provide quantitative evidence to support his theory. Rawson (1952) further developed this concept, considering lake depth to be important primarily for its association with thermal stratification, circulation, and dilution of nutrients. He demonstrated that the biomass of plankton and benthos shared an inverse hyperbolic relationship with mean depth (Figure 4 shows only the former relationship).

It is also possible to relate lake surface to fish productivity. Rounsefell (1946) obtained an inverse relationship between fish yield and lake surface area. This was due to the reduction in fish production per unit area in most large lakes compared to small ones. Small, shallow lakes generally have a high proportion of their total area in the littoral zone. Large lakes are generally deeper than small lakes and contain a lesser proportion of total area in the littoral zone. Because this zone is often the most productive area for both primary and secondary

TABLE 4
Great Lakes Water Volumes

Lake	Volume (km^3)	Mean depth (m)
Superior	12,100	147.4
Michigan	4,920	85.1
Huron	3,540	59.4
Erie	484	18.8
Ontario	1,640	86.5

Data source: Herdendorf (1990a, b).

TABLE 5
Great Lakes Dimensions

Lake	Max length (km)	Max breadth (km)	Max depth (m)
Superior	563	257	405
Michigan	494	190	281
Huron	332	295	229
St. Clair	42	39	6[a]
Erie	388	92	64
Ontario	311	85	244

[a] Deepest sounding outside dredged navigation channel which has project depth of 8.2 m.

Data source: Herdendorf (1990).

TABLE 6
Great Lakes Shoreline Lengths (km)

	Canada		U.S.		Total	
Component	Mainland	Islands	Mainland	Islands	Mainland	Islands
Lake Superior	1390	990	1390	615	2780	1605
St. Mary's River	106	101	47	143	153	244
Lake Michigan	0	0	2250	383	2250	383
Lake Huron	2040	2770	933	414	2973	3184
St. Clair River	48	8	45	0	93	8
Lake St. Clair	114	69	95	135	209	204
Detroit River	98	53	48	63	96	116
Lake Erie	592	47	694	69	1286	116
Niagara River	53	5	58	55	111	60
Lake Ontario	538	80	483	45	1021	125
St. Lawrence River (above Iroquois)	166	253	171	175	337	428

Data source: Herdendorf (1990a, b).

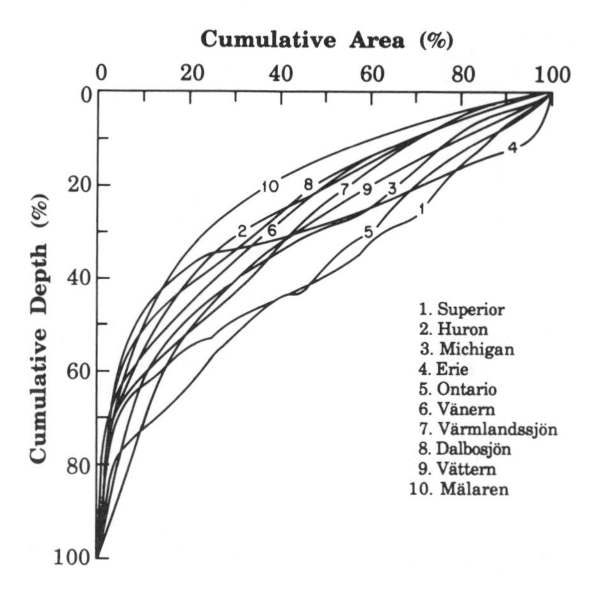

FIGURE 3. Relative hypsographic curves for ten large lakes. (Data source: Ryder, 1982.)

productivity, the average production per unit area decreases as surface area increases (Larkin 1964). Perhaps a more convincing relationship could have been demonstrated by Rounsefell (1946) if he had performed a regression between yield and mean depth rather than yield with area. Carlander (1955) reanalyzed Rounsefell's data and found no correlation between the standing crop of fish (biomass per unit area) and lake area. He also found that biomass per unit area decreased with an increase in maximum depth. Ryder (1982) also suggested that a better relation would have been obtained by substituting mean depth for surface area.

The edaphic factors, e.g., alkalinity, TDS, and total phosphorus, were viewed by Carlander (1955), Moyle (1956), Turner (1960), and Ryder (1965) as also contributing to lake productivity. The importance of nutrients and carbonate alkalinity is reflected mainly through their relationship to phytoplankton growth. These planktonic primary producers serve as a prime food source in the food web. Therefore, any increase in the abundance of such producers could result in increased numbers of secondary and tertiary consumers. Carlander (1955) demonstrated that a significant increase of fish biomass was associated with an increase in carbonate content (measured as alkalinity). Moyle (1956), using data from a number of Minnesota lakes, related fish yield to four independent variables: total phosphorus, total nitrogen, total alkalinity, and carbonate alkalinity. Similarly, Turner (1960) demonstrated a significant positive correlation between carbonate alkalinity and the standing crop of fish, based on a study of 22 Kentucky ponds.

Northcote and Larkin (1956) analyzed interrelations between various physical and chemical variables (e.g., altitude, lake area, mean depth, and dissolved solids) and plankton standing crop, benthic fauna, and fish for 100 lakes in British Columbia. Individual relationships between these variables were not found to be significant. However, a "bio-index" concept developed by combining relative measures of plankton, bottom fauna, and fish abundance in a multiple regression analysis with TDS and mean depth as independent variables demonstrated a significant correlation. The TDS content appeared to be the most important factor in determining the secondary productivity level in these lakes.

Brylinsky and Mann (1973) concluded from statistical analysis of lake and reservoir data (distributed from tropical to polar regions) that variables related to solar energy, e.g., latitude

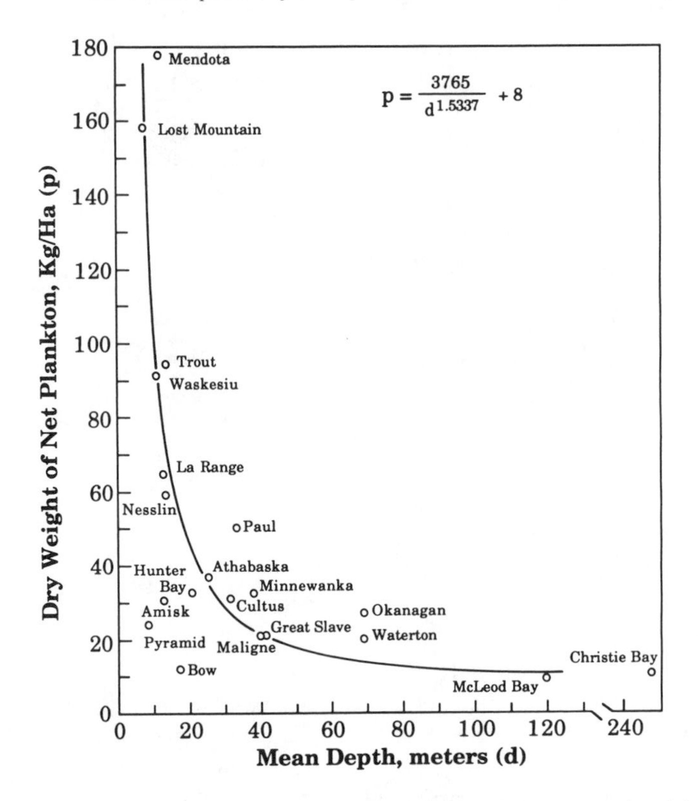

FIGURE 4. Mean depth and the average standing crop of plankton in 20 North American lakes. (Data source: Rawson, 1955.)

and air temperature, have a greater influence on both primary and secondary production than variables related to nutrient concentration. However, in the study of lakes within a narrow range of latitude, e.g., temperate or tropical zones, nutrient-related variables assume greater importance.

Another factor in controlling lake productivity is the circulation pattern. Findenegg (1965) demonstrated that lake stratification may lead to nutrient depletion in the epilimnion during summer; oligomixis (unusual, irregular, and short duration mixing); and meromixis (incomplete circulation of lake water) can lower lake production. The importance of mixing gained further support from the studies by Talling (1969) and Brylinsky and Mann (1973). Talling (1969) emphasized the importance of the annual mixing cycle in governing the productivity of tropical lakes. He explained that in tropic regions, the seasonal variations in temperature and solar radiation were minimal. The growth of phytoplankton in tropical areas therefore, could be enhanced by the vertical mixing that transports accumulated nutrients to the surface. Imboden et al. (1981) stated that long-term lake biological development was influenced by both continuous vertical mixing (i.e., eddy diffusion) and partial or complete turnover.

Schindler (1971) proposed that the supply rate of limiting nutrients to a lake ecosystem is directly proportional to watershed area and inversely proportional to lake volume. Schindler's model predicted the limiting nutrient supply rate, given the area of the watershed, lake surface area, and lake volume. The model, however, is predicated on the fact that the atmospheric input of nutrients is not significant when compared to the runoff input in Schindler's small lakes experimental area.

Productivity estimation involves a large number of variables; therefore, a model describing the relationships between variables influencing productivity is essential. A hierarchical model is most often used. Brylinsky and Mann (1973) developed this type of model based on the

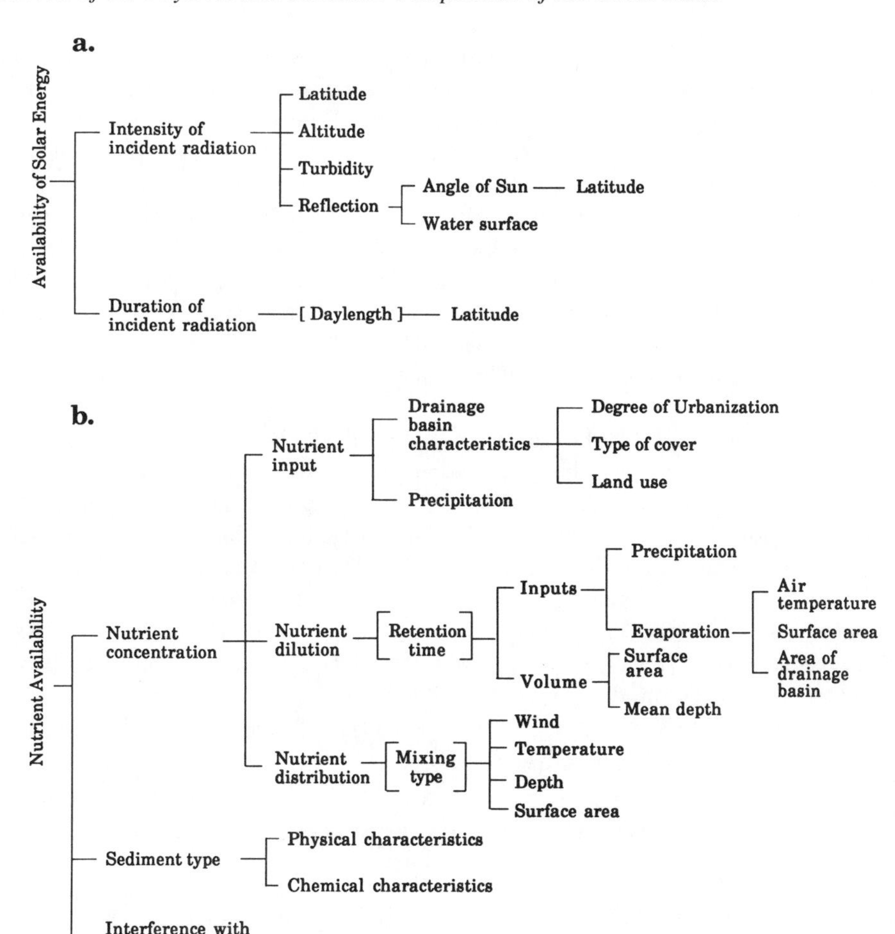

FIGURE 5. Factors influencing phytoplankton productivity in large lakes. (a) Hierarchical model of solar energy availability and (b) hierarchical model of nutrient availability. (Data source: Brylinsky and Mann, 1973.)

regression of individual variables against phytoplankton production. They grouped such variables into those related to solar energy availability and nutrient availability (Figure 5). The energy related variables are subdivided into intensity and duration of incident radiation, both of which include latitude categories.

Secondary production was divided into four major groups: (1) herbivorous zooplankton, (2) carnivorous zooplankton, (3) herbivorous benthos, and (4) carnivorous benthos. Estimates were made by utilizing related biotic and abiotic variables, e.g., phytoplankton chlorophyll, mean depth, mean epilimnion temperature, and duration of stratification. This study introduced the use of categories of controlling factors in predicting lake productivity.

Carlson (1977) used multiple regression to construct a numerical trophic state index (TSI) for lakes which can be calculated by using several parameters, including Secchi disc transparency, chlorophyll, and total phosphorus.

Success in estimating fish production is greater when abiotic factors are combined to produce more complex indices. The greater the number of variables presented in the model, the more details of the environment are represented; therefore, a more accurate prediction may be expected. Ryder (1965) presented a morphoedaphic index (MEI) which was successful when applied to the potential fish yield estimation from large temperate lakes and African lakes and large reservoirs. The MEI can be calculated as the ratio of TDS to mean depth, and thus combines lake chemistry and morphometry data. In Ryder's index, TDS represented morphometric and edaphic factors. If climate is not taken into account, the MEI is applicable only in the narrow range of latitude over which climatic differences are minimal. On the other hand, Schlesinger and Regier (1982) presented a "climatic index" utilizing mean annual air temperature as the only standard for the index. From stepwise regression analyses, the climatic index accounted for 74% of the variability of maximum sustainable fish yield. The correlation of determination was improved after integration of MEI with this model.

Multiple regression (MR) analyses can provide reliable estimates of potential fish yield. Hayes and Anthony (1964) derived a productivity index (PI) for 41 North American ponds and lakes based on fish standing stock, angler harvest, and fish yield. By computing a series of multiple regressions relating fish productivity to morphological and chemical variables, they account for 67% of the fish standing crop variability.

Boonyuen and Herdendorf (1986) substantially agreed with Rawson's (1952; 1955) earlier work and concluded that productivity could be grouped into three categories: climatic, edaphic, and morphometric factors. These factors are represented by latitude, circulation type, and mean depth. A PI was formulated: $PI = LC/M$; where C = circulation type, L = latitude code, $M = \log(\text{mean depth} + 1)$. Throughout the world 41 large lakes (surface area ≥ 500 km) were tested with this model. The regression equation describing the linear relation between fish productivity (P) and fish PI is $P = 1.65\ PI - 13.41$; where $N = 41$, $r^2 = 0.84$, $p \leq 0.0001$. A nonlinear relationship was also obtained from utilizing general linear regression between selected variables and fish productivity in the following equation: $\text{Log}(P) = 0./38 + 0.009\ C^2 + 0.52\ M^{-1} - 1.02$; where $N = 41$, $r^2 = 0.85$, $p \leq 0.0001$. Both the index and model should have application to large freshwater lakes worldwide.

IV. WATER RETENTION

Many factors influence water motions, and all the processes illustrated in Figure 6 can drive and/or mix coastal or nearshore waters. The water retention time is an important value since concentrations and ecological effects of pollutants from point sources can be predicted from such data. Water retention, which is an integration of motions over extended time, sets the framework for all biota. For example, the environments are quite different in coastal waters or lakes in which the retention time varies from hours to years.

A simple model is able to predict the median retention time of surface water, defined as the mixed water above the thermo- and/or the halocline (Figure 7). The model works best where tidal impact is small, such as the Great Lakes. The model does not work along traditional hydrodynamic lines. Instead of focusing on the various processes in one particular basin, it relies on data collected from many coastal areas and, in particular, their varied morphometry. For further information on field work, theories, alternative methods to determine the retention time, statistical methods, reliability of data, etc., see Håkanson et al. (1984, 1986). Two methods of calculating water retention time are presented in the Appendix.

V. KEY VARIABLES AND THEIR ROLE IN SHAPING AND CONTROLLING USE OF HABITAT

A. SUBSTRATE

1. Sediment Texture and Hydraulic Regime Relationships

Differences in sediment texture (particle size) often reflect changes in hydraulic energy, whereby decreasing particle size is a general correlate of diminishing water motion. However,

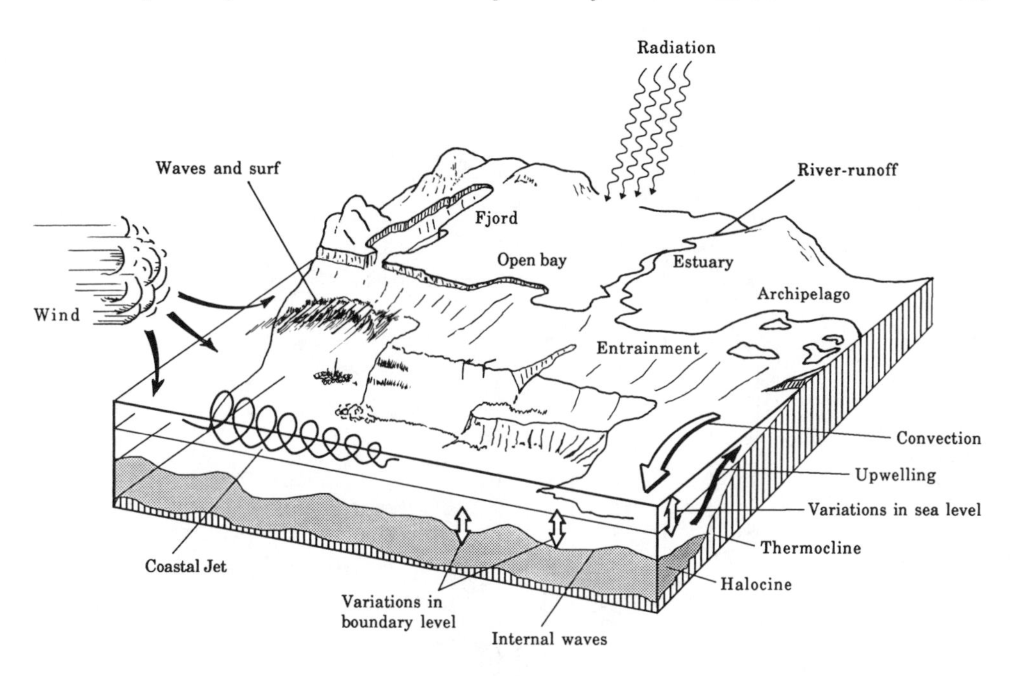

FIGURE 6. Schematic illustration of the complex causal relationships determining the water exchange in coastal areas.

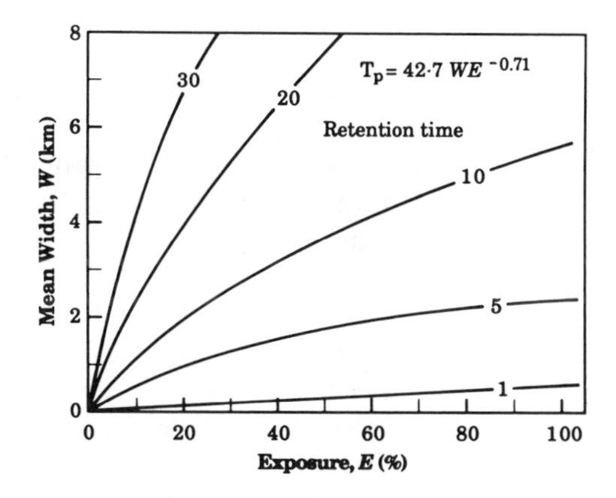

FIGURE 7. The relationship between retention time of surface water (Tp, days), mean coastal width (W, km), and exposure (E, %).

the mean size of a complete sediment sample may not be a good correlate of hydraulic conditions defined by near bed shear velocity, and three reasons explain this:

1. In coarse lag deposits (boulders and cobbles) only the smaller materials can be moved by hydraulic motion and, although the size of particulates in these residual materials can be very large, only the smallest remaining gravel-cobbles are closely related to formative hydraulic energies.
2. Some deposits are of mixed origin in which, for example, materials have been incorporated from both bed and suspended loads, such as the entrapment of muds in coarse gravels. The mean size of a complete sample of this material is unrelated to the formative hydraulic regime.

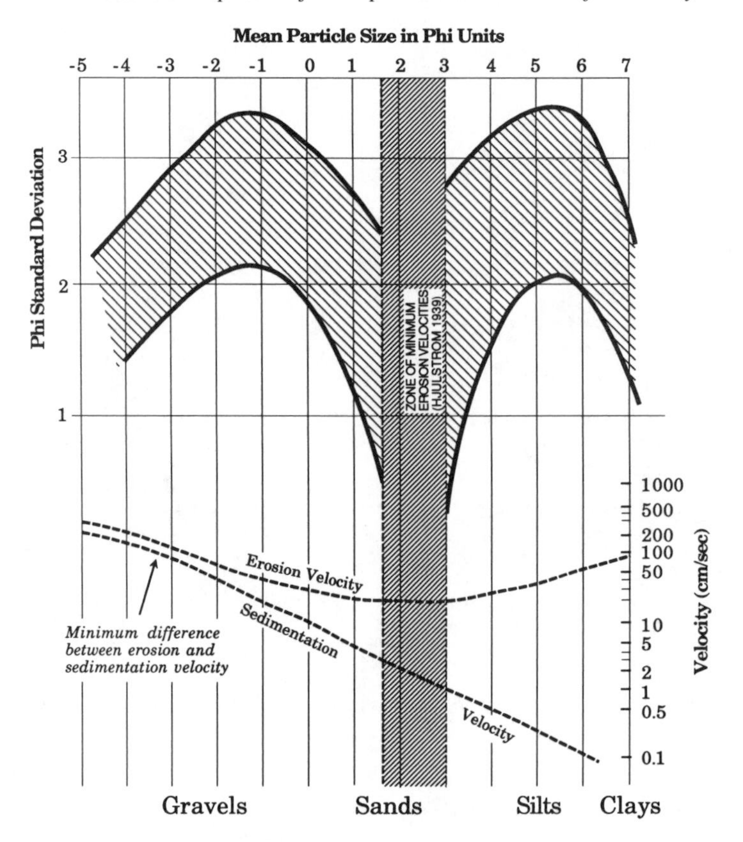

FIGURE 8. Sediment erosion, transport, and deposition velocities. (Data source: Sly et al., 1983.)

3. Due to the complex interactions between electrolytes, particle concentrations, and turbulent motion, the mean size of a large range of silt-clay mixtures may not be consistently related to hydraulic shear velocity.

There is a more or less continuous size spectrum from boulders to clays in which size is typically measured in metric or phi (–log mm/0.301) units. Based on Hjulstrom (1939) and many subsequent studies, researchers have found that minimum hydraulic shear velocities capable of setting particulates in motion will move particles of about 0.3 to 0.13 mm (1.7 to 3.0 phi), and that with increasing velocity both finer and coarser particulates can be moved (Figure 8). Hjulstrom also demonstrated that finer particulates settle from suspensions at progressively lower flow velocities. Because of these characteristics, it is possible to separate water-lain sediments into high (erosional) or low (depositional) energy regimes, depending upon whether their mean size is greater or less than the size equivalent of Hjulstrom's minimum erosion velocity.

Typically, the size relationships of sediments are analyzed by means of statistical moment measures (Sly et al., 1983); the first four such measures are the arithmetic mean, standard deviation, skewness, and kurtosis. The first two are common measures, but the latter two are more specialized:

1. Skewness provides a measure of the separation between the mean and the median; a negative sign indicates a shift toward coarser material and a positive sign indicates a shift toward finer material.
2. Kurtosis provides a measure of peakedness relative to a normal distribution; a positive sign indicates an excess of material in the central part of the distribution (extremely well

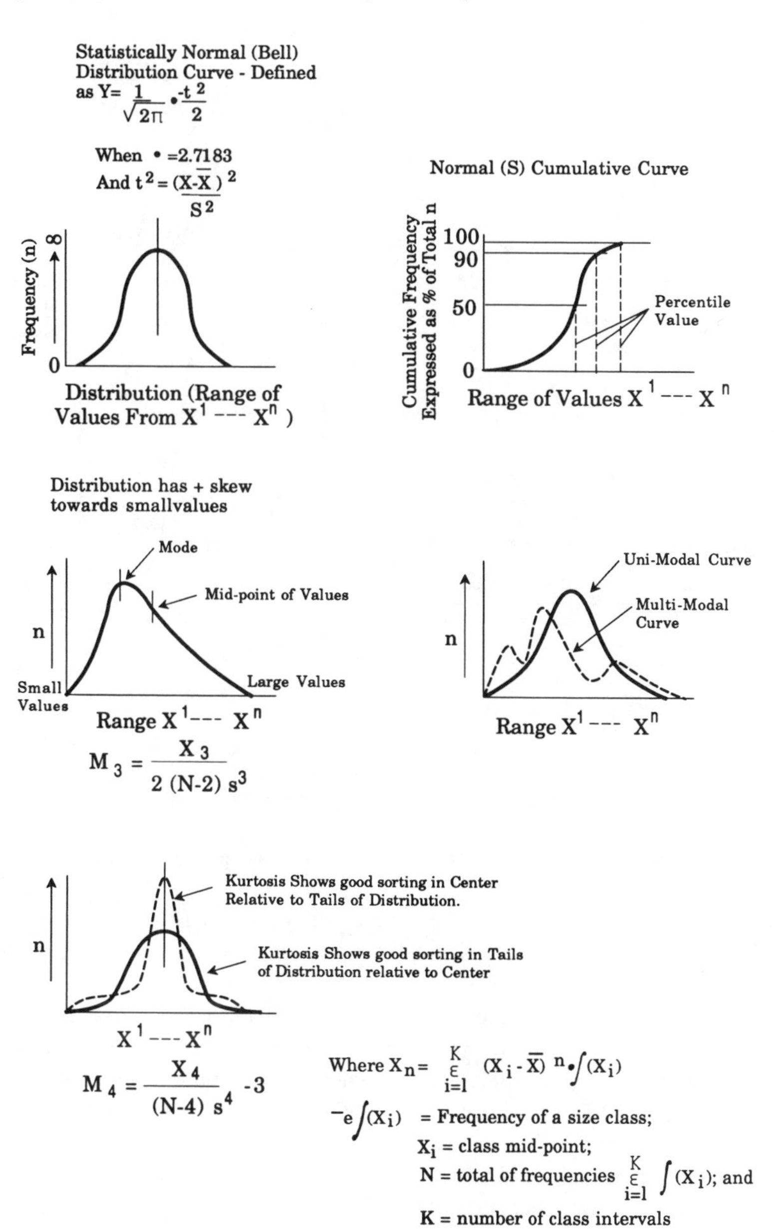

Where $X_n = \sum_{i=1}^{K} (X_i - \bar{X})^n \cdot f(X_i)$

$-e f(X_i)$ = Frequency of a size class;

X_i = class mid-point;

N = total of frequencies $\sum_{i=1}^{K} f(X_i)$; and

K = number of class intervals

FIGURE 9. Derivation of kurtosis statistics and graphic forms.

sorted), and a negative sign indicates that excess material exists in the tails of a distribution (often a mixture of materials from more than one source). The derivation of these statistics and their graphic forms are present in Figure 9.

By using skewness and kurtosis it is possible to plot the entire range of water-lain sediments as a continuous series (Figure 10). Sediments that lie close to the boundary curve are most related to existing hydraulic conditions, and those that plot away from this curve are anomalous. Anomalous sediments, in terms of hydraulic regime, include lag gravels, ice-drop debris, mixed layer or variable regime deposits, and flocculent materials. The use of skewness and kurtosis provides a valuable means of discriminating between materials that are related to

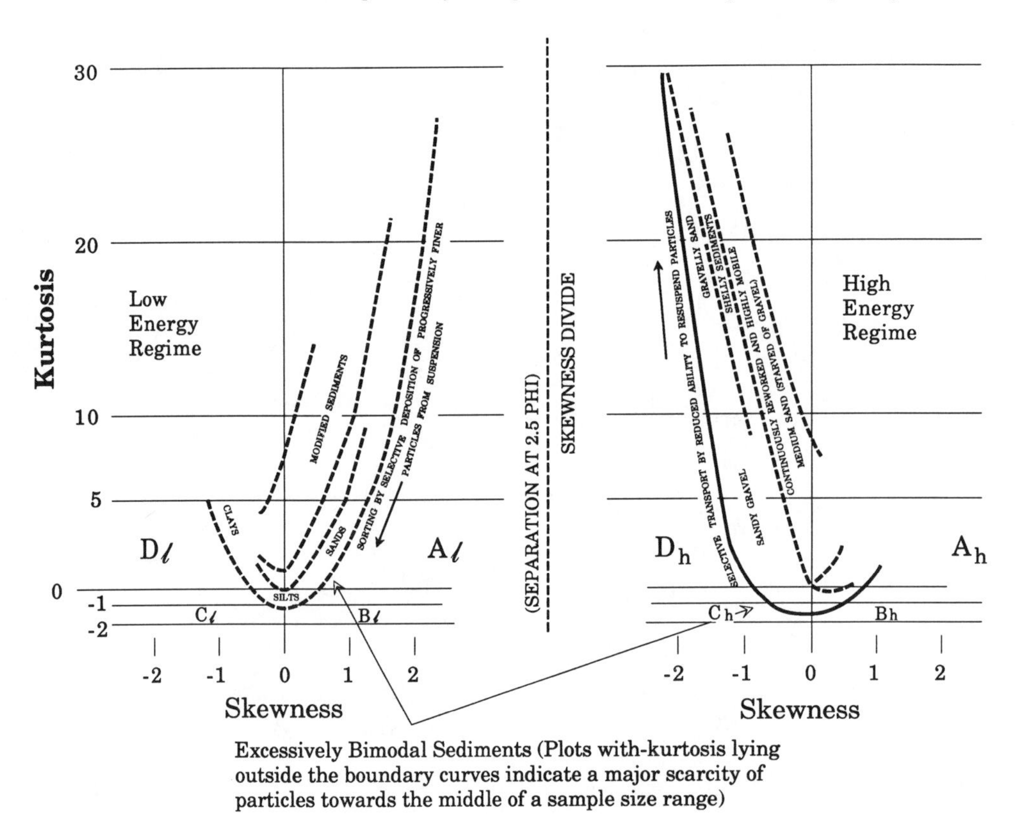

FIGURE 10. Plot of water-lain sediments using skewness and kurtosis statistics. (Data source: Sly et al., 1983.)

hydraulic conditions and those that are not, based upon analysis of complete sediment samples. The technique, however, is not widely understood and presents computational demands in excess of many operators.

Because very high energy regime particulates should not occur in low energy sediments (under equilibrium conditions), and vice versa, it is often possible to take a more simple approach to size analysis (Sly, 1988). This approach is based on the particle-size distribution of the sand size fraction and the relative content of mud (a mixture of silt- and clay-sized particles) in the same sample.

Because the size range of the sand fraction extends from 1 to 0.063 mm or 0 to 4 phi, it contains both high and low energy regime materials that merge at the point of minimum erosion velocity, termed a divide (ca. 0.18 mm or 2.5 phi). This means that the sand fraction alone can be used to separate high and low energy regime sediments by relating mean size to the divide. By plotting mean sand size against standard deviation (Figure 8), empirical data show (1) equilibrium sediments are best sorted near the divide, (2) as mixing with other coarse or fine end member populations occurs, the standard deviation increases, and (3) close to such end members standard deviation again decreases as sand content decreases to minimum values. Thus, although a mixture of fine sand and mud may characterize an equilibrium deposit, a mixture of coarse sand and mud is anomalous. By combining sand size and mud content data, it is possible to define the characteristics of most types of equilibrium and anomalous sediment.

Based on analysis of Lake Ontario data, it seems probable that most bottom samples having a mean size of between 0.5 and 0.22 mm can be related to erosion and transportation. Sediments with a mean size between 0.22 and 0.016 mm become progressively more depo-

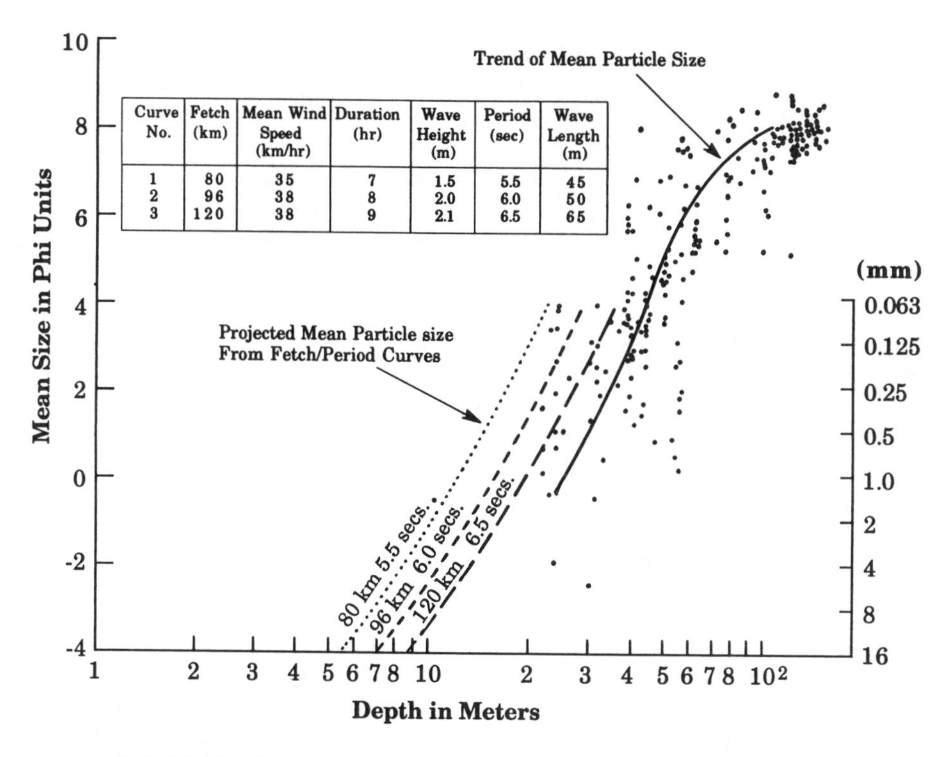

FIGURE 11. Plot of depth vs. mean sediment size. (Data source: Sly et al., 1988.)

sitional and may be influenced by flocculation (many samples are difficult to correlate with formative hydraulic shear velocities). At mean sizes finer than about 0.016 mm the settlement of particles corresponds to Stokean conditions which characterizes the fall of individual particles at very low concentrations.

In addition to providing a basis for process modeling, these definitions make it possible to differentiate between sediments forming under the influence of modern conditions and those of relict origin. Figure 11 illustrates this point; actual sediment samples from Georgian Bay appear to be coarser at a given depth than predicted size curves (1 and 2) derived from hydraulic shear velocities, which were computed from recorded wind data (Sheng and Lick, 1979).

2. Erosion, Transport, and Deposition of Lake Sediment

In the littoral zone, lake sediments are influenced by constantly changing wind-waves, currents, and water levels, and they tend to integrate response to formative processes, often over periods of weeks or even months. Particle-size trends rarely reflect short-term variance unless very thin sediment layers are analyzed. Particle size decreases both laterally and offshore in response to decreasing levels of hydraulic energy. These trends, in turn, reflect general directions of sediment transport. Sediment size is a useful indicator of bed stability, within which transport may occur in more than one direction. Coarse sands and sandy gravels usually occur in areas of intermittent but potentially high energy and trends in their particle size are good indicators of transport. Also, as a result of fetch limitation or the influence of river inflow, these trends are often directionally stable. Sands of about 0.3 to 0.13 mm mean size are most easily moved and can respond to light or moderate wind-wave or current conditions. Because light to moderate wave or current motions are less limited by fetch, these conditions may be generated by winds from several different directions; over time, transport often lacks directional stability. Derivation of transport pathways within distributions of such sands is often impossible. Fine sands are typical of areas of long-term and fairly consistent accumulation, and as such they can be good indicators of transport pathways. As a further complication, however, one must remember that the characteristics of sediment deposits can be greatly influenced by the availability of supply. Not all sizes of material may be available to respond to the range of hydraulic shear velocities that are present.

In lake areas further from shore, the deposition of sediments may be influenced by flocculation when concentrations of suspended material are high. Further, bottom deposits can be much influenced by the flow directions of separate water masses. From these, different types of particulate material may settle to form sediments composed of numerous mixed but exceedingly thin layers. Some intermittent resuspension of these deposits may occur with extreme storm events, with periodic introductions into the offshore environment.

In mid-lake profundal areas, annual resuspension of part of the suspended near bed nepheloid layer can result in temporal migration of particulate fronts (such as evidenced by apparent contaminant creep). This process is associated with sediment focusing and it is generally applicable only to very fine materials. Although trends in particle size are often used as indicators of sediment transport, they are far less useful as indicators of erosion and deposition. For the latter, measurements of profile change (bed elevation) are essential.

3. Substrate Habitat and Significance of Formative Origins

Substrate habitat is an extremely complex component of the ecosystem in which not only the present hydraulic conditions but also the formative history governs use and suitability for different biological species and communities. In this regard, substrate structure and the degree of consolidation play a major part in modifying habitat use. For example, clay surfaces that have been subaerially exposed may form crusts before being covered by later rising waters. In some areas of lake bottom these surfaces provide microrelief and act as an attractant for several benthic species. Clays may be hardened almost to the consistency of rock as a result of overconsolidation by late glacial ice loading, and such materials are indeed used by benthic communities as if they were bedrock surfaces.

Rates of deposition and stability of deposits are other significant factors affecting biological communities. With the exception of artificial dumping, benthic invertebrates are usually sufficiently mobile to compensate for the effects of most natural fluctuations in the rate of material deposition. However, with seasonal storms, exceptional wave action is capable of eroding sediments at water depths much below normally active erosion. Periodically, such events may have a major impact by removing some benthic communities. Further, in areas in which rates of deposition are very low, thin layers of modern sediment may cover quite different and much older material. Thus, the depth at which burrowing organisms can inhabit such substrates may be limited by the layer thickness, a feature that is not apparent from any analysis of surface particle size.

Formative origins have an important influence on the use of coarse substrates. For example, stable lag cobble-gravels with clean and open lattice structures (derived by erosion of glacial till) or similarly sized scree deposits can provide important substrates for spawning fish. However, comparably sized materials in actively forming cobble-gravel beaches are generally unsuitable because they lack stability. Thus, the Great Lakes, having considerable areas of relict sediment as well as modern actively forming deposits, requires that benthic habitat be defined not only by particle size but also by formative origin. At the very least, some estimate of void filling in coarse sediments and the hardness (shear strength) of sands, silts, and clays must be included.

4. Use of Mineral Substrate

Generally, natural substrate diversity is highest nearshore because of the influences of river bays and strong currents in shaping the kinds and amounts of substrate occurring in this zone. Progressing offshore there is a gradient of sediment types and sizes grading to the deepest basins and depositional areas where the fine organic material accumulates (sediment focusing). In many areas, offshore reefs occur which provide some unique rocky and clay bank habitat in areas of substrate high in organic matter. Fish and invertebrate diversities reflect habitat diversity. In the Great Lakes the offshore profundal zone is characterized by only a few common species at certain seasons of the year (i.e., lake trout, deepwater sculpin, burbot, bloater, and occasional rainbow smelt and alewife). Many more species are found in the nearshore zone.

Although bloater and alewife broadcast their eggs, the former offshore and the latter in the nearshore (<30 m) water (Wells, 1966; Tesar and Jude, 1985), many fish species have specific substrate preferences for particular stages in their life histories. As an example, Table 7 shows the habitat characteristics of 33 Lake Erie fish species, as derived from Trautman (1981) and Scott and Crossman (1973). Fish usually have very specific spawning substrate requirements, except for those species that broadcast their eggs. However, even for those species a preferred depth or substrate type may still be required. Many species such as alewife, spottail shiner, and rainbow smelt spawn over sand, rocks, or on *Cladophora*-covered rocks. Johnny darter and slimy sculpin deposit their eggs nearshore on the underside of rocks, logs, or other suitable substrate. Ninespine stickleback build elaborate nests from vegetation and debris while deepwater sculpin presumably make nests on the bottom and guard them during the winter (Westin, 1969). Yellow perch in Lake Michigan are known to spawn in tributaries, connecting waters (Perrone et al., 1983), and in the main lake over rocky habitat, in deep clefts within shoal areas, and over clay banks (Dorr, 1982). Yellow perch can use a range of materials as spawning substrate, including rocks, sticks, and plants, which provide some form of vertical relief. The type and quality of substrate on which eggs are deposited is often crucial to reproductive success. For example, for native lake trout that have been extirpated from all the Great Lakes except Lake Superior, a significant loss of spawning habitat may have occurred due to degradation by increased quantities of organic matter associated with the effects of eutrophication (Dorr et al., 1981a, b; Sly, 1984, 1988). The lack of reproductive success by introduced stocks of lake trout, however, may be due to a number of other possible causes including hatchery effects, contaminants, excessive mortality rates, and genetics (Eshenroder et al., 1984).

Some of the most dramatic changes that have occurred in the Great Lakes have been related to habitat modification. Often, habitat modification has been detrimental to fish because of the destruction of wetlands that serve as important nursery areas, or the location of intakes in areas in which high densities of larval fish occur. However, the Atlantic salmon in Lake Ontario was likely extirpated largely because of the loss of stream spawning habitat associated with deforestation and the construction of dams and wiers during the early stages of European settlement around the lake basin (Christie, 1974). The lower parts of many rivers that flow into the Great Lakes have been modified and made into harbors, and these activities have largely destroyed the deltas and associated wetlands that existed previously. Poe et al. (1986) have documented that species composition and the richness of fish assemblages were higher in the unaltered littoral habitat in Lake St. Clair than in the altered habitat of this lake. A percid-cyprinid-cyprinodontid community dominated in unaltered areas, while a centrachid assemblage dominated in altered areas. On the other hand, artificial substrate has been used by fish as shelter, resting, feeding, and spawning sites. Much of the limited in-lake reproduction by lake trout has been recorded on riprap-protected water intakes and shoreline protection

TABLE 7
Habitat Characteristics of Common Lake Erie Fish Species

Taxon	Habitat characteristics												
	Spawning location		Water depth		Water clarity		Bottom type				Rooted aquatic plants		
Family and sp.	Tribut-aries	Near-shore	Shallow	Deep	Clear	Turbid	Mud	Sand	Rocky	Organic	Absent	Moder-ate	Abund-ant
Acipenseridae													
Acipenser fulvescens	X		X		X			X	X		X		
Lake sturgeon													
Clupeidae													
Alosa pseudoharengus	X	X	X	X	X		X	X	X	X			
Alewife													
Dorosoma cepedianum	X	X	X	X	X	X	X	X	X	X			
Gizzard shad													
Salmonidae													
Coregonous artedii		X	X	X	X		X		X		X		
Cisco													
Salvelinus namaycush		X	X	X	X		X		X		X		
Lake trout													
Osmeridae													
Osmerus mordax	X	X		X	X		X		X		X		
Rainbow smelt													
Esocidae													
Esox lucius	X	X	X	X	X			X	X	X			X
Northern pike													
Esox musquinongy	X	X	X	X	X		X			X			X
Muskellunge													
Cyprinidae													
Carassius auratus	X	X	X		X	X	X	X	X	X			X
Goldfish													
Cyprinus carpio	X	X	X		X	X	X	X	X	X		X	
Carp													

Notemigonus crysoleucas	X		X		X			X		X			X
Golden shiner													
Notropis hudsonius		X	X	X	X			X	X				
Spottail shiner													
Notropis spilopterus	X		X			X	X					X	
Spotfin shiner													
Pimephales promelas	X		X			X	X				X	X	
Fathead minnow													
Pimephales notatus	X		X		X	X	X	X	X	X			
Bluntnose minnow													
Catostonidae													
Carpiodes cyprinus	X	X	X	X	X			X	X				
Quillback													
Costomus commersoni	X		X	X	X	X	X	X	X	X	X	X	X
White sucker													
Ictaluridae													
Ictalurus moles	X	X	X			X	X					X	
Black bullhead													
Ictalurus natalis	X	X	X		X			X	X	X			X
Yellow bullhead													
Ictalurus nebulosus	X	X		X	X			X	X	X		X	
Brown bullhead													
Ictalurus punctatus	X		X		X			X	X		X		
Channel catfish													
Percichthyidae													
Morone chrysops	X		X		X			X	X				
White bass													
Centrarchidae													
Lepomis gibbosus		X	X		X			X		X			X
Pumpkinseed													
Lepomis macrochirus		X	X		X			X		X			X
Bluegill													
Posoxis annularis		X	X		X			X	X	X		X	
White crappie													
Posoxis nigromaculatus		X	X		X			X		X			X
Black crappie													

TABLE 7 (Continued)
Habitat Characteristics of Common Lake Erie Fish Species

Taxon	Habitat characteristics												
	Spawning location		Water depth		Water clarity		Bottom type				Rooted aquatic plants		
Family and sp.	Tribut-aries	Near-shore	Shallow	Deep	Clear	Turbid	Mud	Sand	Rocky	Organic	Absent	Moder-ate	Abund-ant
Micropterus dolomioui Smallmouth bass		X	X		X				X			X	
Micropterus salmoides Largemouth bass	X	X	X		X			X	X	X			X
Percidae													
Perca flavescens Yellow perch		X	X	X	X			X		X			X
Stizostedion cansdense Sauger	X	X	X			X	X				X	X	
Stizostedion v. vitreus Walleye	X	X	X		X			X	X		X		
Sciaenidae													
Aplodinotus grunniens Freshwater drum		X	X	X	X				X				

Data source: Hartley and Herdendorf (1977).

structures (Jude et al., 1981; Wagner, 1980; Peck, 1979). Scuba divers have observed that several fish species (yellow perch, slimy sculpin, spottail shiner, alewife, and johnny darter) spawn on artificial substrates around the intakes at the Cook Generating Plant (Dorr and Jude, 1980; Dorr and Jude, 1986) and the Campbell Generating Plant (Jude et al., 1982) in Lake Michigan. Manny et al. (1985) found that breakwater walls and associated riprap in western Lake Erie (West Harbor, OH) acted as both spawning and nursery areas for fish. Increases in the number of fish eggs and larvae were noted and the areas served as nursery grounds for as many as 20 species, both during and after construction. Secondary production of fish-food organisms was also extensive. Winnell and Jude (1987), Lauritsen and White (1981), and Rutecki et al. (1985) also reported that artificial substrate is an important habitat for fish-food organisms.

Many fish go through several ontogenetic changes in feeding, starting with zooplankton, and later utilizing the macrobenthos or fish. The type and size of benthic populations are often directly related to substrate type, as has been shown by Winnell and Jude (1987) and Johnson et al. (1987). Because larval fish are dependent on the optimal quality and quantity as they transfer from endogenous to exogenous feeding patterns, an inexorable link develops between secondary production and first feeding in larval fish. Further, investigators demonstrated that rocky habitats in the Great Lakes have benthic populations that are more closely related to lotic populations than those found in nearby nonrocky substrate.

5. Use of Plant Substrates

Plant materials can act as spawning substrates, nursery areas, and as a substrate for fish-food organisms. Many studies have focused on the importance of plants as habitat affecting the community structure of fishes, including experimental studies in ponds (Werner and Hall, 1979) and in natural situations (Poe et al., 1986). Their findings have revealed that plant commodities are very important habitats, and that they are a key component in the competitive interactions among species and predator-prey relationships. Many species, including northern pike, yellow perch, spottail shiner, and common carp, broadcast adhesive eggs over macrophytes and algae in marshes and bays (Auer, 1982). Dorr and Jude (1986) have documented the spawning of spottail shiners on *Cladophora*-covered rocks on the intake riprap of a power plant in southern Lake Michigan. Young-of-the-year fish often use vegetation as a nursery area in which they are more protected from predators; often this vegetation supports secondary production and forms a valuable and easily available food source (Chubb and Liston 1986). For example, during the first few months of their life history, yellow perch are pelagic. At this time, they feed primarily on zooplankton and avoid predators by remaining dispersed in the open water zone. At a body length of about 30 mm they become demersal and move into shallow littoral zones for protection and to seek benthic organisms which become increasingly important in their diets (Wong, 1972; Faber, 1967).

B. WATER AND THE EFFECTS OF FORCING FUNCTIONS

1. Wave Action

Wave action follows wind action very closely on Lake Erie (Figure 12) because of the shallowness of the lake. Swells, however, often continue into the day after a storm subsides. The depth of the water and the direction, velocity, duration, and open water fetch of the wind collectively determine the characteristics of waves at a given location. With a fetch of 240 km and a wind velocity of 50 km/h, the maximum wave is developed in 20 h. With this wind velocity and duration, a wave 4 m high and with a 6.5-s period can be developed. Waves of this height break well offshore, but re-formed waves up to 1 m in height may reach the shoreline.

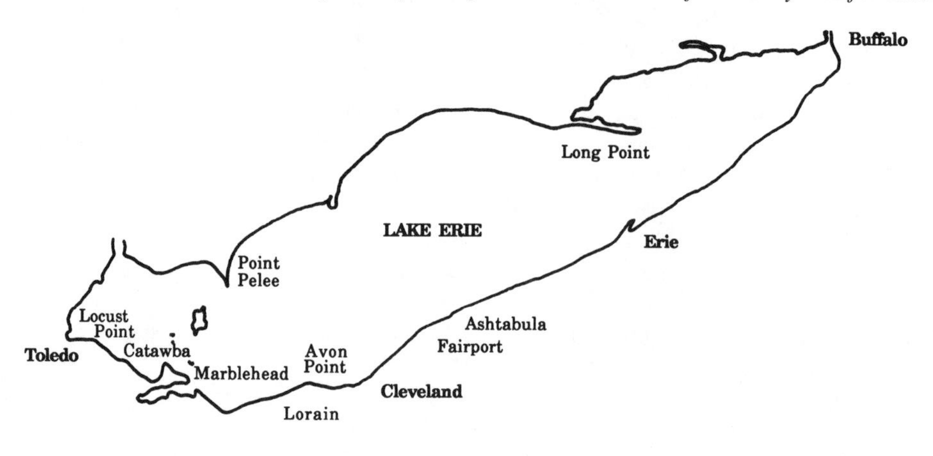

FIGURE 12. Lake Erie geographic locations and communities.

Waves are formed by the orbital motions of water, and motion decreases downward from the surface to negligible values at a depth equal to half the wave length. As a result, only slight net transport of water occurs in the direction of wave progress.

Longshore (littoral) currents are generated by waves in the nearshore zone of Lake Erie and can attain velocities up to 1.2 m s^{-1}. Such currents are capable of eroding and moving beach materials as large as pebbles along the bottom. The movement of sediment by nearshore currents is termed littoral drift. Currents are important agents of erosion, transportation, and deposition of sediments along the shorelines of the Great Lakes. The combination of direction, velocity, duration, and open water fetch of the wind determines the strength of the waves and of the resulting currents. Neutral zones are areas of either divergence or convergence. Divergence occurs where littoral currents flow away from each other and generally cause erosion; convergence occurs where drifts of opposing directions meet, producing accretion or sedimentation.

The relative strengths of predominant drift and the positions of nodal zones can be computed for the shorelines of the Great Lakes by converting wind records to associated wave heights and calculating the wave energy for different directions. The amount of energy utilized in producing littoral drift in one direction is then compared to the opposing energy. The difference gives the theoretical direction and relative strength of the predominant drift.

Areas of convergence and divergence are generally marked by distinct configurations of shoreline: elongated spits are found in areas of convergence and broad, gently rounded headlands are found in areas of divergence. The most pronounced areas of convergence in Lake Erie are on the Canadian shore at Point Pelee and Long Point, Ontario. No comparable areas exist on the south shore. However, areas of divergence are better marked on the Ohio shore than on the north shore: Locust Point, Catawba, Marblehead, and Avon Point. Catawba and Marblehead appear to be preexisting land forms that have controlled the littoral drift, but it is questionable whether Locust Point and Avon Point are antecedent land forms or are products of shore erosion.

2. Circulation and Flow

Owing to the prevailing southwest winds and the orientation of Lake Erie's long axis in the same direction, the dominant surface flow tends to stream toward the east end of the lake, although in some areas, island and channel configurations modify the dominant flow patterns.

Energy is transferred from the wind to the water by frictional drag on the lake surface. Coriolis deflection causes a progressive change of direction with depth known as the Ekman spiral. Theoretically, surface currents deviate 45° to the right of the wind and drag the water immediately below, which also undergoes Coriolis deflection, producing a continuous swing to the right. In addition the wind tends to drive greater volumes of water than are compensated for by discharge from the basin, causing a balancing subsurface return flow in opposition to the surface current. As a result, subsurface currents are often at variance with the surface currents. Therefore, the dominant subsurface flow appears to move toward the west to balance the water economy of the lake. The process described above probably takes place at the bottom of the lake in the winter and in the lower epilimnion water in the summer when the lake is stratified.

Currents in the hypolimnion appear to be largely generated by vertical motions of the thermocline. Parmenter (1929) theorized that during the summer stratification, wind action drives the upper layers of water toward the windward shore of the lake, a process that tends to tilt the thermocline. The depth and angle of inclination of the thermocline are dependent upon the amount of warm epilimnion water above it. When this force is removed or overbalanced by gravity, the thermocline swings back toward a horizontal position and an oscillation (internal seiche) is set up in the cold hypolimnion water as it is released from its deformed shape. Ayers (1962) reasoned that major sinking of water masses can depress the thermocline and upwellings can raise hypolimnion water. Pressures caused by these opposing movements force the displaced hypolimnion water to spread across the basin. Ayers also believed that tilting and oscillations of the thermocline which are initiated by wind or barometric changes can supply energy to generate currents to the deep water. In Lake Erie, for example, these bottom currents can reach velocities >0.3 m s^{-1} and can resuspend bottom sediments.

Water motion, as current, is a physical factor that influences the population levels of both lotic- and lentic-water fishes. At the lake surface, water movements also tend to equilibrate air and water temperatures. Through mixing, water movements tend to homogenize temperature and chemical factors in the epilimnion of the lakes. When turbulence at the surface comes into contact with the water layer beneath, a current in the opposite direction is set up in the lower layer; such currents can produce turbulent internal eddies that lead to a vertical interchange of water particles, known as eddy diffusion (Lagler et al., 1977). In lakes, currents are largely responsible for material transport and associated turbidities (see discussion in subsection 8, Light and Turbidity).

The species composition of fish communities differs markedly with changing current regime and, generally, as in streams, the stronger the current the more depauperate the fish fauna. The development of streamlining and hold fast organs among fishes is an obvious response to current. Thus, even as gradient in streams is an ecological factor for fishes, so are the shape, depth, configuration, and composition of the bottom standing waters (Lagler et al., 1977).

3. Water Masses

Synoptic water surveys have demonstrated that distinct water masses, particularly those created by major inflows, can be mapped as in Lake Erie (Hartley et al., 1966). Conductivity measurements in western Lake Erie show that three distinct masses of Detroit River water enter the lake. Mid-channel flow, which is characterized by water of low conductance, can be traced as far south as the Ohio shore, northward west of the Bass Islands, and into central Lake Erie through Pelee Passage. Temperature and conductivity on inflowing water masses in the central and eastern basins allow the masses to be traced for limited distances once they have entered the lake. Tributary streams along the south shore contribute about 10% of the total inflow into the lake, but the highly conductive water masses from these streams cannot generally be traced more than 1 or 2 km offshore. Harbors such as Lorain, Cleveland, Fairport, Ashtabula, and Buffalo, where large water areas are confined by massive structures, act as effective dispersal or mixing areas for highly conductive water. These structures are apparently impermeable to the passage of water masses; radical differences in conductivity are observed on lake- and shoreward sides. East of inflowing streams the highest conductivity readings are commonly found nearshore, indicating that most of the tributary flow clings to the shoreline rather than moving lakeward. In the late spring and summer, when stratification sets up a thermal barrier within a few kilometers of the shore, most of the tributary discharge tends to stay near the shore and moves predominantly eastward. In the fall and early winter, when the lake becomes nearly isothermal, the cooler tributary water is free to move lakeward and underrun the lake water. Mid-lake water in central and eastern Lake Erie is very uniform in nature and is little affected by north and south shore streams. Minor variations in the dissolved material occur between the epilimnion and hypolimnion water masses. Restricted circulation and an upward flux from interstitial waters and solution effects at sediment/water interface may account for the slightly higher bottom water conductivity.

4. Wind Effects on Fish Habitat Use

Wind is a very important physical force since it is the chief means of generating waves, currents, upwellings, and most seiches. Fish eggs can be adversely affected, as recorded by Clady (1976) and Clady and Hutchinson (1975) for yellow perch in Oneida Lake, and by Busch et al. (1975) for walleye in Lake Erie. Lake Michigan workers have seen lake trout eggs washed up on shore during strong autumn storms. Since most larval fish are planktonic in early life and subject to the vagaries of the currents, water movements can have profound effects on recruitment in fishes. Currents can deposit larval fish in optimal nursery areas in which predation is low and food availability high, or they can be transported to areas in which food supplies are meager, habitat unsuitable, or predators abundant. Lasker (1975) has shown the importance of larval fish being transported by currents into optimal food patches in the ocean. The critical period hypothesis (Hjort, 1914; May, 1974) is closely related to these phenomena.

Upwellings in the Great Lakes can also have adverse impacts on larval fish, as was shown by Heufelder et al. (1982). They demonstrated that during years of above-average upwelling frequency in Lake Michigan, alewife larvae tended to: (1) be found in lower densities, (2) be closer to shore, and (3) have a marked protracted spawning period, presumably as the result of the lower temperatures. Adult fish are influenced strongly by temperature as well, but effects are mostly reflected in movement attraction or avoidance. Emery (1970), however, did record some mortalities of adult sculpins in Georgian Bay during an upwelling. In nearshore Lake Michigan, during a summer upwelling, warmwater species such as yellow perch, alewife, and trout-perch were replaced by coldwater species such as bloaters, lake trout, and to a lesser degree rainbow smelt. Brandt et al. (1980), Crowder and Magnuson (1982), and Brandt (1980) also discuss at length the importance of the thermal regime in shaping fish distributions and ecological interactions.

5. Water Level

Water level changes in the Great Lakes are of short and long period fluctuations. Four basic causes have been found for water level changes: (1) lunar tides, (2) seiches, (3) seasonal fluctuations in watershed runoff, and (4) long-term climatic changes. The important components of level changes and circulation for Lake Erie have been outlined by Herdendorf (1975) and they form the basis for descriptive examples as shown in Table 8. Due to interactions between a variety of hydrologic and weather conditions, the Great Lakes experience significant short- and long-term fluctuations in water levels whose magnitude and timing may be difficult to predict.

Short period fluctuations are due to a tilting of the lake surface by the wind or possibly by atmospheric pressure differentials. Wind tides, seiches, surges, and harbor resonance, which

TABLE 8
Characteristics of Major Currents in Lake Erie

I. Horizontal circulation
   A. Natural flow (hydraulic current)
      1. Is a result of hydraulic gradient from west to east
      2. Has net eastward movement (unidirectional throughout water column)
      3. Has low velocity (maximum estimate: Verber (1952) for western basin, 4.5 cm/s and central basin, 0.9 cm/s; Parmenter (1929) for eastern basin, 3.3 cm/s)
      4. Has no compensating return flow
      5. Has other currents superimposed on it and often masking natural flow
      6. Is important in distribution of dissolved substances (84% introduced at its source)
      7. Is unimportant in transport of suspended material except in restricted channels
   B. Wind-driven currents
      1. Are caused by wind stress on water surface
      2. Are variable in direction
      3. Have high velocity (up to 60 cm/s)
      4. Move large volumes of water in short period of time (wind tide and wind setup)
      5. Have subsurface return flow often associated
      6. Are modified by geostrophic deflection, remnant currents, basin topography, air and water temperatures, and characteristics of the wind
   C. Alongshore (littoral) currents
      1. Are generated by breaking waves in the nearshore zone
      2. Have movement generally parallel to shoreline (controlled by nearshore topography)
      3. Have direction at an angle to wind or wave progress
      4. Have rapid velocity (up to 1.2 m/s)
      5. Are capable of transporting sand- and gravel-sized particles (littoral drift)
      6. Dissipate rapidly when storm subsides
   D. Seiche currents
      1. Are created by standing wave motion of seiches (oscillating waves without progression)
      2. Are degenerated by friction (seldom complete because of modification or rejuvenation)
      3. Have minimum velocity at area of maximum amplitude
      4. Have maximum velocity at nodal zone
      5. Accomplish no net transport of water (balanced by to- and-fro motion)
   E. Hypolimnion currents
      1. Are generated by thermocline depression or elevation as a result of:
         a. Wind setup and ensuing internal seiche
         b. Sinking water masses
         c. Upwelling water masses
      2. Have normally low velocity, but can have velocity as high as 2 ft/s during severe storms and upwelling
      3. Can occur only in summer when lake is stratified (restricted to hypolimnion by thermocline)
      4. Are not present when lake is isothermal
      5. Accomplish no net transport of water
      6. Have high velocities capable of resuspending bottom sediments
   F. Inertial currents
      1. Are related to 18-h inertial rotary period for Lake Erie (Verber, 1966)
      2. Have right-hand acceleration to the existing currents because of Earth's rotation
      3. Have various flow patterns (straight-line, oscillatory, and rotary)
      4. Occur at all depths and in all seasons as well as under ice cover
      5. Have period similar to internal wave periods on thermocline during summer stratification
      6. Have imperfectly known mechanism
II. Horizontal and vertical circulation
   A. Density currents
      1. Are the result of density differences between lake water and inflowing water
      2. Have density differences caused by temperature, and by dissolved solid content and suspended material content differentials
      3. Provide mechanism for rapid distribution of tributary inputs
      4. Underrun lake water when water is cooler and solids-laden (turbidity current)
      5. Can override lake water when warmer if solids content is not too high

TABLE 8 (Continued)
Characteristics of Major Currents in Lake Erie

6. Have movement lakeward with no compensating return flow of same water (thermal bar, Rodgers, 1965)

B. Turbulence
1. Has random motion with horizontal and vertical components
2. Is associated with other types of currents (particularly pronounced with wind-driving and wave-generated currents)
3. Is effective in mixing and dispersing water masses

III. Vertical circulation

A. Temperature gradient currents
1. Are caused by heat transfer (convection cell)
2. Are most important during cooling period (September to January) and warming period (March to June)
3. Are characterized by cooled surface water sinking as warmer water rises to replace it, a process that continues until water column reaches temperature of maximum density (4°C) or by lakeward progression of a thermal bar as the lake is warmed

B. Sinking and upwelling currents
1. When sinking, are caused by convergence of horizontal currents, forcing a downward movement to balance the water level
2. When upwelling, are caused by divergences of horizontal currents, resulting in an upward movement to balance the water level

Data source: Herdendorf (1975).

have periods from a few seconds to several days, are examples of short-term oscillations. Sun and lunar tides are negligible, resulting in a maximum fluctuation of about 3.4 cm for Lake Erie (Verber, 1960). Long period fluctuations are related to volumetric changes of the lake, caused principally by variation in precipitation, evaporation, and runoff. These changes include both seasonal changes and those occurring over a period of several years.

Water levels at the extremities of lakes have a greater fluctuation than at the center where tilting of the surface results in a nodal zone. For example, on Lake Erie high water levels coupled with northeast storms have produced a maximum fluctuation of 3 m above Low Water Datum at Toledo, OH. Conversely, low levels and southwest winds have lowered the level to 2 m below datum. The forced movement of the lake surface is known as wind tide and the amount of rise produced is the wind setup. The resulting free oscillation of the lake surface caused by the inequality of water level is called a seiche. Wind tides are distinguished from seiches by the extreme variation in the period of wind tide duration. Wind tides may pile up water for only a few hours or for as long as 48 h during severe storms.

The major seiches on Lake Erie travel the longitudinal axis of the lake. Seiches along this axis have a period of about 12 to 14 h. A water level gauge at Put-in-Bay on South Bass Island indicates that longitudinal seiches were in effect for about 44% of the time over a 3-year period. Wind records for Sandusky, OH, are in agreement with the frequency of seiche periods; surface winds from the southwest and northeast occur approximately 150 d (42%) of the year (Herdendorf and Braidech, 1972). Seiches with periods shorter than 12 h are generally transverse or oblique to the major axis.

The long-term change in water level in the Great Lakes region is mostly a function of the storage capacity of the lake system. The recent (1971 to 1987) high lake levels can be seen as a result of large inputs of water from surface inflow and direct precipitation. Water input has slightly exceeded output, as outflow and evaporation, for the majority of years for this peroid (Greene, 1987). Similarly, low lake levels occur when surface and subsurface outflow and evaporation exceed inflow and precipitation. This was the case during the low water period between 1955 and 1965.

The size and limited discharge capacities of the Great Lakes have resulted in extreme low and high water level conditions. Levels may prevail for considerable periods of time after the causative factors have changed or disappeared. In the Great Lakes, seasonal fluctuations reflect the annual hydrologic cycle. Rising water levels are seen in the late spring and early summer due to heavier rains and snowmelt. In the late summer and early autumn, lower water levels reflect the effects of evaporation and decreased precipitation and surface runoff.

As a result of their close association with the lake, coastal wetlands are greatly affected by changes in water levels. Depending on the topography of the wetland, changes in water level may be accommodated by lake- or landward shifts in wetland community types. This has been demonstrated in wetland areas with a low gradient slope. However, the extent to which wetland communities shift has been found to be limited (Lyon et al. 1986; Busch and Lewis, 1983).

The upland boundaries of wetlands are generally stable. Typical Great Lakes' wetlands are confined on the upland side by shorelines and beach ridges. Cowardin et al. (1979) defined the upland limit as the boundary between the land that is flooded and those areas that are not flooded during years of normal precipitation or water level. As such, upland boundaries are considered to be fixed and, to a certain extent, their locations are unaffected by changes in water level. Only over a geological time scale does major change occur.

Many of these wetlands are also protected from wave action by a natural barrier such as a beach, spit, or bar, or an artificial barrier such as drainage dikes. The presence of such a barrier also provides an opportunity for runoff to enter the area and remain for a period of time before mixing with open lake water (Greene, 1987). In this way, a barrier can help to prevent wetlands from becoming desiccated at low water levels. Since barriers need to withstand the force of storm waves at high water levels, it is important that they have stability. In addition, they must be able to receive material from rivers or littoral drift to maintain their "permanence" (dynamic equilibrium).

6. Climate and Ice Formation

The effects of climate on fish can range from subtle to catastrophic. Effects of the Pacific *El Nino* are well known, and years of cold winter, such as that experienced in 1977, can cause massive alewife die-offs that impact upon food webs, as for example in Lake Michigan (Jude and Tesar, 1985; Eck and Brown, 1985; Scavia et al., 1986). Similar die-offs have been recorded in Lake Ontario (Bergstedt and O'Gorman, 1983; O'Gorman and Schneider, 1987). Heavy ice cover can reduce primary production in subsequent seasons by limiting deep water overturn and the release of phosphorus from reworked sediment. Meisner et al. (1987) discuss climate changes (global heating) and their impact on many of the habitat variables discussed in ths chapter. Storms can dislodge eggs, increase turbidity, and presumably cause increased mortality among early life stages. Strong year classes of fish are generally observed, especially for yellow perch and walleye, during certain combinations of climate which are optimum for increased survival of incubating eggs and young-of-the-year (Busch et al., 1975; Clady, 1976). In Lake Michigan several of the dominant fish species' growth and survival patterns are related to years of above-average temperatures. In Lake Windemere trapping forced yellow perch populations to very low levels, and consequently this species was unable to rebound until climate factors were near optimum (Craig et al., 1979).

Fluctuating water levels associated with years of drought or high rainfall can cause changes in the physical habitat of fish. High water levels cause increased erosion and subsequent sedimentation which can be detrimental to some species. They also create more nearshore habitat which is probably beneficial to fish that spawn in areas of flooded vegetation and/or use these nearshore zones as nursery areas. Years of drought may adversely impact species that rely on flooded marshes, rivers, and nearshore habitat for spawning and nursery areas.

Robertson and Scavia (1984) provide a useful summary of the climatic characteristics of the Great Lakes region. The Great Lakes are located in the interior of the North American land mass, and lie under the influence of a continental climate. The July mean air temperature on the perimeter of Lake Erie is 21.8°C while the comparable January mean is –3.8°C. For the lakes farther north the corresponding temperatures are lower: Lake Superior's mean perimeter temperature is 18.1°C in July and –11.1°C in January.

Precipitation over the lakes also tends to decrease with increasing latitude. Lake Erie has an annual average precipitation of 880 mm, while Lake Superior experiences only 785 mm (Phillips and McCulloch, 1972; Saulesleja, 1986). Seasonal precipitation is rather evenly distributed, with somewhat larger amounts during summer. Precipitation takes the form of snow during winter and, with the exception of the most southerly areas, all parts of the Great Lakes' basin are permanently snow covered during winter.

Because of their size, the lakes have a pronounced effect on the climate of the adjacent shorelines. They have tremendous capacities for heat storage and so undergo much more limited fluctuations in temperature than occur in the atmosphere. Through air-water exchanges of heat and moisture, the lakes influence the characteristics of overlying air masses, which in turn, affect the temperatures and precipitation in the coastal areas. In such localities, air temperatures are moderated, precipitation is often increased, and humidity and wind speed tend to be greater. The prevailing winds over the Great Lakes are from the west or southwest, and certain areas of their eastern shorelines are notorious for the large amounts of precipitation that they receive (Robertson and Scavia, 1984).

All of the Great Lakes experience at least partial ice cover in winter, but they seldom (if ever) freeze over completely from shore to shore. In a normal winter, only Lake Erie, the most southerly lake, has ice over most of its surface. Of all the Great Lakes, this lake has the smallest heat storage capacity. Lake Erie is shallow and has the smallest volume to surface area ratio of any of the Great Lakes (Assel et al., 1983).

7. Thermal Regime

Temperature is a master variable and affects all life processes of fishes. It is of particular importance to the early life history stage since temperature controls hatching and development of eggs, and the timing of food availability. Detrimental temperatures during any of these critical periods can adversely impact recruitment and growth of fishes. In the Great Lakes, temperature can vary latitudinally, annually, and daily. Latitudinal effects, which are closely related to climate regime, affect species diversity. Lake Superior is the coldest and has the fewest number of species; Lake Erie is the warmest of the Great Lakes and has the most species. Also affected are growth rates, which are related to food availability, and the length of growing season.

Spawning is influenced by the annual temperature cycle, along with the photo regime. As noted earlier, in this region, fish spawn during almost every month of the year. Temperature often determines when spawning and egg deposition occur. However, in the Great Lakes, fluctuations in temperatures during spawning adversely affect many species of fish (Heufelder et al., 1982) and can impact year-class strength and species diversity within the lakes (Shutter et al., 1980).

Daily fluctuations in temperature are most important at the egg incubation stage. Shutter et al. (1980) demonstrated that smallmouth bass required a certain minimum temperature to initiate spawning, and they also found that sudden drop or extreme daily fluctuations of temperature led to decreased survival of progeny. Similar minimum and/or maximum temperature examples exist for other species of fish. Edsall (1970) found that alewife hatched at temperatures <10°C developed nonfunctional jaws.

Evidence exists that most juvenile and adult fish select preferred temperatures (Fry, 1947; Lagler et al., 1977). Thus, the volume and location of space available to fish for feeding and movement change with the thermal structure and density stability of lake waters.

The distribution of fish in the water column is greatly influenced by water temperature (Christie and Regier, 1988). This is particularly important during spawning periods (Table 9). A species that shows a preference for 15°C water may be near the surface in April, whereas by late August it may be 12 to 20 m deep (Figure 13). After the autumn overturn, when the water is isothermal, the same species may be found throughout the water column.

Temperature is a factor of wide and varied significance on food, growth, and development. Generally, within the range of temperatures that can be tolerated by the fish, vital processes are accelerated by warm temperatures and decelerated by cold (Lagler et al., 1977). Temperature extremes or sudden changes are often lethal. Elevated, sublethal temperatures may induce estivation, and cold, hibernation. Within the Great Lakes, temperature may determine the reproductive success of a species as well as its distribution. Within the geographical range of trouts of the family Salmonidae, waters are commonly classified thermally as cold (inhabited by trout) or warm (intolerable to trout). In lakes of the temperate zone, when thermally stratified in midsummer, coldwater fishes such as the trouts (*Salmo*), chars (*Salvelinus*), and whitefishes (*Coregonus*) remain in deep water in or beneath the thermocline, and warmwater fishes are restricted to the shallow upper warmwater layer. When low temperatures bring partial ice cover to lakes, the growth of fish is slowed or stopped. Water temperature can be an important influence in the migration and movement of fishes. As a result of correlating temperature with the ranges of fishes, zoogeographers have developed isothermal theories of fish distribution.

In the Great Lakes, the seasons induce a cycle of physical and chemical changes in the water that are often conditioned by temperature. In the spring, water temperatures and density are vertically homogeneous, making it possible for wind to mix the water to considerable depth and thereby redistribute nutrients and flocculent bottom materials throughout the lake. Oxygen is also redistributed throughout the lake at this time. During much of the summer, full vertical circulation of water and exchange of gases with the atmosphere are restricted by thermal stratification. Until the autumn overturn, only the waters of the epilimnion are mixed and exchange gases with the atmosphere. The hypolimnion remains cold, relatively uniform in temperature, cut off from circulation with the upper waters, and is not involved in gaseous exchange with the atmosphere. Circulation becomes restricted when the lakes freeze over in winter.

8. Light and Turbidity

Light has important implications for predation, feeding, and reproduction. It is one of the major factors that affect final species dominance in an aquatic system. In highly turbid lakes, only those species of fish whose young can effectively feed at low light levels will survive.

Light is also critical in the feeding of many fish, such as the yellow perch, which feed during the day and are dormant at night; species such as the walleye, channel catfish, and slimy sculpin, feed and are more active at night. Light mediated through hormones in fish controls maturation of gonads, while temperature often determines when spawning and egg deposition occurs.

Rooted, suspended, and floating aquatic plants require light for photosynthesis. Light penetration in the Great Lakes is exceedingly variable from lake to lake and from inshore to offshore regions. The principal factors affecting depth of light penetration include suspended phytoplankton and zooplankton, suspended detritus, and mineral particles, water color, and chemical foams (Mackenthun et al., 1964). The lower limit of the euphotic zone is generally considered to be at the level of 1% of incident light. Below this "compensation depth", light

TABLE 9
Spawning Characteristics of Common Lake Erie Fish Species

Taxon family and species	Reproductive characteristics					
			Fecundity			
	Maturity age class	Spawning temp (°C)	Female age or size	Egg production/ female	Spawning season	Longevity (years)
Acipenseridae						
Acipenser fulvescens	XX	12–19	13,608–17,690 g	181,720–188,800	May–Jun.	80+
Lake sturgeon			48,535–52,617 g	652,904–682,640		
Clupeidae						
Alosa pseudoharengua	M–II	22	178 mm	10,000–12,000	Jun.–Jul.	9+
Alewife	F–III					
Dorosoma copedianum	II	19.5	282–297 mm	23,405–96,560	Jun.–Jul.	9+
Gizzard shad			434–452 mm	267,216–350,288		
Salmonidae						
Coregonous artedii	II–III	1.1–5.0	II	16,000–42,500	Nov.–Dec.	13
Cisco			III	14,000–38,600		
Salvelinus namaycush	XIII–XVII			6,000	Sept.–Nov.	41
Lake trout						
Osmeridae						
Osmerus mordax	II–III	10	241 mm	57,910		
Rainbow smelt			185–195 mm	25,102	May	6
Esoclidae						
Esox lucius	M–II–III	8	431–480 mm	22,000	Feb.–Mar.	24
Northern pike	F–II		597 mm	48,950		
Esox musquinongy	M–III	4.5–10	900–1,170 mm	22,092–164,112	April	30
Muskellunge	F–III–IV					
Cyprinidae						
Carassius auratus	100–185 mm			1,400		4
Goldfish						
Cyprinus carpio	M–II–IV	25.2	1,225–1,905 g	72,000–347,000	Apr.–Jun.	16
Carp						

Notemigonus crysoleucas	I–II				May–Aug.	8
Golden shiner		16–27				
Notropis atherinoides	M–II–III	23			Jun.–Aug.	5
Emerald shiner	F–III–IV		III–IV	500–1,500		
Notropis hudsonius	59–84 mm	20	87–127 mm	1,769–4,380	Jun.	5
Spottail shiner						
Notropis spiloptarus	I–II		61–82 mm	316–1,155	Jun.–late Aug.	5
Spotfin shiner						
Pimephales promelas	<I	15.6		800–1,000	May–Aug.	1
Fathead minnow						
Pimephales notatus	<I	>21		200–500	Apr.–Sept.	3
Bluntnose minnow						
Catostonidae						
Carpiodes cyprinus	1,800–2,700 g			4,000–15,000	Apr.–May	3
Quillback						
Catostomus commersoni	M–III–VII	10	406–510 mm	56,000–139,000	Mar.–Apr.	12
	F–III–IX					
White sucker						
Ictaluridae						
Ictalurus moles	III	15.6–23.9	183–224 mm	168–6,820	May–Jun.	9
Black bullhead						
Ictalurus natalis	II–III	15.6–23.9	170–680 g	1,652–6,660	May–Jun.	5
Yellow bullhead						
Ictalurus nebulosus	F–III	15.6–23.9	203–330 mm	2,400–13,800	May–Jun.	6
Brown bullhead						
Ictalurus punctatus	IV–VI	27	406–508 mm	4,200–106,000	Apr.–Aug.	8
Channel catfish						
Percichthyidae						
Morone chrysops	III	19		242,000–933,000	Apr.–May	7
White bass						
Centrarchidae						
Lepomis gibbosus	II	18–21	61–92 mm	600–2,923	Apr.–May	8–10
Pumpkinseed						
Lepomis macrochirus	II–III	19–27		2,360–47,400	May–Aug.	5–10
Bluegill						
Pomoxis annularis	II–III	18 (47)		5,000–30,000	May–Jun.	8
White crappie						

TABLE 9 (Continued)
Spawning Characteristics of Common Lake Erie Fish Species

Taxon family and species	Reproductive characteristics					
			Fecundity			
	Maturity age class	Spawning temp (°C)	Female age or size	Egg production/female	Spawning season	Longevity (years)
Pomoxis nigromaculatus Black crappie	II–III	14–18		11,000–188,000	Mar.-May	8–10
Micropterus dolomioui Largemouth bass	III–VI	13–18		2,000–10,000	May-Jul.	15
Percidae						
Perca flavescens Yellow perch	M–II F–III	16	246 mm	44,000	Mid-Apr.-May	8
Stizostedion canadense Sauger	M–23.7 mm	8.2	305–311 mm	43,000–48,500	Apr.-May	10
Stizostedion v. vitreum Walleye	M–II–III F–III–V	4.5–11.1		48,000–614,000	Mar.-May	13
Sciaenidae						
Aplodinotus grunniens Freshwater drum	M–III–VII F–V–VI	21.0		100,000–500,000	Spring	9

Data source: Hartley and Herdendorf (1977).

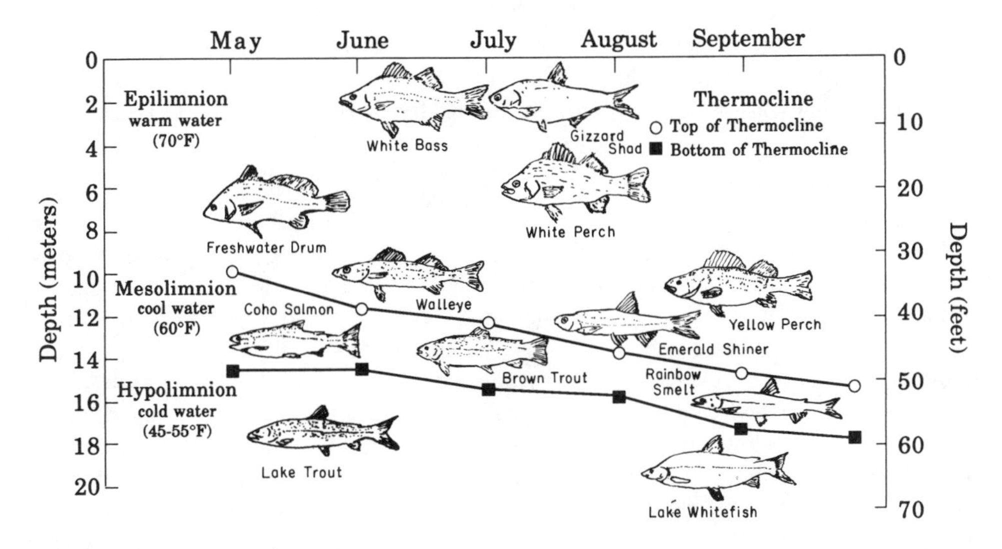

FIGURE 13. Distribution of fish in central Lake Erie in relation to summer temperature profile. (Data source: Ohio Sea Grant.)

is inadequate for photosynthesis to overcome respiration. In winter, the presence of ice with snow cover further limits the amount of already low incident light energy that reaches the water. Clear ice has a light absorptive quality similar to that of water (Neess and Bunge, 1957). A 5-cm snowcover over 40 cm of clear ice permits only 1% of incident light to reach the water (Mackenthun and McNabb, 1961).

Light affects fish directly through vision, but there are many indirect effects as well. Coloration of the integument at any given time is a direct function of the spectral characteristics and intensity of light. Light triggers and directs migrations and movements, and influences the rate and pattern of growth. Light also determines the kinds and amounts of food available for fish and is, of course, the direct energy source for the first photosynthetic link at the base of the food web of all fishes (Lagler et al., 1977).

Many species of Great Lakes fish are sight feeders and thus require sufficient light to illuminate their prey. The design of an aquatic visual system, such as the eyes of sight-feeding fish, must meet more stringent requirements than the eyes of most terrestrial vertebrates because of photic conditions peculiar to the underwater realm. Light is colored by the selective filtering action of water and dissolved substances. It is scattered and refracted by suspended particles, making distant objects appear faint and blurred, and it is normally dim. During evolution, these factors shaped the color vision, pattern vision, and visual sensitivity of fishes to match the variety of the light environments they inhabit (Northmore et al., 1978).

Turbidity, which is an expression of the optical property of water that causes light rays to be scattered and absorbed rather than transmitted in a straight line, results from a variety of suspended particulate matter. Fine particulate inorganic and organic materials in suspension can limit the penetration of sunlight, thus restricting the growth of submerged hydrophytes as well as planktonic algae. In turn, by limiting growth of aquatic plants, food chains are interrupted, and this can ultimately result in a sparsity of fish life. Also, Lagler (1956) states that natural inorganic agents causing turbidity can harm fish directly, such as through impaired respiration, if present in high concentrations:

0–25 ppm	No effect on fisheries
25–80 ppm	Somewhat reduced fish yield
80–400 ppm	Not likely to support good freshwater fisheries
>400 ppm	Only poor fisheries at best

Locally, high turbidity may be caused by the presence of a particular fish species, such as carp, which disturb the bottom sediments during feeding and spawning activities. This occurs commonly in some coastal marshes, embayments, and freshwater estuaries of the Great Lakes.

C. THE CHEMICAL MEDIUM

1. Dissolved Oxygen

Oxygen enters the water by absorption from the atmosphere or by plant photosynthesis. That derived from the atmosphere enters by direct diffusion or by surface water agitation (wind and waves). The annual oxygen production (via photosynthesis) curve closely follows the solar radiation curve in western Lake Erie (Verduin, 1956). At all depths, respiration and decomposition of organic matter consume oxygen and liberate carbon dioxide. Throughout the period of summer stratification, a gradual decrease occurs in dissolved oxygen. After the oxygen is depleted below the thermocline (such as in the central basin hypolimnion of Lake Erie), anaerobic decomposition continues with the evolution of carbon dioxide, methane, and hydrogen sulfide. Thus, the thermocline acts as a transition zone for dissolved oxygen as well as temperature. Below the thermocline, the water is cooler, light is reduced, photosynthesis decreases, and if sufficient oxygen is present, some species of coldwater fish abound.

The availability of oxygen to fish is not constant. The solubility of oxygen in water varies with temperature and is greater in cold water than in warm. As the temperature of lake water rises during spring and early summer, the amount of dissolved oxygen the water can hold decreases. Since fish are ecotootherms, i.e., their body temperature rises with increasing water temperature and their metabolic rate approximately doubles with each increase in temperature of 10°C, warmwater fish experience a particularly high oxygen demand (Eddy and Underhill, 1976). The demand occurs at a time when dissolved oxygen generally is least available to them. Lagler (1956) states that the amount of dissolved oxygen present during summer stratification is an important indicator of environmental suitability, particularly for coldwater fish that are restricted to the hypolimnion. If such hypolimnion waters do not contain at least 4 to 5 mg/l of oxygen, they cannot support coldwater fish assemblages (e.g., lake trout). Dissolved oxygen is a limiting factor in some Great Lakes, in some harbors and backwater areas, and in inflowing polluted rivers.

Tonn and Magnuson (1982) found that winter-dissolved oxygen and aquatic macrophyte diversity were the chief environmental factors affecting community structure in a series of Wisconsin lakes. Effects on juvenile and adult fish are probably only minimal; fish move to less stressful thermal habitats. However, growth and survival may be affected if fish must move from preferred areas because of low dissolved oxygen concentrations to areas with adequate oxygen concentrations, but where food or predators may be major limiting factors.

Fish have evolved various mechanisms to avoid potentially adverse conditions on or near the substrate interface, such as low dissolved oxygen and the buildup of toxic compounds (i.e., hydrogen sulfide, ammonia, and carbon dioxide, CO_2). These strategies include (1) pelagic eggs and larvae, (2) stalked eggs attached to sand grains or other substrate, an accordion-like matrix which suspends eggs off the bottom, (3) short incubation periods, (4) guarding of eggs, (5) building nests of debris, (6) ascending rivers and streams to lay eggs in current swept areas, or (7) laying eggs on wave-swept reefs in the open lake.

2. Alkalinity and pH

A total alkalinity of about 40 mg/l appears to be a natural separation between hard- and softwater lakes. Hardwater lakes, such as the Great Lakes in which the bicarbonate content is high, store CO_2 to a higher degree than softwater lakes, although the amount of free CO_2 present in the surface waters is often greater in softwater lakes. Bicarbonates indirectly furnish a large amount of CO_2 for plant growth in hardwater lakes. The following alkalinity scale has been derived by Moyle (1949) for natural lakes in Minnesota:

Alkalinity (mg/l)	Plant and fish productivity
20	Low
20–40	Low to medium
40–90	Medium to high
>90	High

Lagler (1956) contends that CO_2 is perhaps the best single criterion of environmental suitability for fish. High concentrations of free CO_2, which are in themselves toxic to fish, are usually accompanied by low values of dissolved oxygen. Free CO_2, in excess of 20 ppm, is regarded as harmful to fish, although lower values may be equally harmful in water of low oxygen content (3 to 5 mg/l). Eddy and Underhill (1976) point out that contrary to expectations, fish do not respond to high concentrations of CO_2 in the water but only to low oxygen concentrations.

Neel et al. (1961) found that pH values >8.0 are produced by a photosynthetic rate that demands more CO_2 than the quantities furnished by respiration and decomposition, whereas pH values <8.0 indicate the failure of photosynthesis to totally utilize the amounts of CO_2 so produced. Fish are most commonly found in water with a pH range between 5.0 and 9.0. Although some resistant fish species can tolerate pH values below 4.0 and well above 10, such extreme conditions associated with industrial or atmospheric pollution are hazardous to fish life in waters that are not naturally at such extreme levels. Most fish in the Great Lakes can tolerate both the wide ranges (6.5 to 9.5) and rapid, but linked changes of acidity typical of these bodies of water.

3. Nutrients

The biological productivity in the Great Lakes is basically dependent upon the autotrophic community, which in turn is controlled by the availability of nutrients. Table 10 lists the common nutrients generally considered necessary for plant growth. The processes by which nutrients enter and leave a lake are complex. Figure 14 simply illustrates the major nutrient pathways in a lake system. Once in the lake, nutrients are subjected to various removal and/or recycling functions. Chemical reactions, sediment uptake and subsequent burial, and biological utilization tend to remove large quantities of nutrients. These processes may be reversed, however, through biochemical action. The conversion of inorganic nutrients into biomass within a zone of high algae productivity may be reversed in the profundal zone having active decomposition or variable chemical conditions (such as oxic-anoxic). These activities often result in uneven nutrient distributions when the lake is stratified (Tobin and Youger, 1977).

The Great Lakes are of recent geologic origin, their basins being largely excavated by the ice sheets of the Pleistocene glacial period. The ice sheets undoubtedly were guided to a considerable extent by the existing topographic features and tended to deepen river valley floors, especially where they were composed of relatively soft rocks. All of the lake basins lie on or near the southern edge of the Canadian shield (Precambrian), a large mass of very old metamorphic and igneous rocks. These rocks are very resistant to weathering, and thus the lakes, or parts of the lakes that lie in the Precambrian area, have low concentrations of dissolved solids. Those lakes or parts of lakes that lie outside this area are in basins composed of younger sedimentary rocks that are more easily eroded; thus, these lakes have higher concentrations of total dissolved materials (Table 11). The nutrient concentrations, and thus the biological productivity of these lakes, are greatly influenced by the composition of the

TABLE 10
Nutrient Requirements for Algal Growth in Aquatic Systems

Element[a]	Symbol	Some common forms in water[a,b]	Min. requirements[c]	Examples of natural sources[a,d]	Examples of man-made sources[e-g]
Aluminum	Al	Al^{3+}, $AISO_4$, AIO^2, (salts of aluminum)	Probably trace quantities	Clay minerals, silicate rock minerals	Domestic sewage, industrial wastes, mine drainage
Boron	B	B, H_3BO_3	100 μg/l	Evaporite deposits, igneous rock minerals, volcanic gases	Cleaning aids, detergents, industrial wastes, irrigation, sewage
Calcium	Ca	Ca^{2+}, $CaCO_3$, $CaSO_4$	20 mg/l	Igneous rock minerals, rainwater, sedimentary rocks, soil	Industrial wastes (metallurgy, steelmaking), treatment plant wastes
Carbon	C	CO_2, CO_3, HCO_3, II_2CO_3, $CaCO_3$	Quantities always sufficient in surrounding medium	Atmosphere, organic compounds and decay products, rainwater, soil	Industrial wastes (carbonation, metallurgy, pulp and paper, soda, and steelmaking), domestic sewage
Chlorine	Cl	Cl^{-1} (oxides of chlorine)	Trace quantities	Evaporite deposits, igneous rock minerals, ocean water, rainwater, sedimentary rocks, volcanic gases	Chlorinated hydrocarbon process, cleaning aids, industrial wastes (petroleum and refining), irrigation, salt mining
Cobalt	Co	Co	500 μg/l	Coal ash, soil, ultramafic rocks	Manufacturing wastes (tools and instruments), metallurgy
Copper	Cu	Cu^{3+}, Cu, $CuSO_4$	6.0 μg/l	Crustal rocks, groundwater, marine animals	Industrial wastes (fabrication of pipes, refining, smelting), manufacturing wastes (electrical, foods), mill tailings, mine wastes, ore dumps, treatment plant wastes
Hydrogen	H	H^+, H_2S, H_2O, HCO_3, H_2CO_3, OH	Quantities always sufficient in surrounding medium	Atmosphere, oxidation processes, rainwater, volcanic activity	Industrial wastes (hydrocarbon process), oils
Iron	Fe	Fe^{2+}, Fe^{3+}, $FeSO_4$, $Fe(OH)_2$	0.65–6000 μg/l	Groundwater, igneous rock minerals, iron minerals, organic decomposition, soil	Acid drainage from mines, industrial wastes (steelmaking), iron ore mining, manufacturing wastes, oxides of iron metals (car bodies, refrigerators)
Magnesium	Mg	Mg^{2+}, $MgSO_4$	Trace quantities	Igneous rock minerals, groundwater, rainwater, sedimentary rocks	Irrigation, manufacturing wastes (transportation vehicles)

Manganese	Mn	Mn^{2+}, MnO^2	5.0 µg/l	Groundwater, plants, rocks, soil, tree leaves	Acid drainage from coal mines, industrial wastes (steelmaking)
Molybdenum	Mo	Mo, MoO_4	Trace quantities	Groundwater, rocks, soil	Industrial wastes (electrical devices, metallurgy, steelmaking), manufacturing wastes (alloys)
Nitrogen	N	N, NO_2, NO_3, organic nitrogen, NH_3	Trace quantities to 5.3 mg/l	Atmosphere, bacterial and plant fixation, limestone, rainwater, soil	Agricultural wastes (feedlots, fertilizers), domestic sewage, industrial wastes, storm drainage
Oxygen	O	O_2, H_2O, oxides	Quantities always sufficient in surrounding medium	Atmosphere, oxidation processes, photosynthesis, rainwater	Industry (metallurgy)
Phosphorus	P	P^{5+}, PO_4, HPO_3, organic phosphorus	0.002–0.09 mg/l	Groundwater, igneous and marine sediments, rainwater, soil, waterfowl	Agricultural wastes (feedlots, fertilizers), domestic sewage (detergents), industrial wastes
Potassium	K	K^+ (sales of potassium)	Trace quantities	Evaporite deposits, igneous rock minerals, plant ash, sedimentary rocks	Agricultural wastes (feedlots, fertilizers), industrial wastes (preservatives, pulp ash)
Silicon	Si	Si^{4+}, SiO_2	0.5–0.8 mg/l	Diatom shells, igneous rock minerals, metamorphic rocks	Domestic sewage, industrial wastes
Sodium	Na	Na^+, Na salts (NaCl, $NaCO_3$)	5.0 mg/l	Groundwater, igneous rock minerals, ocean water, soil	Industrial wastes (paper and pulp, rubber, soda, water softeners), manufacturing wastes (dyes and drugs)
Sulfur	S	SO_2, HS, H_2S, SO_4	5.0 mg/l	Animal and plant decomposition, igneous rocks, rainwater, sedimentary rocks, springs, volcanic activity	Agricultural wastes (fertilizers), industrial wastes (fuels, paper and pulp)
Vanadium	V	V^{2+}, V^{3+}, V^{4+}, V^{5+} (salts and oxides of vanadium)	Trace quantities	Groundwater, plant ash	Industrial wastes
Zinc	Zn	Zn^{2+} (salts of zinc), ZnO_2	10–100 µg/l	Igneous and carbonate rock minerals	Industrial wastes (piping, refining), mine wastes

Note: The minimum nutrient requirements of algae in the aquatic environment are difficult to determine, and this uncertainty is shown by the wide range of concentrations in the table (source: Britton et al. 1975). "Trace" quantities generally refer to concentrations <1 mg/l, and more exact concentration requirements for these elements have not been determined. "Quantities always sufficient in surrounding medium" refers to those elements that are never below minimum concentrations so as to limit algal growth.

[a] Hem (1970),[b] McKee and Wolf (1971),[c] Greeson (1971),[d] Reid (1961),[e] Gurnham (1965),[f] Nebergall et al. (1963),[g] Sawyer and McCarty (1967).

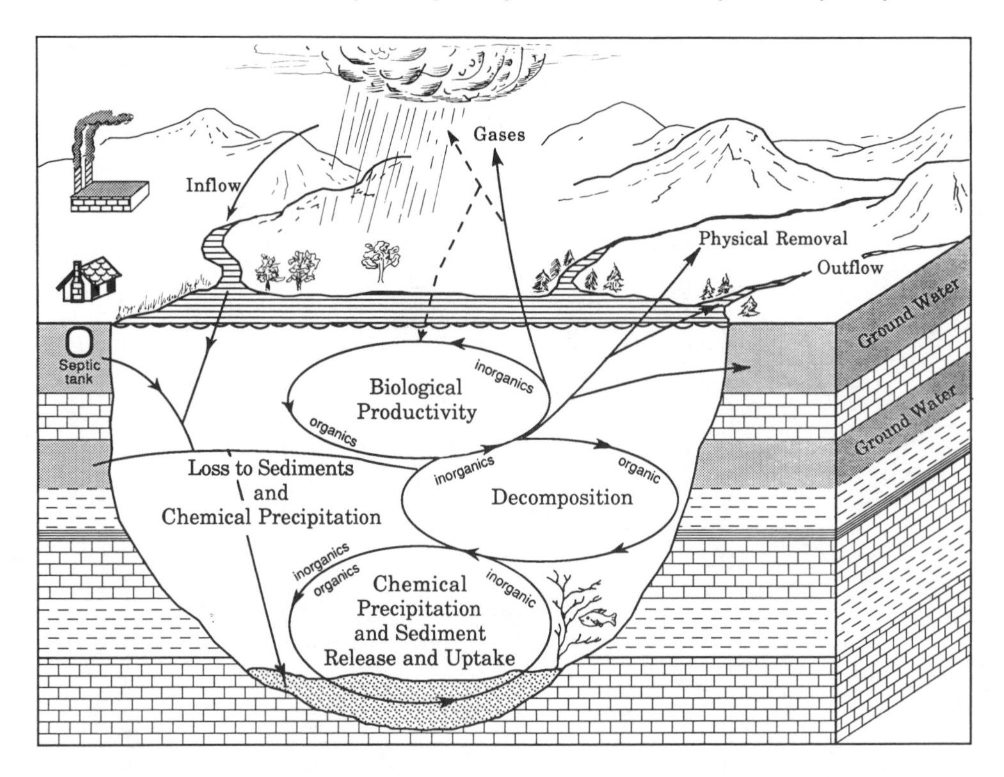

FIGURE 14. Nutrient pathways in a lake system. Nutrients entering the lake from the left may be permanently or temporarily trapped in the benthic zone or may be incorporated within the lake's biochemical cycles (arrows). Avenues of nutrient removal from the lake are shown at the right. (Data source: Tobin and Youger, 1977.)

rocks in their drainage basins. Those in the Precambrian area have very low concentrations and are markedly oligotrophic, while the lakes in sedimentary areas tend to have higher nutrient concentrations and levels of production (Robertson and Scavia, 1984).

Moyle (1956) found a positive relationship between the total phosphorus concentration in Minnesota lakes and their standing crop of fish:

Phosphorus (mg/l)	Standing crop of fish (kg/ha)
0.034	101
0.058	168
0.126	414

The nutrient concentration of a lake is related to edaphic (soils) characteristics in the drainage basin, anthroprogenic inputs, and atmospheric deposition. The Great Lakes vary considerably in their concentrations of phosphorus, nitrogen, and silicon. Thus, the fertility of these lakes in terms of fish production also vary considerably (Table 12).

Silica is a key nutrient and its availability is usually a limiting factor in diatom production. Phosphorus is usually the limiting nutrient for primary production during the stratified period in the Great Lakes. In a few areas, especially where large phosphorus loads occur due to municipal effluents and/or agricultural runoff, this may not always be true. In such areas available nitrogen or even light, as controlled by turbidity caused by large amounts of organic and inorganic particles, may play a limiting role. Increased phosphorus concentrations arising from human activities have caused accelerated "cultural eutrophication" in several of the lakes, especially Erie, Ontario, and Michigan. Blue-green algae are typical of highly eutrophic

TABLE 11
Relation between Average TDS Concentration and Type of Rocks Underlying the Drainage Basins of the Great Lakes

Lake	TDS (mg l^{-1})	Type of drainage basin
Superior	60	Precambrian rocks
Huron	115	Mixed Precambrian and sedimentary rock
Michigan	155	Sedimentary rocks
Erie (west)	165	Sedimentary rocks
Erie (central)	185	Sedimentary rocks
Erie (east)	190	Sedimentary rocks
Ontario	190	Sedimentary rocks

Data source: Patalas (1975).

TABLE 12
Comparative Commercial Fish Harvests for the Great Lakes 1880–1980

	Tons/mi² of lake surface					
Year	Superior	Michigan	Huron	St. Clair	Erie	Ontario
1880	0.007	0.52	0.25	2.88	1.54	0.46
1890	0.13	0.59	0.53	3.28	3.71	0.50
1900	0.15	0.84	0.53	1.11	3.48	0.36
1910	0.20	1.06	0.45	1.51	2.68	0.27
1920	0.20	0.45	0.39	1.56	2.47	0.35
1930	0.31	0.60	0.48	0.49	2.13	0.31
1940	0.38	0.51	0.32	0.98	1.65	0.29
1950	0.24	0.61	0.21	0.68	2.06	0.16
1960	0.26	0.55	0.22	0.81	2.54	0.15
1970	0.13	1.19	0.10	0.09	2.08	0.21
1980	0.13	0.51	0.13	0.35	2.63	0.15
Mean	0.20	0.68	0.33	1.25	2.45	0.29
SD	±0.09	±0.24	±0.15	±1.01	±0.68	±0.12
SE	±0.03	±0.07	±0.05	±0.30	±0.30	±0.04

Data source: Herdendorf (1983).

conditions and certain species have the ability to fix atmospheric nitrogen, thus utilizing a large proportion of available phosophorus and frequently being an important component of algal blooms.

VI. RECOMMENDATIONS

A habitat classification system is proposed based on three major levels that describe the physical habitat: horizontally, vertically, and by substrate. A series of habitat quality factors is invoked to describe limiting and forcing variables which are most important to fish and which determine their distribution and behavior. Theoretically, this system should be able to describe the habitat of any substrate with water over it in the Great Lakes.

Level 1 factors describe the horizontal spatial scale in a Great Lake; moving offshore these include the coastal zone (rivers, wetlands, connecting lakes, and water bodies), the littoral or

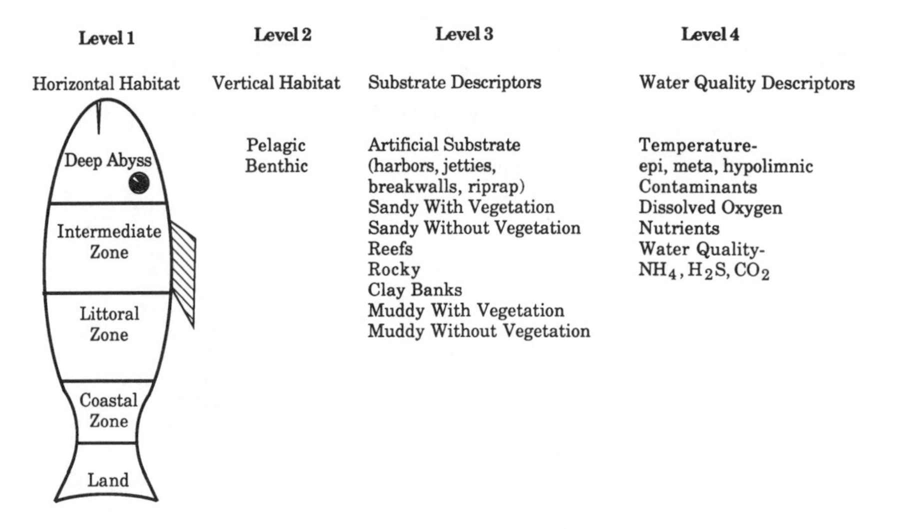

FIGURE 15. Proposed simple habitat descriptors for classification of Great Lakes habitat based on use by fish, especially early life history stages.

nearshore zone, the intermediate zone, and the deep abyss or offshore zone. Level 2 factors describe the vertical habitat and consist of two variables, pelagic and benthic. Level 3 characterizes the substrate. Descriptors may include artificial substrate (harbors, jetties, breakwalls, intake riprap, etc.), sandy with vegetation, sandy without vegetation, reefs, rocky with algae, rocky without algae, clay banks, muddy with vegetation, muddy without vegetation. The classification system should also attempt to provide a description of habitat quality, for example, temperature (epi-, meta-, and hypolimnion), contaminants, dissolved oxygen, nutrients, water quality (ammonia, CO_2, and hydrogen sulfide). A brief schemata (Figure 15) displays these ideas graphically. An example for a particular area of Saginaw Bay might be littoral zone, benthic, muddy with vegetation, temperature (epilimnion), impacted by contaminants and nutrients. The next step would be to assign some fish community assemblages to specific areas of the Great Lakes so classified. This system is suggested as a starting point and must include many other aspects, including the temporal factor, which would accommodate seasonal changes in some of the habitat descriptors (for example, temperature), and hence the fish utilizing the habitat.

APPENDIX

METHOD 1

Water velocity in section areas was determined for 21 Swedish coastal areas. The mean value was 2.92 (cm/s); the median, 1.98 (cm/s); and the standard deviation, 2.07 (cm/s), and can be used to determine a first estimate of the surface water retention time (Tp) as

$$Tp = (1 / 2) * (V / At) \quad (1)$$

where V = the water volume (km^3) and At = the section area (km^2).

When the Tp values from Equation 1 are compared to the empirical T values, the degree of explanation (r^2) is 0.64. Thus, the retention time may be determined from two standard morphometrical parameters (V and At), which are easy to calculate from bathymetric maps. The degree of explanation, however, is not very high.

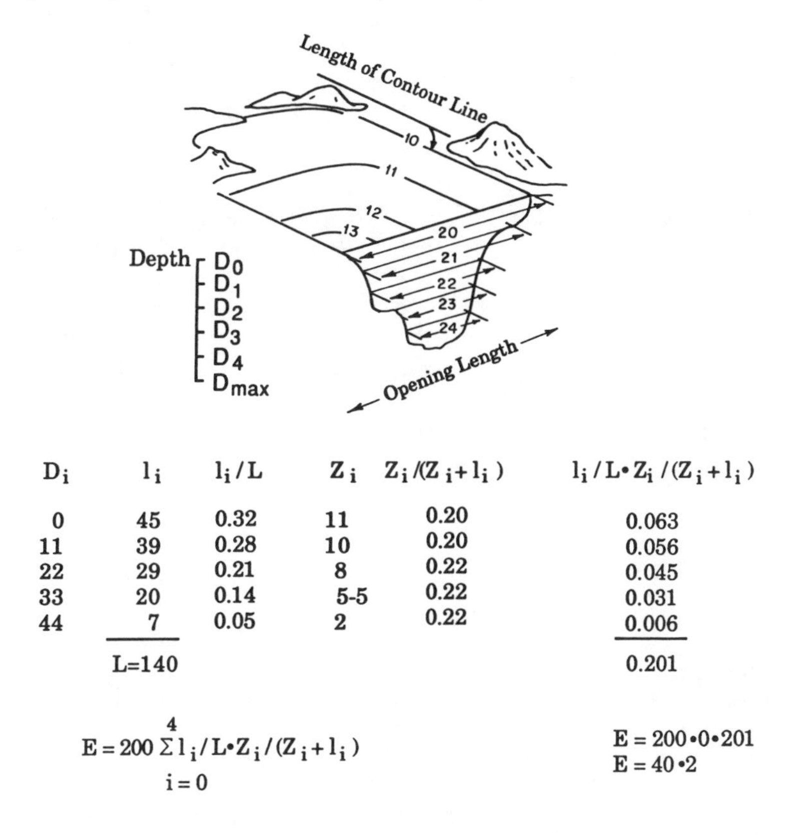

D_i	l_i	l_i/L	Z_i	$Z_i/(Z_i+l_i)$	$l_i/L \cdot Z_i/(Z_i+l_i)$
0	45	0.32	11	0.20	0.063
11	39	0.28	10	0.20	0.056
22	29	0.21	8	0.22	0.045
33	20	0.14	5-5	0.22	0.031
44	7	0.05	2	0.22	0.006
	L=140				0.201

$$E = 200 \sum_{i=0}^{4} l_i/L \cdot Z_i/(Z_i+l_i)$$

$$E = 200 \cdot 0 \cdot 201$$
$$E = 40 \cdot 2$$

FIGURE 16. Definition of the topographical openness (or the exposure) from the relationship between the length of opening lines (zi) at 4 water depths and the length of the corresponding contour lines (li). The formula gives E = 100% for a straight coastline and E values <100% for bays.

METHOD 2*

A better equation (in terms of higher statistical degree of explanation) is given by

$$Tp = 42.7 * W * E^{-0.71} \tag{2}$$

where 42.7 and –0.71 are empirical constants; Tp = the retention time (days) for the surface water; W = the mean coastal width (km); E = the topographical openness (or exposure, %). The equation is graphically shown in Figure 13 and gives a high degree of explanation (r^2 = 0.86). One should note that many morphometric parameters cannot be considered "independent" from a statistical viewpoint. That is not of primary concern here; the crucial point is that the various parameters should reveal different topographical form elements of the coast and that the required equation should be built on parameters with a minimum of interdependence. The mean coastal width (W) is defined as the ratio between total coastal area (A) and the length of the coastline (lo); the exposure is defined from other premises (Figure 16). One must also note that

1. The higher the E, the shorter the retention time and vice versa. This is quite logical.
2. The larger the W, the longer the retention time. This is also reasonable, as W is a size-dependent parameter (dimension in kilometers).

* The systems for calculating water retention time in both methods were derived from Håkanson et al. (1984, 1986).

3. That a dimension analysis of Equation 2 reveals that the constant 42.7 has the dimension time/length and could be linked to an exchange velocity, and that the equation can be physically coupled to a basic mass balance equation.
4. The equation cannot be used in river mouth areas or in areas dominated by strong tides; precise limits cannot be given at present.
5. The equation can only be used for the bioproductive season (April to October).
6. The equation gives a characteristic value (a median) and not a frequency distribution.
7. The absolute differences between predicted and empirical values increase with T, but that the percentage error is quite constant for different T values.
8. The main benefit with Equation 2, as compared to Equation 1, is that the exposure explains a statistically significant part of the variation in water velocity in the section area.
9. The stability of Equation 2 has been tested and the result is most satisfactory; r^2 is about 0.9 and k1, k2, and k3 are very close to 43, 110, and –0.7, whatever test method applied.

REFERENCES

Assel, R. A., F. H. Quinn, G. A. Leshkevich, and S. J. Bolsenga. 1983. Great Lakes Ice Atlas. NOAA, Great Lakes Environmental Research Laboratory, Ann Arbor, MI.

Auer, N. A. Ed. 1982. Identification of Larval Fish of the Great Lakes Basin with Emphasis on the Lake Michigan Drainage. Spec. Publ. 82–3. Great Lakes Fisheries Commission, Ann Arbor, MI.

Ayers, J. C. 1962. Great Lakes Waters, Their Circulation and Chemical Characteristics. Publ. 71. AAAS Great Lakes Basin Symp. Chicago, 1954. pp. 71–89.

Bergstedt, R. and R. O'Gorman. 1983. Lake Ontario Fish Stocks: Surveillance and Status of Fish Populations. Progress in Fishery Research 1983. Great Lakes Fish. Lab., U.S. Fish Wildl. Serv., Ann Arbor, MI. pp. 15–16.

Boonyuen, S. and C. E. Herdendorf. 1986. Fish Productivity in Large Lakes: A Numerical Model for Estimating Annual Rates. Sea Grant Tech. Bull. No. 3. Ohio State University, Columbus.

Boyce, F. M. 1974. Some aspects of Great Lakes physics of importance to biological and chemical processes. *J. Fish. Res. Bd. Can.* 31(5):689–730.

Brandt, S. B. 1980. Spatial segregation of adult and young-of-the-year alewives across a thermocline in Lake Michigan. *Trans. Am. Fish. Soc.* 109:469–478.

Brandt, S. B., J. J. Magnuson, and L. B. Crowder. 1980. Thermal habitat partitioning by fishes in Lake Michigan. *Can. J. Fish. Aquat. Sci.* 37:1557–1564.

Britton, L. J., R. C. Averett, and R. F. Ferreira. 1975. An Introduction to Processes, Problems, and Management of Urban Lakes. U.S. Geological Survey Circ. 601-K. USGS, Washington, D.C.

Brylinsky, M. and K. H. Mann. 1973. An analysis of factors governing productivity in lakes and reservoirs. *Limnol. Oceanogr.* 18:1–14.

Busch, W.-D. N., R. L. Scholl, and W. L. Hartman. 1975. Environmental factors affecting the strength of walleye (*Stizostedion v. vitreum*) year-classes in western Lake Erie 1960–70. *J. Fish. Res. Bd. Can.* 32:1733–1740.

Busch, W.-D. N. and L. M. Lewis. 1983. Responses of wetland vegetation to water level variations in Lake Ontario. in Lake and Reservoir Management, Proc. 3rd Annu. Conf. N. American Lake Management Society.

Camanzo, J., C. Rice, D. Jude, and D. Rossmann. 1987. Organic priority pollutants in nearshore fish from 14 Lake Michigan tributaries and embayments, 1983. *J. Great Lakes Res.* 13:296–309.

Carlander, K. D. 1955. The standing crops of fish in lakes. *J. Fish. Res. Bd. Can.* 12:543–570.

Carlson, R. E. 1977. A trophic state index of lakes. *Limnol. Oceanogr.* 22:361–369.

Christie, G. C. and Regier, H. A. 1988. Measures of optimal thermal habitat and their relationship to yields for four commercial fish species. *Can. J. Fish. Aquat. Sci.* 45: 301–314.

Christie, W. J. 1974. Changes in the fish species composition of the Great Lakes. *J. Fish. Res. Bd. Can.* 31:827–854.

Chubb, S. and C. Liston. 1986. Density and distribution of larval fishes in Pentwater Marsh, a coastal wetland on Lake Michigan. *J. Great Lakes Res.* 12:332–343.

Clady, D. D. 1976. Influence of temperature and wind on the survival of early stages of yellow perch (*Perca flavescens*). *J. Fish. Res. Bd. Can.* 33:1887–1893.

Clady, M. and B. Hutchinson. 1975. Effect of high winds on eggs of yellow perch (*Perca flavescens*), in Oneida Lake, New York. *Trans. Am. Fish. Soc.* 104:524–525.

Cowardin, L., V. Carter, F. Golet, and E. LaRoe. 1979. Classification of Wetlands and Deepwater Habitats of the United States. U.S. Fish Wildl. Serv. Pub. FWS/OBS-79/31. U.S. Government Printing Office, Washington, D.C.

Craig, J., C. Kipling, E. LeCren, and J. McCormack. 1979. Estimates of the numbers, biomass and year-class strengths of perch *Perca fluviatilis* L. in Windemere from 1967 to 1977 and some comparisons with earlier years. *J. Anim. Ecol.* 48:315–325.

Crowder, L. B. and J. J. Magnuson. 1982. Thermal habitat shifts by fishes at the thermocline in Lake Michigan. *Can. J. Fish. Aquat. Sci.* 39:1046–1050.

Doan, K. H. 1942. Some meteorological and limnological conditions as factors in the abundance of certain fishes in Lake Erie. *Ecol. Monogr.* 12:293–314.

Dorr, J. A., III. 1982. Substrate and Other Environmental Factors in Reproduction of the Yellow Perch (*Perca flavescens*). Ph.D. thesis. University of Michigan, Ann Arbor.

Dorr, J. A., III and D. J. Jude. 1980. SCUBA assessment of abundance, spawning, and behavior of fishes in southeastern Lake Michigan near the Donald C. Cook Nuclear Plant, 1975–1978. *Mich. Acad.* 12:345–364.

Dorr, J. A., III and D. J. Jude. 1986. Diver Assessment of the Inshore Southeastern Lake Michigan Environment Near the D.C. Cook Nuclear Plant, 1973–1982. Spec. Rep. No. 120, Great Lakes Research Division, University of Michigan, Ann Arbor.

Dorr, J. A., III, D. J. Jude, G. Heufelder, S. Klinger, G. Noguchi, T. Rutecki, and P. Schneeberger. 1981a. Preliminary Investigation of Spawning Habitat Conditions and Reproduction of Lake Trout in Eastern Lake Michigan near Port Sheldon, Michigan. Michigan Sea Grant Publ. No. Michu-81–213. University of Michigan, Ann Arbor.

Dorr, J. A., III, D. V. O'Connor, N. R. Foster, and D. J. Jude. 1981b. Substrate conditions and abundance of lake trout eggs in a traditional spawning area in southeastern Lake Michigan. *N. Am. J. Fish. Manage.* 1:165–172.

Eadie, J. and A. Keast. 1984. Resource heterogeneity and fish species diversity in lakes. *Can. J. Zool.* 62:1689–1695.

Eck, G. and E. Brown, Jr. 1985. Lake Michigan's capacity to support lake trout and other salmonines: an estimate based on the status of prey populations in the 1970s. *Can. J. Fish. Aquat. Sci.* 42:449–454.

Eddy, S. and J. C. Underhill. 1976. *Northern Fishes.* University of Minnesota Press, Minneapolis.

Edsall, T. A. 1970. The effect of temperature on the rate of development and survival of alewife eggs and larvae. *Trans. Am. Fish. Soc.* 99:376–380.

Edsall, C. and M. Mac. 1982. Comparative hatching success of lake trout eggs in Lake Michigan water and well water. *Prog. Fish. Cult.* 44:47–48.

Eisenreich, S. and G. Hollod. 1979. Accumulation of polychlorinated biphenyls (PCBs) in surficial Lake Superior sediments. Atmospheric deposition. *Environ. Sci. Technol.* 13:569–573.

Emery, A. R. 1970. Fish and crayfish mortalities due to an internal seiche in Georgian Bay, Lake Huron. *J. Fish. Res. Bd. Can.* 27:1165–1168.

Eshenroder, R., T. Poe, and C. Olver. 1984. Strategies for Rehabilitation of Lake Trout in the Great Lakes. Tech. Rep. No. 40. Great Lakes Fisheries Commission. Ann Arbor, MI.

Faber, D. 1967. Limnetic larval fish in northern Wisconsin lakes. *J. Fish Res. Bd. Can.* 24:927–937.

Findenegg, I. 1965. Factors controlling primary productivity especially with regard to water replenishment, stratification, and mixing, in *Primary Productivity in Aquatic Environment.* C. R. Goldman, Ed. pp. 106–119.

Fry, F. E. J. 1947. Effects of the Environment on Animal Activity, University of Toronto Press. No. 55, Ontario Fisheries Research Lab. Toronto.

Golterman, H. L., P. G. Sly, and R. L. Thomas. 1983. Study of the Relationship between Water Quality and Sediment Transport. Tech. Papers in Hydrology No. 26, UNESCO, Paris.

Greene, R. G. 1987. Effects of Lake Erie Water Levels on Wetlands as Measured from Aerial Photographs: Pointe Mouillee, Michigan. M.S. thesis. Ohio State University, Columbus.

Greeson, P. E. 1971. The Limnology of Oneida Lake with Emphasis on Factors Contributing to Algal Blooms. U.S. Geological Survey open-file report. USGS, Washington, D.C.

Gurnham, F. C. 1965. *Industrial Wastewater Control.* Academic Press, New York.

Håkanson, L. 1981. *A Manual of Lake Morphometry.* Springer-Verlag, Berlin.

Håkanson, L., I. Kulinski, and H. Kvarnas. 1984. Water dynamics and bottom dynamics in Swedish coastal areas (in Swedish). Swedish Environmental Protection Board (SNV) PM 1905, Solna, Sweden. p. 228.

Håkanson, L., H. Kvarnas, and B. Karlsson. 1986. Coastal morphometry as regulator of water exchange — a Swedish example. *Estuar. Coast. Shelf Sci.* 23:1–15.

Hartley, R. P., C. E. Herdendorf, and M. Keller. 1966. Synoptic Survey of Water Properties in the Western Basin of Lake Erie. Div. Geol. Surv. Rep. Invest. 58. Ohio Department of Natural Resources, Columbus.

Hartley, S. M. and C. E. Herdendorf. 1977. Spawning Ecology of Lake Erie Fishes. CLEAR Tech. Rep. No. 62. Ohio State University, Columbus.

Hayes, F. R. 1957. On the variation in bottom fauna and fish yield in relation to trophic level and lake dimensions. *J. Fish. Res. Bd. Can.* 14:1–32.

Hayes, F. R. and E. H. Anthony. 1964. Production capacity of North America lakes as related to the quantity and the trophic level of fish, the lake dimensions, and the water chemistry. *Trans. Am. Fish. Soc.* 93:53–57.

Hem, J. D. 1970. Study and Interpretation of the Chemical Characteristics of Natural Water, 2nd ed. Water-Supply Pap. U.S. Geological Survey, 1473. Washington, D.C.

Herdendorf, C. E. 1975. Shoreline changes of Lakes Erie and Ontario with species reference to currents, sediment transport and shore erosion. *Bull. Buffalo Soc. Nat. Sci.* 25(3):43–76.

Herdendorf, C. E. 1983. Our changing fish species history in the Great Lakes. *Inland Seas* 39(4):276–286.

Herdendorf, C. E. 1987. The Ecology of the Coastal Marshes of Western Lake Erie: A Community Profile. Biol. Rep. 85 (7.9). U.S. Fish and Wildlife Service. Washington, D.C.

Herdendorf, C. E. 1990a. Distribution of the world's large lakes. in Large Lakes: Ecological Structure and Function. M. M. Tilzer and C. Serruya, Eds. Springer-Verlag, Berlin. pp. 3–38.

Herdendorf, C. E. 1990b. Great Lakes estuaries. *Estuaries* 13(4): 493–503.

Herdendorf, C. E. and L. L. Braidech. 1972. Physical Characteristics of the Reef Area of Western Lake Erie. Ohio Div. Geol. Surv. Rep. Invest. 83. Columbus.

Heufelder, G. R., D. J. Jude, and F. J. Tesar. 1982. Upwelling effects on local abundance and distribution of larval alewife (*Alosa pseudoharengus*) in eastern Lake Michigan. *Can. J. Fish. Aquat. Sci.* 39:1531–1537.

Hjort, J. 1914. Fluctuations in the great fisheries of northern Europe viewed in the light of biological research. *Rapp. P.-V. Reun. Cons. Int. Explor. Mer.* 20:1–228.

Hjulstrom, F. 1939. Transportation of detritus by moving water. in *Recent Marine Sediments.* Symp. Am. Assoc. Petrol. Geol., R. D. Trask, Ed. AAPG, Tulsa. pp. 5–31.

Hocutt, C. H. and E. O. Wiley. 1986. *The Zoogeography of North American Freshwater Fishes.* John Wiley & Sons, New York.

Hokanson, K. 1977. Temperature requirements of some percids and adaptations to the seasonal temperature cycle. *J. Fish. Res. Bd. Can.* 34:1524–1550.

Hunt, A. R. 1885. On the influence of wave-currents on the fauna inhabiting shallow seas. *J. Linn. Soc. London (Zool.).* 18:262–274.

Hutchinson, G. E. 1957. *A Treatise on Limnology.* Vol. 1. *Geography, Physics and Chemistry.* John Wiley & Sons, New York.

Imboden, D. M., V. Lemmin, T. Joller, K. H. Fisher, and W. Weiss. 1981. Lake mixing and trophic state. *Verh. Int. Ver. Limnol.* 21:115–119.

Janssen, J. 1978. Feeding behavior repertoire of the alewife, *Alosa pseudoharengus*, and the ciscoes *Coregonus hoyi* and *C. artedii. J. Fish. Res. Bd. Can.* 35:249–253.

Johnson, M., C. McNeil, and S. George. 1987. Benthic macroinvertebrate associations in relation to environmental factors in Georgian Bay. *J. Great Lakes Res.* 13:310–327.

Jude, D. J., S. A. Klinger, and M. D. Enk. 1981. Evidence of natural reproduction by planted lake trout in Lake Michigan. *J. Great Lakes Res.* 7:57–61.

Jude, D. J., C. P. Madenjian, P. J. Schneeberger, H. T. Tin, P. J. Mansfield, T. L. Rutecki, G. Noguchi, and G. R. Heufelder. 1982. Adult, Juvenile and Larval Fish Populations in the Vicinity of the J. H. Campbell Power Plant, 1981, with Special Reference to the Effectiveness of Wedge-Wire Intake Screens in Reducing Entrainment and Impingement of Fish. Spec. Rep. No. 96. Great Lakes Res. Div. University of Michigan, Ann Arbor, MI.

Jude, D. J. and F. J. Tesar. 1985. Recent changes in the inshore forage fish of Lake Michigan. *Can. J. Fish. Aquat. Sci.* 42:1154–1157.

Jude, D. J., F. J. Tesar, T. J. Tomlinson, T. J. Miller, N. J. Thurber, G. G. Godun, and J. A. Dorr III. 1979. Inshore Lake Michigan Fish Populations Near the Donald C. Cook Nuclear Power Plant During Preoperational Years — 1973, 1974. Spec. Rep. No. 71. Great Lakes Res. Div. University of Michigan, Ann Arbor.

Krecker, F. H. and L. Y. Lancaster. 1933. Bottom shore fauna of western Lake Erie: a population study to a depth of six feet. *Ecology* 14(2):79–93.

Lagler, K. F. 1956. *Freshwater Fishery Biology.* Wm. C. Brown, Dubuque, IA.

Lagler, K. F., J. E. Bardach, R. R. Miller, and D. R. M. Passino. 1977. *Ichthyology,* 2nd ed. John Wiley & Sons, New York.

Langlois, T. H. 1954. *The Western End of Lake Erie and Its Ecology.* J.W. Edwards Publishing, Ann Arbor, MI.

Larkin, P. A. 1964. Canadian lakes. *Verh. Int. Ver. Limnol.* 15:76–90.

Lasker, R. 1975. Field criteria for survival of anchovy larvae: the relation between inshore chlorophyll maximum layers and successful first feeding. *U.S. Fish. Bull.* 73:453–462.

Lauritsen, D. D. and D. S. White. 1981. Comparative Studies of the Zoobenthos of a Natural and a Man-Made Rocky Habitat on the Eastern Shore of Lake Michigan. Spec. Rep. No. 74, Great Lakes Res. Div. University of Michigan, Ann Arbor.

Leach, J. H., M. G. Johnson, R. M. Kelso, J. Hartmann, W. Naumann, and B. Entz. 1977. Responses of percid fishes and their habitats to eutrophication. *J. Fish. Res. Bd. Can.* 34:1964–1971.

Lyon, J., R. Drobvney, and C. Olsen. 1986. Effects of Lake Michigan water levels on wetland soil chemistry and distribution of plants in the Strait of Mackinac. *J. Great Lakes Res.* 12(3):175–183.

Mackenthun, K. M. and C. D. McNabb. 1961. Stabilization pond studies in Wisconsin. *J. Water Pollut. Contr. Fed.* 33(12):1234–1251.

Mackenthun, K. M., W. N. Ingram, and R. Porges. 1964. Limnological Aspects of Recreational Lakes. Div. Water Supply & Pollution Control. U.S. Public Health Service, Washington, D.C.

Manny, B., D. Schloesser, C. Brown, and J. French, III. 1985. Ecological Effects of Rubble-Mound Breakwater Construction and Channel Dredging at West Harbor, Ohio (Western Lake Erie). Tech. Rep. EL-85-10, prepared by U.S. Fish and Wildlife Service for U.S. Army Engineer Waterways Experiment Station, Vicksburg, MS.

Mansfield, P. J., D. J. Jude, D. T. Michaud, D. C. Brazo, and J. Gulvas. 1983. Distribution and abundance of larval burbot and deepwater sculpin in Lake Michigan. *Trans. Am. Fish. Soc.* 112:162–172.

Mansfield, P. 1984. Reproduction by Lake Michigan fishes in a tributary stream. *Trans. Am. Fish. Soc.* 113:231–237.

May, R. 1974. Larval mortality in marine fishes and the critical period concept, in *The Early Life History of Fish,* J. H. S. Blaxter, Ed. Springer-Verlag, New York. pp. 3–19.

McKee, J. E. and H. W. Wolf. 1971. Water Quality Criteria, 2nd ed. Publ. No. 3-A. California Water Resources Control Board, Sacramento.

Meisner, J. D., J. L. Goodier, H. A. Regier, B. J. Shuler, and W. J. Christie. 1987. An assement of the effects of climate warming on Great Lakes basin fishes. *J. Great Lakes Res.* 13:340–352.

Methot D. D., Jr. and D. Kramer. 1979. Growth of northern anchovy, *Engaulis mordax,* larvae in the sea. *U.S. Nat. Mar. Fish. Ser. Bull.* 77:413–423.

Moyle, J. B. 1949. Some indices of lake productivity. *Trans. Am. Fish. Soc.* 76:322–334.

Moyle, J. B. 1956. Relationships between the chemistry of Minnesota surface waters and wildlife management. *J. Wildl. Manage.* 20(3):302–320.

Nebergall, W. J., F. C. Schmidt, and H. F. Holtzclaw. 1963. *General Chemistry,* 2nd ed. D. C. Heath, Boston.

Neel, J. K., J. H. McDermott, and C. A. Monday, Jr. 1961. Experimental lagooning of raw sewage at Fayette, Missouri. *J. Water Pollut. Contr. Fed.* 33(6):603–641.

Neess, J. C. and W. W. Bunge. 1957. An unpublished manuscript of E. A. Birge on the temperature of Lake Mendota. II. *Trans. Wis. Acad. Sci. Arts Lett.* 46:31–89.

Northcote, T. G. and P. A. Larkin. 1956. Indices of productivity in British Columbia lakes. *J. Fish. Res. Bd. Can.*. 13:515–540.

Northmore, D., F. C. Volkman, and D. Yager. 1978. Vision in fishes: color and pattern. in *The Behavior of Fish and Other Aquatic Animals.* D. I. Mostofsky, Ed. Academic Press, New York. pp. 79–126.

O'Gorman, R. and C. Schneider. 1987. Dynamics of alewives in Lake Ontario following a mass mortality. *Trans. Am. Fish. Soc.* 115:1–14.

Parmenter, R. 1929. Hydrology of Lake Erie. Preliminary report on the cooperative survey of Lake Erie. *Buffalo Soc. Nat. Sci. Bull.* 14(3):25–50.

Patalas, K. 1975. The crustacean plankton communities of fourteen North American great lakes. *Verh. Int. Ver. Limnol.* 19:504–511.

Peck, J. 1979. Utilization of Traditional Spawning Reefs by Hatchery Lake Trout in the Upper Great Lakes. Fish. Res. Rep. No. 1871. Michigan Department of Natural Resources, Fisheries Division. Lansing.

Perrone, M. P., Jr., P. J. Schneeberger, and D. J. Jude. 1983. Distribution of larval yellow perch (*Perca flavescens*) in nearshore waters of southeastern Lake Michigan. *J. Great Lakes Res.* 9:517–522.

Phillips, D. W. and J. A. W. McCulloch. 1972. The Climate of the Great Lakes Basin. Atmospheric Environ. Climatol. Study No. 20, Environment Canada, Toronto.

Poe, T., C. Hatcher, C. Brown, and D. Schloesser. 1986. Comparison of species composition and richness of fish assemblages in altered and unaltered littoral habitats. *J. Freshwater Ecol.* 3:525–536.

Rawson, D. S. 1939. Some physical and chemical factors in the metabolism of lakes. AAAS Sci. Publ. No. 10. American Association for the Advancement of Science, Washington, D.C. pp. 9–26.

Rawson, D. S. 1952. Mean depth and the fish production of large lakes. *Ecology* 33:513–521.

Rawson, D. S. 1955. Morphometry as a dominant factor in the productivity of large lakes. *Verh. Int. Ver. Limnol.* 12:164–175.

Reid, G. K. 1961. *Ecology of Inland Waters and Estuaries.* Reinhold Publishing, New York.

Rice, J., L. Crowder, and F. Binkowski. 1985. Evaluating otolith analysis for bloater (*Coregonus hoyi*): do otoliths ring true. *Trans. Am. Fish. Soc.* 114:532–539.

Robertson, A. and D. Scavia. 1984. North American Great Lakes, in *Lakes and Reservoirs.* F. B. Taub, Ed. Elsevier, Amsterdam. pp. 135–176.

Rodgers, G. K. 1965. The Thermal Bar in the Laurentian Great Lakes. Proc. 8th Conf. Great Lakes Res., Great Lakes Res. Div., University of Michigan, Lansing. pp. 358–363.

Rounsefell, G. A. 1946. Fish production in lakes as a guide for estimating production in proposed reservoirs. *Copeia* 1:29–40.

Rutecki, T. L., J. A. Dorr, III, and D. J. Jude. 1985. Preliminary analysis of colonization and succession of selected algae, invertebrates, and fish on two artificial reefs in inshore southeastern Lake Michigan, in *Artificial Reefs: Marine and Freshwater Applications.* F. M. D'Itri, Ed. Lewis Publishers, Chelsea, MI. pp. 459–489.

Ryder, R. A. 1965. A method for estimating the potential fish production of north temperate lakes. *Trans. Am. Fish. Soc.* 94:214–218.

Ryder, R. A. 1982. The morphoedaphic index — use, abuse and fundamental concepts. *Trans. Am. Fish. Soc.* 111:154–164.

Ryder, R. A., S. R. Kerr, K. H. Loftus, and H. A. Regier. 1974. The morphoedaphic index, a fish yield estimator. Review and evaluation. *J. Fish. Res. Bd. Can.* 31:663–688.

Saulesleja, A. 1986. Great Lakes Climatological Atlas. Atmospheric Environment Service, Environment Canada, Ottawa.

Sawyer, C. N. and P. L. McCarty. 1967. *Chemistry for Sanitary Engineers,* 2nd ed. McGraw-Hill, New York.

Scavia, D., G. Fahnenstiel, M. Evans, D. Jude, and J. Lehman. 1986. Influence of salmonine predation and weather on long-term water quality trends in Lake Michigan. *Can. J. Fish. Aquat. Sci.* 43:435–443.

Schindler, D. W. 1971. A hypothesis to explain differences and similarities among lakes in the experimental lake area, northern Ontario. *J. Fish. Res. Bd. Can.* 28:295–301.

Schlesinger, D. A. and H. A. Regier. 1982. Climatic and morphoedaphic indices of fish yield from natural lakes. *Trans. Am. Fish. Soc.* 111:141–150.

Scott, W. B. and E. J. Crossman. 1973. Freshwater Fishes of Canada. Fisheries Research Board of Canada, Ontario, Bull. No. 184.

Sheng, P. Y. and W. Lick. 1979. The transport and resuspension of sediments in a shallow lake. *J. Geophys. Res.* 84(C4):1809–1826.

Shuler, B. J., J. A. MacLean, F. E. J. Fry, and H. A. Regler. 1980. Stochastic simulation of temperature effects on first-year survival of smallmouth bass. *Trans. Am. Fish. Soc.* 109:1–34.

Sly, P. G. 1984. Habitat. in Strategies for Rehabilitation of Lake Trout in the Great Lakes. Tech. Rep. No. 40. R. Eshenroder, T. Poe, and C. Olver, Eds. Great Lakes Fisheries Commission, Lansing, MI.

Sly, P. G. 1988. Dispersal: all we need to know, mean of sand fraction and percent mud. *Hydrobiologia* 176–177:111–124.

Sly, P. G., R. L. Thomas, and B. R. Pelletier. 1983. Interpretation of moment measures derived from water-lain sediments. *Sedimentology* 30:219–233.

Smith, S. H. 1968. Species succession and fishery exploitation in the Great Lakes. *J. Fish. Res. Bd. Can.* 25:667–693.

Smith, S. H. 1970. Species interactions of the alewife in the Great Lakes. *Trans. Am. Fish. Soc.* 99:754–765.

Smith, S. H. 1972a. Factors of ecological succession in oligotrophic fish communities of the Laurentian Great Lakes. *J. Fish. Res. Bd. Can.* 29:717–730.

Smith, S. H. 1972b. The future of salmonid communities in the Laurentian Great Lakes. *J. Fish. Res. Bd. Can.* 29:951–957.

Talling, J. F. 1969. The incidence of vertical mixing and some biological and chemical consequences of tropical African lakes. *Verh. Int. Ver. Limnol.* 17:988–1012.

Tesar, F. J. and D. J. Jude. 1985. Adult and Juvenile Fish Populations of Inshore Southeastern Lake Michigan Near the Cook Nuclear Power Plant During 1973–82. Spec. Rep. No. 106. Great Lakes Res. Div. University of Michigan, Ann Arbor.

Thienemann, A. 1927. Der Bau des Seebeckens in Seiner Bedetung fur den Ablauf des Lebens. *Verh. Zool. Bot.* 77:87–91.

Tobin, R. L. and J. D. Youger. 1977. Limnology of Selected Lakes in Ohio — 1975. U.S. Geological Survey Water-Resour. Invest. 77–105. USGS, Washington, D.C.

Tonn, W. and J. Magnuson. 1982. Patterns in the species composition and richness of fish assemblages in northern Wisconsin Lakes. *Ecology* 63:1149–1166.

Trautman, M. B. 1981. *The Fishes of Ohio*, 2nd ed. Ohio State University Press, Columbus.

Turner, W. R. 1960. Standing crops of fishes in Kentucky farm ponds. *Trans. Am. Fish. Soc.* 89:333–337.

Verber, J. L. 1952. Currents in Lake Erie. Ohio Dept. Nat. Res., Div. Shore Erosion Dept., Columbus.

Verber, J. L. 1960. Long and Short Period Oscillations in Lake Erie. Ohio Dept. Nat. Res., Div. Shore Erosion Rep., Columbus.

Verber, J. L. 1966. Inertial Currents in the Great Lakes. Proc. 9th Conf. Great Lakes Res., Great Lakes Res. Div., University of Michigan, Ann Arbor. pp. 375–379.

Verduin, J. 1956. Energy fixation and utilization by natural communities in western Lake Erie. *Ecology* 37:40–50.

Wagner, W. 1980. Reproduction of Planted Lake Trout in Lake Michigan. Fish. Res. Rep. No. 1885. Michigan Department of Natural Resources, Fisheries Division. Lansing, MI.

Wells, L. 1966. Seasonal and depth distribution of larval bloaters (*Coregonus hoyi*) in southeastern Lake Michigan. *Trans. Am. Fish. Soc.* 95:388–396.

Wells, L. 1968. Seasonal depth distribution of fish in southeastern Lake Michigan. *Fish. Bull.* 67:1–15.

Wells, L. 1973. Distribution of Fish Fry in the Nearshore Waters of Southeastern and East-Central Lake Michigan, May-August 1972. Admin. Rep. Great Lakes Fish. Lab., Ann Arbor, MI.

Werner, E. and D. Hall. 1979. Foraging efficiency and habitat switching in competing sunfishes. *Ecology* 60:256–264.

Werner, E., D. Hall, and M. Werner. 1978. Littoral zone fish communities in two Florida lakes and a comparison with Michigan lakes. *Environ. Biol. Fish.* 3(3):163–172.

Westin, L. 1969. The mode of fertilization, parental behavior and time of egg development in fourhorn sculpin, *Myxocephalus quadricornis* (L.). *Inst. Freshwater Res. Drottingholm Rep.* 49:175–182.

Wetzel, R. G. 1983. *Limnology*. 2nd ed., W.B. Saunders, Philadelphia.

Winnell, M. and D. Jude. 1984. Associations among chironomidae and sandy substrates in nearshore Lake Michigan. *Can. J. Fish. Aquat. Sci.* 41:174–179.

Winnell, M. and D. Jude. 1987. Benthic community structure and composition among rocky habitats in the Great Lakes and Keuka Lake, New York. *J. Great Lakes Res.* 13:3–17.

Wong, B. 1972. Growth, Feeding, and Distribution of Yellow Perch Fry, *Perca fluviatilis flavescens* (Mitchell), During Their First Summer in West Blue Lake. M.S. thesis, University of Manitoba, Winnipeg, Canada.

Chapter 7

APPLICATION OF BIOLOGICAL MEASURES TO CLASSIFICATION OF AQUATIC HABITATS IN THE LAURENTIAN GREAT LAKES

W. G. Duffy, R. T. Oglesby, and J. A. Reckahn

TABLE OF CONTENTS

ABSTRACT

A review of the literature was undertaken to determine if commonly used biological measures and emerging ecological theories could be used in classifying the aquatic habitats of the Great Lakes and other large lacustrine ecosystems. Existing classification schemes based on phytoplankton most often rely on indicator species and are aimed at the trophic status of a lake. Schemes based on zoobenthic data also rely heavily on indicator species and thus have been used primarily in trophic evaluations. The distribution of aquatic macrophytes has been used to classify zones in lakes that might then be used as surrogates of animal habitat. Information on fish communities has not been widely applied in lake classification. However, classification schemes based on fishes, such as the Index of Biotic Integrity or the classification of fishes into reproductive guilds may be useful in delineating habitats. Biological measures that might lend themselves to habitat classification include organism size and health, as expressed by morphological attributes. Defining the boundaries of fish populations might also assist in habitat delineation and classification. Knowledge of biotic diversity, considered together with the nature of species composing a community, is useful in assessing habitat degradation. Finally, among emerging ecosystem organization theories, particle-size distribution theory apparently holds promise for predicting standing stock or production in pelagic zones.

I. INTRODUCTION

The task of developing a habitat classification system for the Great Lakes or any large lacustrine ecosystem is formidable. The Great Lakes cover a vast area, contain numerous environments that vary physically and chemically, fluctuate seasonally, and are inhabited by a greater number of species than are usually found in typical inland lakes. However, it is perhaps time that the empirical descriptive data gathered on fishes and other biota in the Great Lakes be reduced into a logical scheme of classification. The predictive value of such a classification system in managing large lakes would be immense if the system was founded on causal relations between the various components of the biota and the composition of their habitats (Brinkhurst, 1974). Although habitat is most often considered a property of the physical and chemical environment, it includes all parts of the ecosystem that must be present for a species to exist (Hutchinson, 1967). Biotic communities, therefore, must be considered as integral parts of any habitat classification system (Carpenter and Lodge, 1986).

Our present goal is to identify biological measures having potential application to aquatic habitat classification in the Great Lakes. We review biologically based classification systems developed for lacustrine ecosystems and other biological measures that appear to be amenable to habitat classification. The review has been guided by the primary goal of the Classification and Inventory of Great Lakes Aquatic Habitat (CIGLAH) workshop — to develop a classification system that will help improve the management of Great Lakes fishes. Treatment is selective and admittedly biased in favor of fishes. However, we have endeavored to review all levels of biological organization while emphasizing those we feel merit further consideration.

II. BASES OF SOME EXISTING LAKE CLASSIFICATION SCHEMES

A. PHYTOPLANKTON

The classification of phytoplankton associations has been based primarily on the concept of indicator organisms, in which either dominant or rare species are the indicators. Dominant species are those having relatively broad tolerances to environmental conditions, or those occupying a large niche (Hutchinson, 1975; Ryder and Edwards, 1985). Although populations of dominant species may fluctuate temporally, they remain important over the seasonal cycle. Rare indicator species are those whose environmental tolerances tend to be narrower than those of dominant species.

Each of the two classifications — dominant and rare indicator species approaches — provides information about a lake or habitat, but at different levels of organization. Characterization of a phytoplankton community based on dominant indicator species gives a general picture of the community, whereas a characterization based on rare indicator species gives more specific information about the nature of the community (Hutchinson, 1975). Such specific information might include pH, relative light penetration, or conductivity, all of which can be rather easily measured by other means (Table 1).

B. AQUATIC MACROPHYTES

Many schemes have been devised for the classification of freshwater plants; detailed reviews were provided by Sculthorpe (1967), Hutchinson (1975), and Spence (1982). Systems developed for classifying freshwater plants vary in the detail of treatment and the emphasis placed on plant associations. Two main approaches have been taken (Hutchinson, 1975). The first, the "continental European approach", has been applied primarily in Europe; the second, the "Clements-Pearsall approach", has been applied primarily in North America. The plant association is considered the basic vegetational unit in the continental European approach. A plant association is viewed as relatively discrete in its occurrence, and indicates the presence of certain environmental conditions.

An association is distinguished by dominant and rare species that exhibit fidelity to a given association. Overall appearance and the life forms of important species, relative to seasonal cycles, are used in distinguishing associations.

The North American or Clements-Pearsall approach, emphasizes community succession rather than discrete associations of plants. The zones of plants in a lake are viewed as successional stages that progress from the deeper edge of the euphotic zone through the various littoral zones and into the terrestrial community (Hutchinson, 1975). As in the continental European approach, certain environmental conditions, such as sediment particle size or percent of organic material, are associated with basic classification units.

Aquatic macrophyte communities have also been used to classify wetlands. A wetland classification system now widely applied in North America (Cowardin et al., 1979) uses information on plant communities, hydrology, and soils to classify wetlands into a variety of categories. It is hierarchical and may be applied in a variety of environmental settings. McKee et al. (Chapter 4) reviewed wetland classification systems in detail.

C. ZOOBENTHOS

Zoobenthos communities have been used extensively in aquatic habitat classification, first in Europe during the early 1900s, then in North America and elsewhere (see Brinkhurst, 1974 for a review of benthic classification). The earliest classification scheme based on zoobenthos was referred to as the "Saprobien system". This system was developed in response to organic pollution problems and attempted to classify habitats on the basis of observed changes in dominant organisms in relation to organic content of the sediment and depletion of oxygen (Brinkhurst, 1974).

Later classification systems attempted to relate lake trophic status to indicator species, or to zoobenthos community composition. These systems based on trophic indicators have been repeatedly applied to the Great Lakes (Brinkhurst et al., 1968; Hiltunen, 1969; Howmiller and Beeton, 1971; Nalepa and Thomas, 1976; Lauritsen et al., 1985; Winnel and White, 1985).

During the International Field Year for the Great Lakes, Nalepa and Thomas (1976) conducted a lakewide survey of the Lake Ontario zoobenthos. They used indicator species

TABLE 1
Summary of Biological Information of Use in a Classification System for Great Lakes Aquatic Habitats

Element of classification	Organization level	Property(ies) upon which classification is based	Data requirements	Difficulty and expense in obtaining data	Current usefulness for Great Lakes
Taxa					
Phytoplankton	Species	Dominance/rarity	Low	Low	Low
Aquatic macrophytes	1. Species	Association	Moderate	Moderate	Moderate
	2. Species	Succession	Moderate	Moderate	Moderate
Zoobenthos	Communities	Relation of composition to pollution, trophic status	High	High	Mod.-low
Fishes	1. Communities	Relation to physical environment	Low-moderate	Low	Moderate
	2. Stocks	Stock density	Moderate	Moderate	Moderate
Organismal attributes					
Size	General	Fish length and/or weight	Low-moderate	Low	Low
Presence/absence	Species	Occurrence	Low	Low	Low
Population attributes					
Dynamics	Populations	Recruitment, growth, and mortality	High	High	High
Niche occupancy	Community	Niche utilization and relations between niches	Variable	Variable	?
Community attributes					
Indicator groups	Guilds, species	Occurrence	Moderate	Moderate	Moderate
Diversity indices	Species	Weighted relative abundances	High	High	Low
Particle-size distribution	General	Relation of particle size to trophic role	Low	Low	High
Trophic levels	Trophic levels	Nutrient supply, habitat availability, and biological indicators	Moderate-high	Moderate-high	Moderate
Food webs	Ecosystem	Interdependence of species	High	High	Moderate

(*Limnodrilus hoffmeisteri, Tubifex tubifex, Stylodrilus heringianus,* as well as other species) to evaluate the trophic status of different parts of the lake. They also related their data to various trophic indices that were based on either the density of oligochaetes (Wright and Tidd, 1933; Mozley and Alley, 1973), the proportion of the total sample represented by oligochaetes (Goodnight and Whitley, 1960), or the proportion of the oligochaete fauna represented by the pollution tolerant *L. hoffmeisteri* (Brinkhurst, 1967). In comparing trophic evaluations made after applying each of these methods to their data, Nalepa and Thomas (1976) concluded that little was gained that was not already obvious from using the indicator species approach. Although categorizations made from indicator species and trophic indices were generally in agreement, discrepancies were noted among some of the indices.

The trophic status of southeastern Lake Michigan was recently evaluated by Winnel and White (1985), using a 9-year zoobenthic data set. The assignment of trophic status based on the Chironomidae was compared to that from the Oligochaeta. Results from both the chironomid and oligochaete fauna were in agreement.

Problems associated with sampling and taxonomy of the zoobenthos have retarded its use in trophic classification in the Great Lakes. Acceptance of the results from indicator species studies depends strongly on the taxonomic expertise of the investigator (Brinkhurst, 1974), and the taxonomy of some groups is as yet incompletely known (Winnel and White, 1985). Furthermore, if extensive data sets are not available, variability of the zoobenthic community composition associated with depth and seasons complicate interpretation of results (Nalepa and Thomas, 1976; Winnel and White, 1985).

D. FISHES

Fish communities have not been as widely used as some other biotic components of lakes to classify lake trophic status or narrowly circumscribed environmental conditions. Exceptions include Balon's (1975) classification of fishes into reproductive guilds (see Section V). Also, the Index of Biotic Integrity (Karr et al., 1986), developed for midwestern U.S. streams, is based on fish community data.

The separation of warm-, cool-, and coldwater fishes on the basis of annual temperature regimes in which they exist is perhaps the simplest classification system for fish communities used in North America. As Trandahl (1978) acknowledged, these communities are not rigorously defined. However, warmwater fish communities are generally considered to be diverse, often centrarchid-dominated, assemblages of fishes in southern lakes. In contrast, coldwater fish communities are synonymous with salmonid-dominated fauna in northern lakes. Coolwater fish communities, then, are those whose distribution by temperature preference lies between the extremes (Trandahl, 1978).

An index of fish community health considered to be time and personnel efficient is the proportional stock density index of Anderson (1976) and Anderson and Weithman (1978). This index is based on the concept that a healthy fish community is one which is balanced, with a species size distribution intermediate between many small and few large fish. It is calculated as the proportion of "quality" size fish relative to "stock" size fish. Anderson (1976) used length as a measure of size and defined quality and stock sizes relative to world-record lengths for the species.

The Index of Biotic Integrity, unlike many "pollution indices", utilizes fish community information in assessing water quality (Karr et al., 1986). It is calculated from 12 measures of a fish community that can be lumped under 3 categories: species richness and composition, trophic composition, and abundance and condition of fishes. It was developed for streams in the midwestern U.S. and requires a different suite of measurements if applied elsewhere.

III. SOME BIOLOGICAL APPRAISAL SCHEMES

A. SIZE

Organism size, age, growth rate, reproductive capacity, health, and presence or absence are routinely recorded biological variables that are sensitive to environmental factors (Green, 1979) and thus amenable to use in habitat classification. Furthermore, the relative ease of obtaining these measures has promoted their collection so that both current and historical data sets exist. Size, in particular, is one of the most obvious features of an organism (Roff, 1986) and would appear to be an attractive measure for habitat assessment because it integrates information over several levels of biological organization.

The size of an organism is ultimately limited by physical factors of the environment (Roff, 1986). It is recognized that temperature regulates the growth of both plants and poikilothermic animals. Maximum size in both fish (Craig, 1985) and aquatic insects (Sweeney and Vannote, 1978) is achieved at relatively low temperatures, whereas rates of growth increase with increasing temperature up to a certain point. Other physical-chemical characteristics that influence growth are light, substrate, and dissolved oxygen concentration. However, these factors, along with depth and water current, have a greater influence on distribution than on growth (Brinkhurst, 1974; Barton and Hynes, 1978).

Although physical variables control the ultimate size of an organism, certain biological factors may control the ultimate realized size (Spencer and King, 1984; Moyle et al., 1986). Size selective predation has been demonstrated for such divergent fishes as the planktivorous alewife, *Alosa pseudoharengus*, and predatory sea lamprey, *Petromyzon marina* (Wells, 1970; Cochran, 1985). Parasite infestations may also increase with body size in fishes, depending on feeding behavior (Amin, 1985). However, fish and other aquatic organisms are adaptively flexible and their growth differs in different environments (Craig, 1985).

Growth in fishes is indeterminant and fecundity, in general, varies with the cube of length (Roff, 1986). Furthermore, population rate of increase (r) may be calculated from information on survivorship and fecundity, and these two measures in turn are related to size (Figure 1). Since size is related to environmental factors, it may be helpful in assessing the ability of a habitat to support populations. The population consequences of body size were described in greater detail by Roff (1986), along with examples of how one might predict population responses to decreasing food resources.

B. HEALTH

The health of an organism might be another biological measure that could be used to access habitat suitability. Harbors and other areas with severely degraded environmental conditions are often characterized as having high proportions of fishes in poor health (Baumann et al., 1982; Maccubbin et al., 1985). Genetic anomalies in aquatic insects have also been associated with areas of contaminated sediments (Warwick et al., 1987).

C. PRESENCE-ABSENCE

Finally, the presence or absence of species is a frequently recorded biological variable that has obvious application to classification. However, since many of the biological classification systems previously described use this measure, it is not considered further here.

IV. POPULATION MEASURES

Populations are the most important element to fishery managers in any ecosystem classification scheme. Although some introduced salmonids may roam lakewide during their life

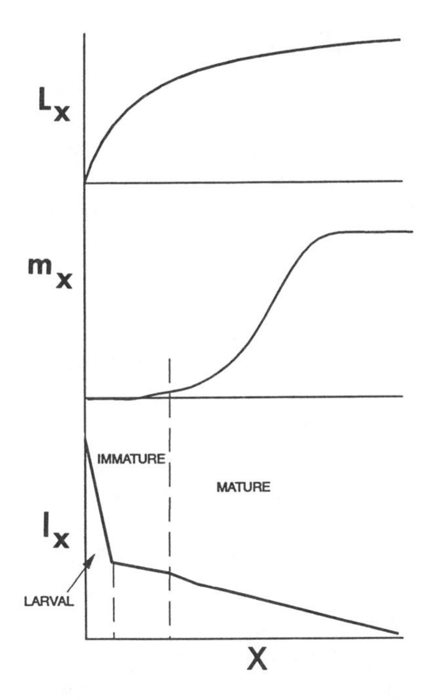

FIGURE 1. Generalized life history parameters for fish. Length (L_x) increases to an asymptot with increasing age (x) and reproductive potential or fecundity (m_x) increases to an asymptot with age after maturity. Survivorship (l_x) decreases rapidly in the larval stage and more gradually thereafter. (Modified from Roff, 1986.)

history, most fish populations are characterized by home ranges within specific portions of each lake. Discrete stocks that may or may not be genetically distinct often have even more restricted distributions. For example, a distinct stock of lake whitefish (*Coregonus clupeaformis*) are confined within the northern basin of South Bay in Manitoulin Island, a site of longstanding fisheries investigations (Fry, 1947a; Regier, 1966; Reckahn, 1970, 1986; Ihssen et al., 1981; Casselman et al., 1981). Thus, the functional orientation in addressing fisheries problems is by species within the boundaries of a population or stock. Regulations of harvest are, however, frequently delimited by statistical districts (Smith et al., 1961) or fishing areas (Lawrie and Rahrer, 1972).

The "four primary factors" controlling the production of exploited fish populations are recruitment, growth, exploitation ("capture"), and "natural" death (Russell, 1931; Beverton and Holt, 1957). The first two are weight increments and the last two are weight decrements. The mathematical descriptions of these factors have achieved monumental proportions and Ricker (1975) provided the most convenient entrance to this broad topic.

A. POPULATION BOUNDARIES

One of the problems initially encountered in population management is the identification of spatial boundaries of populations and stocks (Kutkuhn, 1981). A series of techniques has been devised to address this problem: (1) statistical analysis of morphometric and meristic characters are used to delineate stocks, (2) biochemical, serological, electrophoretic, and mitochondrial DNA:RNA sequencing methods, along with physical characteristics and descriptions of chromosomes have all been used to delineate stocks and their genetic characteristics, and (3) physical marks (fin clipping) and tags of a wide variety have also been used. Tagging has the advantage of providing additional information on growth and exploitation. It may also be used to evaluate the effectiveness of sampling gear (Fry, 1953; Reckahn, 1988). Finally, descriptions of the spatial boundaries of fish stocks may often be provided by fishermen (Koelz, 1929). Smith (1968) and Goodier (1981) used this approach to delineate the spawning sites of lake trout (*Salvelinus namaycush*), in Lakes Huron and Superior. These and other methods of determining the boundaries of populations and stocks have been documented in a special symposium stock (Berst and Simon, 1981).

Fish population boundaries may expand or contract with changes in abundance (Beverton et al., 1984). Furthermore, environmental factors affecting abundance may not be immediately expressed in populations. The importance of long data series to monitor changes in stocks cannot be exaggerated.

B. RECRUITMENT

Variability in recruitment to a fishery caused by environmental fluctuations continues to be a source of uncertainty in fisheries management (Sissenwine, 1984). This was recognized early in the history of fisheries science (Hjort, 1914). The establishment of index fishing techniques such as the Leman-Haaks line for plaice (*Plueronectes platessa*), a set of index trawling recruitment stations off the coast of Holland, was begun in 1906 to monitor variability in recruitment (Beverton and Holt, 1957). Index fishing was applied to the Great Lakes by Hartman (1980). Index fishing and other sampling techniques are attempts to monitor annual fluctuations in spawning success so that highly variable recruitment of important fish can be sampled early in their life history. The goal of index netting is to successfully predict fluctuations in fish abundance. With adequate predictions, quota regulations or other harvest controls can be imposed to provide protection to brood stocks.

A key concept in fisheries conservation is that of "ensuring the perpetuation of the fishery resource" (Loftus et al., 1980) which must have precedence over any allocation of excess fish production. Without adequate brood stock, the fishery fails and far too many examples of such failures are found in the history of the Great Lakes fisheries. Perhaps the best-documented example of this is the sequential loss of lake whitefish populations after the introduction of deepwater trap nets in Lakes Huron and Michigan (van Oosten et al., 1946). It is the exceptionally strong year-classes that must be managed carefully, because it is the source of excess harvestable yields (Cooper and Polgar, 1981). More importantly, these dominant year-classes provide the long-term reproductive potential to carry a stock through adverse years when unfavorable weather or man-induced stresses prevent the development of other strong year-classes.

C. GROWTH

The growth of fishes and the environmental factors regulating growth have been intensively investigated (Brett, 1979; Hoar et al., 1979; Ricker, 1979; Craig, 1985). Important abiotic factors regulating growth of freshwater fishes are temperature, light, and oxygen. Important biotic factors are food availability, size, and competition. Many interactions between these factors may be generated by combinations of controlling, masking, directing, and limiting factors (Fry, 1947b; Brett, 1979).

D. EXPLOITATION

Healey (1975, 1980) presented comparisons of growth and recruitment of lake trout and lake whitefish in relation to exploitation rates over a wide range of stocks. These comparisons come the closest to being a classification of stocks at the population level.

Although we have not included the broad topic of the effects of industrial wastes, toxic chemicals, and the accumulations of chlorinated hydrocarbons on fish populations, it is worth noting that one of the most impressive rehabilitating responses of a fish population stems from the mercury contamination of walleyes (*Stizostedion vitreum*) in the western basin of Lake Erie. A moratorium on walleye harvests for 5 years due to mercury contamination caused an immense buildup in walleye abundance. The abundance represents what can be achieved when fish harvests are stopped, but it was public concern about environmental contaminants, rather than fishery management efforts, that brought about the moratorium.

E. MORTALITY

The two population decrements of exploitation and "natural" death need to be expanded in the Great Lakes to include a third factor, mortality, induced by the sea lamprey (*Petromyzon marinus*), as a category distinct from natural mortality. The impact of this invader in the upper Great Lakes has been the subject of much research (Smith, 1980). Continuing research on mortality rates of lake whitefish has indicated that in northern Lake Huron 25 to 27% of the annual mortality of adults could be attributed to sea lampreys. In juvenile lake whitefish, lamprey-induced mortality may have reached >40% at peak attack rates (Reckahn, manuscript in preparation). Laboratory studies (Parker and Lennon, 1957; Farmer and Beamish, 1973) and extensive field studies (e.g., Reckahn, manuscript in preparation) have shown that lake trout are preferentially attacked. Thus, in a classification scheme lake trout rank as the species most vulnerable to attack by sea lampreys. This may be exceedingly important where large river systems that are difficult to treat with lampricides continue to produce large numbers of lampreys. For example, the St. Mary's and Cheboygan rivers produce exceptionally large lamprey spawning runs (see trapping records in *Annual Reports to the Great Lakes Fishery Commission*).

F. COMMUNITY COMPOSITION

Some fish populations are restrained by temperature requirements to warm shallow waters (warm stenotherms) and others to deep cold waters (cold stenotherms); still others are more flexible in their requirements and move between these areas (eurythermal). These are called coolwater fishes and were the subject of a symposium (Kendall, 1978; Trandahl, 1978). Other classification schemes have been built on similar thermal attributes.

The importance of thermal effects associated with electrical generating plants located on the shores of the Great Lakes has been reported on by several investigators (Brown, 1974; Coutant, 1977; Houston, 1982). For Great Lakes fish, the most comprehensive compilation of temperature relation to date was made by Christie (1985). The need to predict species composition at potential electrical generating plant sites led Christie (1982) to develop a forecasting method based on a multiple set of environmental characteristics, including thermal preferences. Five sets of characteristics are useful for predictive purposes: (1) geographical distribution (14 zones cover Ontario); (2) adult habitat preferences (lotic vs. lentic environments with subcategories under each, e.g., depth distribution; (3) spawning behavior (choice of lotic or lentic waters, season, temperature, depth, substrate, egg dispersal, and fecundity); (4) egg characteristics (size, adhesiveness, duration of incubation); and (5) young-of-the-year development (size at time to age 1). This approach proved highly useful in predicting the species composition actually found at the Mission Island site in Thunder Bay, Lake Superior, where it was tested (Christie, 1982). Although the method has not been developed as a classification scheme, its accurate projections in an initial trial indicate that more extensive testing is warranted.

Even a brief review of temperature effects on fish must include the concepts of thermal zones of activity and tolerance developed by Fry (1947b) and Brett (1952) which have been central to many studies. Their basic concepts have been frequently elaborated upon. Giattina and Garton (1982) provided a general graphical model that summarizes and extends their thermal concepts (Figure 2). These concepts have been used for many years by scientists of Ontario Hydro to define the zones of activity (Figure 3) within which different species will be found (Griffiths, 1979). This can be considered a two-dimensional example of the n-

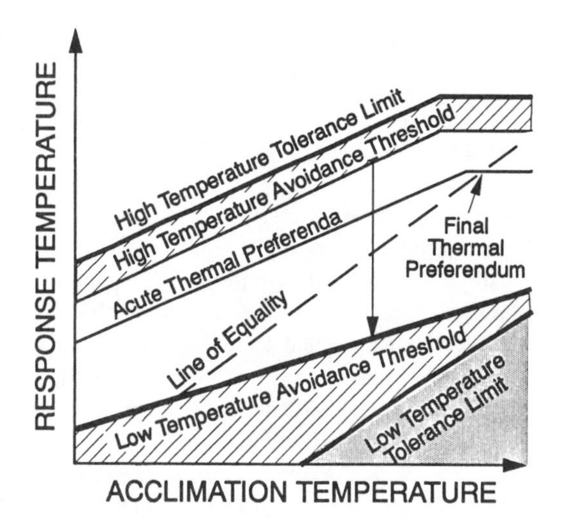

FIGURE 2. Graphic model of thermal limits of fish behavior. (After Giattina and Garton, 1982.)

dimensional niche theory developed by Hutchinson (1957) and Kerr (1976). Most recently, Christie and Regier (1988) have used thermal habitat zones as a guide to the relative production of some fish species in the Great Lakes.

V. COMMUNITY MEASURES

A. GUILD-INDICATOR SPECIES

Habitat is generally considered to be part of an ecological community. The terms plankton community, benthic community, and percid community convey more information than simply the plant or animal populations characteristic of them. The concept of community also includes interactions among organisms and interactions between organisms and their environment (Whittaker, 1975). Therefore, information describing biological communities should be valuable in habitat classification.

Community ecology has been the subject of intense study during the past few decades. As in many active branches of science, empirical studies in community ecology have generally not kept pace with developments in theory. Consequently, though intellectually appealing, some concepts such as the niche, optimal foraging, or island biogeography remain vague or difficult to apply to natural resource management. However, several measures of community information, the guild-indicator species concept, and species diversity may prove useful in the classification of habitats.

An indicator species is a plant or animal that is closely associated with specific environmental factors (Block et al., 1987). Indicator species are used as an index of environmental quality (see existing classifications schemes for phytoplankton, zoobenthos, and fish above).

A guild is considered to be a group of species that exploits the same class of resources in a similar way (Root, 1967). Although associating in the same class, the actual resources used and the means by which they are obtained may differ. A guild indicator species is simply a species which is a member of a guild and acts as an indicator for all other species in the guild (Severinghaus, 1981; Block et al., 1987). In theory, the response of the indicator species to environmental change reflects the response of the other species in the guild. Species composing a guild, as well as the geographic range over which the environmental response is to be measured, must be carefully selected if predictable responses are to be expected (Severinghaus, 1981; Mannan et al., 1984; Block et al., 1987).

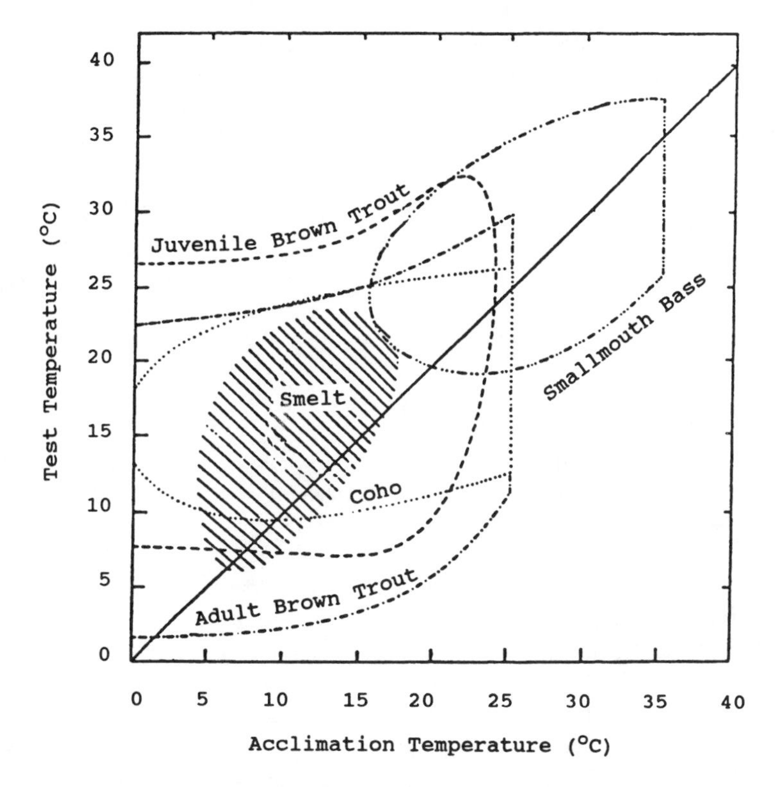

FIGURE 3. Thermal envelopes of some common species in the Great Lakes enclosing swimming performance isopleths for 75% of peak swimming speed. (Modified from Griffiths, 1979.)

The resources most often selected to define ecological guilds are food, cover, and space (Hairston, 1984; Block et al., 1987). However, reproduction and adaptations to early ontogeny have been used to define guilds in fishes (Balon, 1975; Balon et al., 1977). In extending a hypothesis developed by Kryzhanovsky (1948) that adult biology, including feeding, was a reflection of spawning and early development, Balon (1975) reasoned that feeding style was an incomplete determinant of an ecological guild. In his classification system, Balon (1975) separated guilds on the basis of form and function at intervals of early development, on preferred spawning grounds, and on features of reproductive behavior. He felt that predators and oxygen concentration were of primary importance in determining survival and that spawning behavior and spawning grounds (habitat) determined the availability of oxygen and protection from predators.

In the classification system of Balon's (1975), 32 guilds were identified, which included the living fish taxa of the world, for a total of about 20,000 species. The indigenous species of freshwater fish in Canada were identified as belonging to 14 guilds.

B. DIVERSITY INDICES

Interest in diversity indices and their application in ecology was popularized by Margalef (1958). For a time, the use of diversity indices gained particular favor among biologists who were trying to assess the effects of pollution on aquatic communities. These applications stemmed in part from an understanding that pollution reduced diversity in biological communities. The argument that diversity indices were a dimensionless measure of the relative importance of each species and were independent of sample size (Wilhm and Dorris, 1968) also stimulated their use.

More recently, the use of various indices to express diversity in biological communities has been criticized. Unlike most measures used by biologists, which are direct, diversity indices are variables derived from two or more independent measures (Green, 1979). Green (1979) argued that a flaw in many diversity indices was the improper selection of variables used in the index. He suggested that the total number of species present (species richness) was preferable as a measure of diversity over the many other indices developed. Biological diversity continues to be an important concept in community ecology, although it is no longer applied as widely.

VI. ECOSYSTEM MEASURES

A. DIVERSITY AND STABILITY

Some measure of how much biological information, i.e., species richness and abundance, a system contains is obviously an important descriptor of that system. Indices expressing this "diversity" have played a prominent role in both theoretical ecology and environmental management during the past several decades.

Similarly, the abilities to resist change, to recover from a disturbed state, and other processes that reflect "stability" are essential elements in holistic analyses of ecosystems and are of particular concern to those considering the response of a system to stress. Since stability is inherently more difficult to define and measure than diversity, attention given to stability has generally lagged behind that of diversity.

Diversity and stability have been strongly linked in the ecological literature for at least 2 decades. Margalef's (1958, 1968) thoughtful essays did much to stimulate the theorists. Initially, there was widespread acceptance of the simplistic assumption that stability increased as a function of diversity. As a consequence, diversity indices were included in U.S. regulations as a requisite part of environmental impact statements, and their importance in denoting system stability was repeated with regularity in the ecological texts of the 1970s.

Meanwhile, ecologists were busy defining diversity and stability and taking a harder look at the linkage between the two. By the mid-1970s it was clear that no simplistic, general relation existed between diversity and stability. Orians (1975), Odum (1975), Margalef (1981), and Pimm (1984) emphasized that if the diversity-stability concept is to have any utility, the questions that are addressed must be relatively narrowly defined. In meeting this restriction, the user must decide what form of diversity index is appropriate (Pielou, 1975) and what particular characteristic of stability (resilience, persistence, resistance, variability) should be used. If the contemplated use is to define the effect of some stress on the system, the stress and the response to it must be defined in specific spatial and temporal terms.

Even when the above conditions are met, the question of whether the resultant formulation will result in a useful product is debatable. Empirical evidence of success is thus far lacking. Prudence would suggest that much basic science remains to be done before diversity and stability can offer practical results for those seeking to define, monitor, and understand the workings of biological systems. One of the more promising testing grounds for this concept appears to be in studies of community succession. Evans et al. (1987) provided an overview of the diversity-stability concept and certain aspects of its potential applicability to the Great Lakes.

B. PARTICLE-SIZE DISTRIBUTION

Particle-size distribution theory and the role it might play in managing the Great Lakes was reviewed by Dickie et al. (1988) and Evans et al. (1988). The following summary is taken largely from Evans et al. (1987).

The particle-size distribution theory states that concentrations of particles in pelagic systems conform to regular patterns when grouped in size categories defined at order-of-magnitude (logarithmic) intervals. Relations can be "normalized" for particular systems and the properties of resulting curves compared between systems. Curve slope describes the efficiency with which small particles are converted to larger ones.

Borgmann (1983) carried the theory a step further by combining physiological conversion efficiency and predator-prey size ratio into a particle-size conversion efficiency coefficient. Using this coefficient, knowledge of production at any one trophic level can be used to calculate production at any other level.

Particle-size theory contains the simplifying assumption that biomass remains stable at each trophic level. Such homeostasis may be reasonably approximated for large pelagial systems over relatively long periods, e.g., years to decades, but destabilizing influences such as temperature changes, seasonal fishing, and anthropogenic stress cause departures from the normal. It is this fact that makes the particle-size theory attractive as a tool for defining and monitoring habitat, especially if further research permits the factoring out of cyclic changes, such as seasonal patterns of light and temperature.

The particle-size distribution theory appears to be founded on solid logic, since it has performed well when predicted long-term average fishery yields have been compared with actual values for pelagic fisheries. This performance suggests that it might be useful in predicting sustainable harvest of size classes of fish.

It is doubtful if particle-size theory will be applicable to littoral habitats due to the difficulty of defining size categories in that environment.

C. TROPHIC LEVELS

Trophic levels and their interactions constitute a cornerstone concept in ecology (Lindeman, 1942). They provide organizational structure for research on ecosystem functional properties such as nutrient and energy flux and on structural aspects such as particle-size distribution. The concept is simple and elegant, but its application is fraught with difficulties.

It is now widely accepted that nutrient supply determines total production and that habitat availability and biological interactions (including evolutionary and successional processes) are responsible for community structure (e.g., Evans et al., 1987). The principal research on trophodynamic relations has focused on production rather than on structure, particle-size distribution and ecological succession being the exceptions. One of the chief reasons for this focus on production is the difficulty in the assignment of organisms to trophic levels. Only with autotrophs is a relatively clear distinction possible, and even here minor problems are encountered (e.g., euglenoids).

If total production is controlled by nutrient supply to a system, it should be possible to define a quantitative linkage between nutrient input or productivity and biomass of phytoplankton and comparable properties of the fish community (insofar as they can be defined according to trophic level). This linkage assumes total production of fish is controlled by nutrient inputs and that phytoplankton clearly constitute the sole component of the lowest trophic level in pelagial environments. The difficulty in assigning trophic status to nonautotrophs suggests that developing a similar relationship for intermediate trophic levels is less promising.

Like the other ecosystem properties already reviewed, hypothesized linkages are dependent upon assumptions of steady state. One assumes that trophic compartments maintain a constant size relative to one another and that transfer rates of material and energy between compartments are constant.

D. FOOD WEBS

Traditional approaches to food webs have focused on some particular component, such as an insect pest that needs to be controlled or a fish population that needs to be harvested and the functional relations of other forms of life in the community to this component. When viewed from the standpoint of the cycling of materials, it is obvious that an entire ecosystem is a food web. Food webs constructed with a target organism in mind are simplified versions that attempt to portray the most important interrelations. The accuracy with which "importance" can be defined is problematical and certainly varies markedly from situation to situation.

Even if major elements of a food web are properly identified, their quantification is extremely difficult. Defining food-web structure is highly data intensive and transfer coefficients between components are seldom known or are difficult to develop. Feedback loops and rates of change within compartments are usually not even considered.

Evans et al. (1988) cited recent examples of the use of the food-web concept to calculate the sizes of stocks necessary to support system output and predation by other fish. This approach appears to be useful in managing some oceanic stocks of fish, but is so data intensive that it is unlikely to be widely used.

VII. CONCLUSIONS

Much information on Great Lakes fishes and their habitats has been gathered and published in the literature. Goodyear et al. (1982) summarized a voluminous amount of material on spawning and nursery habitats of Great Lakes fishes. The identification of larval fish in the Great Lakes area has been aided by the publication of an atlas describing 148 species from 28 families (Auer, 1982). Relations between Great Lakes fishes and water temperature have been published by Christie (1982) and updated by Wismer and Christie (1987) to include information on 116 species, and by Christie and Regier (1988). Nepszy (1988) compiled and edited the works of Alex Dechtiarenko on parasites of fishes from Lakes Superior, Huron, Erie, and Ontario. Finally, commercial fishery harvest data so valuable in monitoring population trends have been compiled (Baldwin et al., 1979). The data contained in these and many other publications form the underpinnings of our understanding of fish habitat associations in the Great Lakes. These data should be used extensively in developing any habitat classification scheme for the Great Lakes.

None of the biological measures considered stand alone as a classification scheme for the Great Lakes. We categorized 13 broad biological measures considered as having either a high, moderate, or low degree of usefulness in classifying Great Lakes aquatic habitats. Five measures — zoobenthic community structure, organism health, population dynamics, diversity indices, and food webs — were considered to be highly useful. Four of these (organism health being the exception) also have high data requirements and are expensive or difficult to obtain. Organism health is a useful measure and data are not expensive or difficult to gather, but it is most applicable in separating severely degraded habitats from those less degraded. Even though data are sometimes difficult and expensive to obtain, each of the five measures we categorized as being most useful in habitat classification are routinely gathered in the Great Lakes. The collection of these data and transfer of information from the results of data analysis should be better coordinated to maximize their usefulness.

Several of the measures we examined that were not considered highly useful in classifying Great Lakes aquatic habitats are also topics of current research. Future developments in theory or empirical methods could make these measures more useful in classification. For example, particle-size theory may not be developed to a point where application is advisable. However, the information returned from particle-size theory, relative to resources expended, could be immense.

REFERENCES

Amin, O. M. 1985. The relationship between the size of some salmonid fishes and the intensity of the acanthocephalan infections. *Can. J. Zool.* 63:924–927.

Anderson, R. O. 1976. Management of small warmwater impoundments. *Fisheries* 1(6):5–7.

Anderson, R. O. and A. S. Weithman. 1978. The concept of balance for coolwater fish populations. *Trans. Am. Fish. Soc.* 11(Spec. Publ.):371–381.

Balon, E. K. 1975. Reproductive guilds of fishes: a proposal and definition. *J. Fish. Res. Bd. Can.* 32:821–864.

Balon, E. K., W. T. Momot, and H. A. Regier. 1977. Reproductive guilds of percids: results of the paleogeographical history and ecological succession. *J. Fish. Res. Bd. Can.* 34:1910–1921.

Barton, D. R. and H. B. N. Hynes. 1978. Wave-zone macrobenthos of the exposed Canadian shores of the St. Lawrence Great Lakes. *J. Great Lakes Res.* 4:27–45.

Baumann, P. C., W. D. Smith, and M. Ribick. 1982. Hepatic tumor rates and poly-nuclear aromatic hydrocarbon levels in two populations of brown bullhead (*Ictalurus nebulosus*), in *Polynuclear Aromatic Hydrocarbons: 60th Int. Symp. Phys. Biol. Chem.* M. W. Cooke, A. J. Dennis, and G. L. Fisher, Eds. Batelle Press. Columbus, OH. pp. 93–102.

Berst, A. H. and R. C. Simon. 1981. Introduction to the Proceeding of the 1980 Stock Concept International Symposium (STOCS). *Can. J. Fish. Aquat. Sci.* 12:1457–1458.

Beverton, R. J. H. and S. J. Holt. 1957. On The Dynamics Of Exploited Fish Populations. Fishery Invest. Series II Vol. 19. Ministry of Agriculture, Fisheries, and Food, London.

Beverton, R. J. H. (Rapporteur of 12 man working group). 1984. Dynamics of single species. in Exploitation of Marine Communities, R. M. May, Ed. Life Sci. Res. Rep. 32, Dahlem Konferenzen, Berlin.

Block, W. M., L. A. Brennan, and R. J. Gutierrez. 1987. Evaluation of guild-indicator species for use in resource management. *Environ. Manage.* 11(2):265–269.

Borgmann, U. 1983. Effect of somatic growth and reproduction on biomass transfer of pelagic food webs as calculated from particle-size-conversion efficiency. *Can. J. Fish. Aquat. Sci.* 40:2010–2018.

Brett, J. R. 1952. Temperature tolerance in young Pacific salmon genus *Onchorhynchus. J. Fish. Res. Bd. Can.* 9:265–323.

Brett, J. R. 1979. Environmental factors and growth. in *Fish Physiology.* Vol. 8. W. S. Hoar, D. J. Randall, and J. R. Brett, Eds. Academic Press. New York. pp. 599–675.

Brinkhurst, R. O. 1967. The distribution of aquatic oligochaetes in Saginaw Bay, Lake Huron. *Limnol. Oceanogr.* 12:137–143.

Brinkhurst, R. O. 1974. *The Benthos of Lakes.* Macmillan. London.

Brinkhurst, R. O., A. L. Hamilton, and H. B. Herrington. 1968. Components of the Bottom Fauna of the St. Lawrence, Great Lakes. Preliminary Rep. No. 33. University of Toronto Great Lakes Institute, Toronto.

Brown, H. W. 1974. *Handbook of the Effects of Temperature on Some North American Fishes.* American Electric Power Service Corp., Canton, OH.

Carlander, K. D. 1977. Life History Data on Centrachid Fishes of the United States and Canada, *Handbook of Freshwater Fishery Biology,* Vol. 2, Iowa State University Press. Ames.

Carpenter, S. R. and D. M. Lodge. 1986. Effects of submersed macrophytes on ecosystem processes. *Aquat. Bot.* 26:341–370.

Casselman, J. M., J. J. Collins, E. J. Crossman, P. E. Ihssen, and G. R. Spangler. 1981. Lake whitefish (*Coregonus clupeaformis*) stocks of the Ontario waters of Lake Huron. *Can. J. Fish. Aquat. Sci.* 38:1772–1789.

Christie, A. E. 1982. Fish Species of Ontario, A Summary of Some Characteristics and Their Validity to Predict Presence at a Given Site. Rep. No. 82287. Environmental Studies and Assessments Dept., Ontario Hydro. Toronto.

Christie, A. E. 1985. Temperature Relationships of Great Lakes Fishes: A Data Compilation. Rep. No. 85276. Environmental Studies and Assessments Dept. Ontario Hydro. Toronto.

Christie, G. C. and H. A. Regier. 1988. Measures of optimal thermal habitat and their relationship to yields for four commercial fish species. *Can. J. Aquat. Sci.* 45:301–314.

Cochran, P. A. 1985. Size-selective attack by parasitic lampreys: Consideration of alternate null hypotheses. *Oecologia,* 67:137–141.

Cooper, J. C. and T. T. Polgar. 1981. Recognition of year-class dominance in striped bass management. *Trans. Am. Fish. Soc.* 110:180–187.

Coutant, C. C. 1977. Compilation of temperature preference data. *J. Fish. Res. Bd. Can.* 34:739–745.

Craig, J. F. 1985. Aging in fish. *Can. J. Zool.* 63:1–8.

Dickie, L. M., S. R. Kerr, and P. Schwinghamer. 1988. An ecological approach to fisheries assessment. *Can. J. Fish. Aquat. Sci.* 44(Suppl.2):68–74.

Evans, D. O., B. A. Henderson, N. J. Bax, T. R. Marshall, R. T. Oglesby, and W. J. Cristie, 1988. Concepts and methods of community ecology applied to freshwater fishery management. *Can. J. Fish. Aquat. Sci.* 44(Suppl. 2):448–470.

Farmer, G. J. and F. W. H. Beamish, 1973. Sea lamprey (*Petromyzon marinus*) predation on freshwater teleosts. *J. Fish. Res. Bd. Can.* 30:601–605.

Fry, E. E. J. 1947a. The South Bay experiment. *Sylva (Toronto)*, 3:15–25.

Fry, E. E. J. 1947b. Effects of environment on animal activity. *Univ. Toronto Stud. Biol. Ser.* 55:1–62.

Fry, E. E. J. 1953. The 1944 year class of lake trout in South Bay, Lake Huron. *Trans. Am. Fish. Soc.* 82:178–192.

Giattina, J. D. and R. R. Garton, 1982. Graphical model of thermoregulatory behavior by fishes with a new measure of eurythermality. *Can. J. Fish. Aquat. Sci.* 39:524–528.

Goodier, J. L. 1981. Native lake trout (*Salvelinus namaycush*) stocks in the Canadian waters of Lake Superior prior to 1955. *Can. J. Fish. Aquat. Sci.* 38:1724–1737.

Goodnight, C. J. and L. S. Whitley, 1960. Oligochaetes as indicators of pollution, in *Proc. 15th Annu. Waste Conf.*, Purdue University, Lafayette, IN. pp. 139–142.

Green, R. H. 1979. *Sampling Design and Statistical Methods for Environmental Biologists.* John Wiley & Sons. New York.

Griffiths, J. S. 1979. Performance Response of Great Lakes Fishes to Unstable Thermal Regimes — A Review and Synthesis. Res. Div. Rep. No. 79-367-K. Ontario Hydro. Toronto.

Hairston, N. G., Sr. 1984. Inferences and experimental results in guild structure, in *Ecological Communities: Conceptual Issues and the Evidence.* D. Simberloff, D. R. Strong, Jr., L. A. Abele, and A. B. Thistle, Eds. Princeton University Press. Princeton, NJ. pp. 19–27.

Hartman, W. 1980. Fish-stock assessment in the Great Lakes, in *Biological Monitoring of Fish,* C. H. Hocutt and J. R. Stauffer, Jr., Eds. Lexington Books. Lexington, MA. pp. 119–147.

Healey, M. C. 1975. Dynamics of exploited whitefish populations and their management with special reference to the Northwest Territories. *J. Fish. Res. Bd. Can.* 32:427–448.

Healey, M. C. 1978a. Fecundity changes in exploited populations of lake whitefish (*Coregonus clupeaformis*) and lake trout (*Salvelinus namaycush*). *J. Fish. Res. Bd. Can.* 35:945–950.

Healey, M. C. 1978b. The dynamics of exploited lake trout populations and implications for management. *J. Wildl. Manage.* 42:307–328.

Healey, M. C. 1980. Growth and recruitment in experimentally exploited lake whitefish (*Coregonus clupeaformis*) populations. *Can. J. Fish. Aquat. Sci.* 37:255–267.

Hiltunen, J. K. 1969. The benthic macrofauna of Lake Ontario. *Great Lakes Fish. Comm. Tech. Rep.* 14:39–50.

Hjort, J. 1914. Fluctuations in the great fisheries of northern Europe viewed in the light of biological research. *Rapp. P-V Cons. Perm. Int. Explor. Mer.* 20:238.

Hoar, W. S., D. J. A. Randall, and J. R. Brett, 1979. *Fish Physiology: Bioenergetics and Growth.* Vol. 8. Academic Press. New York.

Houston, A. H. 1982. Thermal Effects Upon Fishes. Rep. NRCC No. 18566 of the Environmental Secretariat. National Research Council of Canada. Ottawa.

Howmiller, R. P. and A. M. Beeton, 1971. Biological evaluation of environmental quality, Green Bay, Lake Michigan. *J. Water Pollut. Contr. Fed.*, 43:123–133.

Hutchinson, G. E. 1957. Concluding remarks. *Cold Spring Harbor Symp. Quant. Biol.* 22:415–427.

Hutchinson, G. E. 1967. *A Treatise on Limnology.* Vol. 2. *Introduction to Lake Biology and the Limnoplankton.* John Wiley & Sons. New York.

Hutchinson, G. E. 1975. *A Treatise on Limnology.* Vol. 3. *Limnological Botany.* John Wiley & Sons. New York.

Ihssen, D., O. Evans, W. J. Christie, J. A. Reckahn, and R. L. DesJardine, 1981. Life history, morphology, and electrophoretic characteristics of five allopatric stocks of lake whitefish (*Coregonus clupeaformis*) in the Great Lakes region. *Can. J. Fish. Aquat. Sci.* 38:1790–1807.

Karr, J. R., K. D. Fausch, P. L. Angermeir, P. R. Yant, and I. J. Schlosser, 1986. Assessing Biological Integrity in Running Waters: A Method and Its Rationale. Spec. Publ. No. 5. Illinois Natural History Survey, Champaign.

Kendall, R. L. Ed. 1978. *Selected Coolwater Fishes of North America.* American Fisheries Society. Washington, D.C.

Kerr, S. R. 1976. Ecological analysis and the Fry paradigm. *J. Fish. Res. Bd. Can.* 33:329–332.

Koelz, W. 1929. Coregonid fishes of the Great Lakes. *Bull. U.S. Bur. Fish.* 63:297–643.

Kutkuhn, J. H. 1981. Stock definition as a necessary basis for cooperative management of Great Lakes fish resources. *Can. J. Fish. Aquat. Sci.* 38:1476–1478.

Lauritsen, K. D., S. C. Mozlery, and D. S. White, 1985. Distribution of oligochaetes in Lake Michigan and comments on their use as indices of pollution. *J. Great Lakes Res.* 11:67–76.

Lawrie, A. H. and J. F. Rahrer, 1972. Lake Superior: effects of exploitation and introductions on the salmonid community. *J. Fish. Res. Bd. Can.* 29:765–776.

Lindeman, R. C. 1942. The trophic-dynamic aspect of ecology. *Ecology* 23:399–410.

Loftus, K. H. and H. A. Regier, 1972. Introduction to the proceedings of the 1971 symposium on salmonid communities in oligotrophic lakes. *J. Fish. Res. Bd. Can.* 29:613–616.

Loftus, K. H., A. S. Holder, and H. A. Regier, 1980. A necessary new strategy for allocating Ontario's fishery resources, in *Allocation of Fishery Resources.* J. H. Grover, Ed. UN/FAO. Rome. pp. 255–264.

Maccubbin, A. E., P. Black, L. Trzeciak, and J. J. Black, 1985. Evidence for polynuclear aromatic hydrocarbons in the diet of bottom-feeding fish. *Bull. Environ. Contam. Toxicol.* 34:876–882.

Mannan, R. W., J. L Morrison, and E. C. Meslow, 1984. Comment: the use of guilds in forest bird management. *Wildl. Soc. Bull.* 12:426–430.

Margalef, R. 1958. Information theory in ecology. *Gen. Syst.* 3:36–71.

Margalef, R. 1968. *Perspectives in Ecological Theory.* The University of Chicago Press. Chicago.

Margalef, R. 1981. Stress in ecosystems: a future approach, in *Stress Effects on Natural Systems.* G. W. Barrett and R. Rosenberg, Eds. John Wiley & Sons. New York. pp. 281–289.

McKee, P. M., T. R. Batterson, T. E. Dahl, V. Glooschenko, E. Jaworski, J. B. Pearce, C. N. Raphael, and T. H. Whillans, 1989. Great Lakes aquatic habitat classification, based on wetland classification systems. This volume.

Moyle, P. B., H. W. Li, and B. A. Barton, 1986. The Frankenstein effect: impact of introduced fishes is North America, in *Proc. Symp. Role of Fish Culture in Fisheries Management.* R. Stroud, Ed. American Fisheries Society. Bethesda MD. pp. 415–425.

Mozley, S. C. and W. P. Alley, 1973. Distribution of benthic invertebrates in the south end of Lake Michigan, in *Proc. 16th Conf. Great Lakes Res.* International Association of Great Lakes Research. Ann Arbor, MI. pp. 87–96.

Nalepa, T. F. and N. A. Thomas. 1976. Distribution of macrobenthic species in Lake Ontario in relation to sources of pollution and sediment parameters. *J. Great Lakes Res.* 2:150–163.

Nepszy, S. J., Ed. 1988. Parasites of Fishes in the Canadian Waters of the Great Lakes. Tech. Rep. No. 51, Great Lakes Fisheries Commission, Ann Arbor, MI.

Odum, E. P. 1975. Diversity as a function of energy flow, in *Unifying Concepts in Ecology.* W. H. van Dobben and R. H. Lowe-McConnell, Eds. Dr. W. Junk. The Hague. pp. 11–14.

Orians, G. H. 1975. Diversity, maturity and stability in ecosystems, in *Unifying Concepts in Ecology.* W. H. van Dobben and R. H. Lowe-McConnell, Eds. Dr. W. Junk. The Hague. pp. 139–150.

Parker, P. S. and R. E. Lennon, 1957. Biology of the Sea Lamprey in its Parasitic Phase. Res. Rep. 44. U.S. Fish and Wildlife Service. Washington, D.C.

Pielou, E. C. 1975. *Ecological Diversity.* John Wiley & Sons. New York.

Pimm, S. L. 1984. The complexity and stability of ecosystems. *Nature* 307:321–326.

Reckahn, J. A. 1970. Ecology of young lake whitefish (*Coregonus clupeaformis*) in South Bay, Manitoulin Island, Lake Huron. in *Biology of Coregonid Fishes.* C. C. Lindsey and C. S. Woods, Eds. University of Manitoba Press. Winnipeg. pp. 437–460.

Reckahn, J. A. 1986. Long-term cyclical trends in growth of lake whitefish in South Bay, Lake Huron. *Trans. Am. Fish. Soc.* 115:787–804.

Reckahn, J. A. 1988. Lake whitefish mortality rates in relation to sea lamprey abundance and other species in South Bay, Manitoulin Island, Lake Huron. Manuscript in preparation.

Regier, H. A. 1966. A perspective on research on the dynamics of fish populations in the Great Lakes. *Prog. Fish-Cult.* 28:318.

Ricker, W. E. 1975. Computation and Interpretation of Biological Statistics of Fish Populations. Bull. No. 191. Fisheries and Marine Service. Ottawa.

Ricker, W. E. 1979. Growth rates and models. in *Fish Physiology: Bioenergetics and Growth.* Vol. 3. W. S. Hoar, D. J. Randall, and J. R. Brett, Eds. Academic Press. New York. pp. 677–786.

Roff, D. A. 1986. Predicting body size with life history models. *Bioscience* 36:316–323.

Root, R. B. 1967. The niche exploitation patterns of the blue-gray gnatcatcher. *Ecol. Monogr.* 37:317–350.

Russell, E. S. 1931. Some theoretical considerations on the "overfishing" problem. *J. Cons. Cons. Int. Explor. Mer.* 6:3–27.

Russell, E. S. 1942. *The Overfishing Problem.* Cambridge University Press. London.

Ryder, R. A. and C. J. Edwards, Eds. 1985. A Conceptual Approach for the Application of Biological Indicators of Ecosystem Quality in the Great Lakes Basin. Report to the Science Advisory Board, International Joint Commission, Windsor, Ontario.

Sculthorpe, C. D. 1967. *The Biology of Aquatic Vascular Plants.* St. Martin's Press. New York.

Severinghaus, W. D. 1981. Guild theory as a mechanism for assessing environmental impact. *Environ. Manage.* 5:187–190.

Sissenwine, M. P. 1984. Why do fish populations vary? in *Exploitation of Marine Communities. Dahlem Konferenzen.* R. M. May, Ed. Springer-Verlag, Berlin. pp. 59–94.

Smith, B. R. 1980. Introduction to the proceedings of the 1979 Sea Lamprey International Symposium (SLIS). *Can. J. Fish. Aquat. Sci.* 37:1585–1587.

Smith, J. B. 1968. Former Lake Trout Spawning Grounds in Lake Huron. Res. Sect. Rep. (Fish.) 68. Ontario Department of Lands and Forests. Toronto.

Smith, S. H., H. J. Buettner, and R. Hile, 1961. Fishery statistical districts of the Great Lakes. Tech. Rep. No. 2. Great Lakes Fisheries Commission. Ann Arbor, MI.

Spence, D. H. N. 1982. The zonation of plants in freshwater lakes. *Adv. Ecol. Res.* 12:37–125.

Spencer, C. N. and D. L. King, 1984. Role of fish in regulation of plant and animal communities in eutrophic ponds. *Can. J. Fish. Aquat. Sci.* 41:1851–1855.

Sweeney, B. W. and R. L. Vannote, 1978. Size variation and the distribution of hemimetabolous aquatic insects: two thermal equilibrium hypotheses. *Science* 200:444–446.

Trandahl, A. 1978. Preface. in *Selected Coolwater Fishes of North America*. Spec. Publ. No. 11. R. L. Kendall, Ed. American Fisheries Society. Washington, D.C.

van Oosten, J., R. Hile, and F. Jobes, 1946. The whitefish fishery of Lakes Huron and Michigan with special reference to the deep-trap-net fishery. *Fish. Bull. Fish. Wildl. Serv.* 50:297–394.

Warwick, W. F., J. Fitchko, P. M. McKee, D. R. Hart, and A. J. Burt, 1987. The incidence of deformities in *Chironomus* spp. from Port Hope Harbour, Lake Ontario. *J. Great Lakes Res.* 13:88–92.

Wells, L. 1970. Effects of alewife predation on zooplankton populations in Lake Michigan. *Limnol. Oceanogr.* 15:556–565.

Whittaker, R. H. 1975. *Communities and Ecosystems.* 2nd ed. Macmillan. New York.

Wilhm, J. L. and T. C. Dorris, 1968. Biological parameters for water quality criteria. *Bioscience* 18:477–481.

Winnell, M. H. and D. S. White, 1985. Trophic status of southeastern Lake Michigan based on the Chironomidae (Diptera). *J. Great Lakes Res.* 11:540–548.

Wismer, D. A. and A. E. Christie, 1987. Temperature Relationships of Great Lakes Fishes: A Data Compilation. Spec. Publ. No. 87–3. Great Lakes Fisheries Commission. Ann Arbor, MI.

Wright, S. and W. M. Tidd, 1933. Summary of limnological investigations in western Lake Erie in 1929 and 1930. *Trans. Am. Fish. Soc.* 63:271–285.

Chapter 8

STATE-OF-THE-ART TECHNIQUES FOR INVENTORY OF GREAT LAKES AQUATIC HABITATS AND RESOURCES

T. A. Edsall, R. H. Brock, R. P. Bukata, J. J. Dawson, and F. J. Horvath

TABLE OF CONTENTS

PART A: OVERVIEW

T. A. Edsall

This section of the Classification and Inventory of Great Lakes Aquatic Habitat report was prepared as a series of individually authored contributions that describe, in various levels of detail, state-of-the-art techniques that can be used alone or in combination to inventory aquatic habitats and resources in the Laurentian Great Lakes system. No attempt was made to review and evaluate techniques that are used routinely in limnological and fisheries surveys and inventories because it was felt that users of this document would be familiar with them.

The first contribution (Dawson) evaluates the utility of hydroacoustic systems to inventory pelagic habitats in the Great Lakes and contiguous waters and to provide quantitative data on the use of those habitats by fish. The second contribution (Edsall) focuses on the use of side-scan sonar to inventory and map geophysical features on the water body floor. Small manned submarines and remotely operated submersibles carrying video and photographic systems capable of providing high resolution records of water column and substrate features are also discussed; these vehicles can provide "ground truth" information to supplement side-scan sonar records or can be used independently for habitat and resource inventory. The third contribution (Bukata) explores applications of remote aerial sensing techniques for the inventory of Great Lakes aquatic habitats and adjacent shorelines. The applications that are discussed are ranked on a scale ranging from "feasible and simple" to those that are more difficult and nonfeasible. Survey and control needs for inventory-related navigation and mapping are considered in the fourth contribution (Brock) and systems and techniques that can provide accurate positioning data in real time or for database construction are discussed. The final contribution of the set (Horvath) examines the potential role of Geographic Information Systems (GISs) and artificial intelligence systems in inventory and classification of Great Lakes aquatic habitats, and in the analysis of habitat data for use by the research community, resource managers, and other decision makers whose activities affect the resource.

Collectively, the inventory techniques described in the following five contributions can be used to generate a wide variety of physical, chemical, and biological data that would be difficult or impossible to obtain by other means. These data, if structured according to a functional habitat classification system and manipulated with other available habitat data in a GIS, could contribute significantly to the development of an objective basis for management of aquatic habitat in the Great Lakes — a major goal of this workshop.

PART B: HYDROACOUSTIC ASSESSMENT OF FISH POPULATIONS IN GREAT LAKES AQUATIC HABITATS

J. J. Dawson

ABSTRACT

Hydroacoustic techniques for assessing fish populations have become more powerful due to advances in microcomputers, microcircuitry, signal processing, and digital design. These techniques, used successfully throughout the world, supply quantitative estimates of fish density and acoustic size distributions. Hydroacoustic assessment methods are currently being used in the Great Lakes system. An effort is underway to classify Great Lakes aquatic habitats and this section discusses the suitability of hydroacoustics for assessing the fish populations in each of these habitats.

I. INTRODUCTION

The use of quantitative hydroacoustics in the Great Lakes system has increased in the last decade. The techniques themselves have been developed and improved, and the study objectives have become more complex and quantitative. For some habitats in the Great Lakes system, hydroacoustic techniques are ideally suited; in others, the strengths of the techniques cannot be fully utilized.

II. HYDROACOUSTICS: THE TECHNIQUE

A. EVOLUTION

Early uses of acoustic instruments in biological research were limited to presence or absence determinations. If dots appeared on a chart, fish were present. A natural extension of these measurements was to count the dots to estimate the number of fish. At this point, biologists found that the number of dots could be increased by turning up the gain of their fish finder. They also realized that they needed a water volume estimate to expand the dot counts into an estimate of density. In turning to the electrical engineer and the physicist for help, the biologist found that hydroacoustic assessment had become a multidisciplinary science, one for which most biologists were not well trained. Modern hydroacoustic research requires that the user have a reasonable understanding of the physics of underwater sound, of electronic instrumentation and computers, and of sample design and statistics.

B. PRINCIPLES

To understand the strengths and weaknesses of hydroacoustic techniques in the many habitats present in the Great Lakes ecosystem, a brief discussion of terms and concepts is necessary. Echo sounders transmit a sound pulse through a roughly conical volume of water. Typical (ultrasonic) frequencies used in hydroacoustics range from 20 kHz to 1 MHz (a hertz is 1 c/s). Higher frequencies can detect smaller organisms. In the marine environment, higher frequency energies are absorbed rapidly by the salts in the water, limiting detection ranges. In freshwater, all of the above frequencies are available to the user.

When the transmitted sound pulse encounters something of a different density than the surrounding water, an echo is reflected. The echo sounder receives this echo and amplifies it. The width (time duration) of the pulse received should be essentially the same as the width of the pulse transmitted for an individual fish. If more than one fish is in the sound beam at about the same range from the transducer, the echos arrive back at the echo sounder at about the same time, resulting in a multiple target.

C. INSTRUMENTATION REQUIREMENTS

The output of a scientific echo sounder is a series of voltage pulses. To calibrate the echo sounder, the voltage out of the receiver is measured for a range of sound pressures at the face of the transducer (the underwater transmitting and receiving device), and the sound pressure transmitted from the transducer into the water is measured. Electronic transmission stability is a fundamental necessity for a scientific echo sounder.

Due to the spreading of the sound cone, the voltage that returns from a given fish decreases as that fish goes deeper. A scientific echo sounder removes this depth dependency with an electronic circuit called time varied gain (TVG). The accuracy and stability of this TVG circuit is critical, and must be equivalent to the TVG required by the acoustic theory. Any temptation to use nonscientific sounders that provide only a "pretty picture" should be avoided if quantitative results are desired, since they do not have accurate TVG circuits.

D. TRANSDUCER DIRECTIVITY

Transducers convert the voltage pulse from the echo sounder's transmitter into a sound pressure pulse, and send this energy into the water. The conical volume through which this pulse travels has an acoustic axis that is perpendicular to and coaxial with the face of the transducer. When a fish on this axis reflects an echo, the returning sound pressure is received by the transducer and is amplified by the echo sounder's receiver. The voltage that comes out of the receiver is a measure of the acoustic size (reflectivity) of that fish, and is related to its physical size (length or weight). The "cone of sound" is not uniform in intensity. The sound intensity is highest on the acoustic axis and decreases away from the axis. As a result, the echo voltage that comes out of the echo sounder's receiver decreases as a fish moves away from the center of the cone. Since most fish are not on the acoustic axis, the voltages that come out of the echo sounder's receiver do not represent the acoustic size distribution. Attempts to statistically remove this "transducer directivity" effect become troublesome. A better method is to have the scientific echo sounder automatically remove this confounding effect. Two manufacturers currently provide such equipment. One uses the dual-beam technique (Ehrenberg, 1984), the other uses the split-beam technique (Ehrenberg, 1982). Both methods effectively "move" each fish echo onto the acoustic axis. Although the split-beam technique is marginally superior from a theoretical standpoint, the implementation of the technique is very difficult. In the only commercially available split-beam unit, any theoretical advantage over the dual-beam is lost due to these difficulties. Hereafter, in this chapter, only the dual-beam technique is mentioned.

E. OUTPUT PRODUCTS

One of the primary products from the acoustic survey techniques is estimation of fish density. If relatively few multiple targets are encountered, density may be calculated by counting echoes and estimating the corresponding sample volume. If multiple targets (such as schools of fish) are found, a technique called echo integration is used. The output of an echo integrator (relative density) is the accumulation of all energy reflected from the ensonified

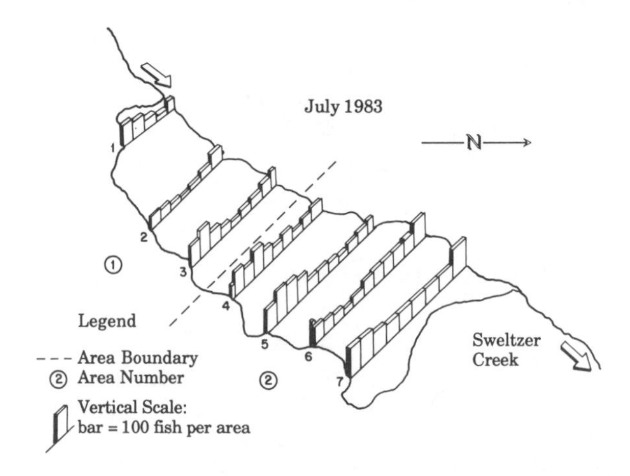

FIGURE 1. Distribution of fish per unit surface area.

volume. If the average energy per fish (target strength) is known, the echo integrator output can be scaled to units of absolute density, e.g., number of grams per cubic meter (Burczynski, 1982). This estimate of the average energy per fish comes from the dual-beam process. In Figure 1, densities per unit surface area are plotted by segment along each of seven transects. These data from Burczynski and Johnson (1986) show a shoreward orientation of sockeye salmon smolts in Cultus Lake, British Columbia. Density mapping provides a useful tool for examining the utilization of habitat by a fish population.

The acoustic size of a fish, called its target strength, has been related to fish length by several researchers (Love, 1971; McCartney and Stubbs, 1971; Nakken and Olsen, 1977). The resulting regressions relating target strength to length are quite variable and at times are poor fits for certain species and sizes. One of the better fitting relationships is by Love (1971):

$$TS = 19.1 \log(l) - 0.9 \log(F) - 62.0$$

where L = fish length in centimeters and F = echo sounder frequency in kilohertz.

The estimation of the acoustic size distribution of a fish population allows the comparison of acoustic data with physical data (length frequency data from trawl catch, for example). The results of this type of comparison, made by Burczynski and Johnson (1986) using Love's equation, are shown in Figure 2. The solid histogram bars represent the translated catch data and show narrower distribution.

III. EVALUATION OF THE TECHNIQUE

A. SAMPLING POWER

One of the strengths of acoustic sampling is the very great sampling volume associated with the sound cone. Assessments of the mid-water fish habitats utilize this strength, since the volume of the cone increases with depth. For transducers mounted near the surface and aimed downward, the sampling volume for near-surface populations or in shallow water is much lower. Any shipboard sampling device might be susceptible to boat avoidance by fish. Organisms that are located within about 0.5 m of reflecting boundaries (e.g., surface, bottom, thermocline, etc.) are difficult to assess with hydroacoustic tools. Standing timber, bottom vegetation, gas bubbles seeping from the sediments, and surface turbulence due to wind and waves are factors that usually limit the sampling power of the technique.

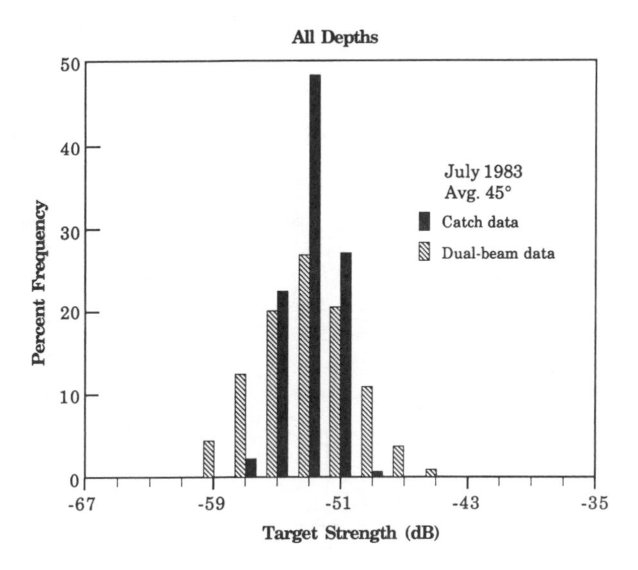

FIGURE 2. Comparison of the target strength distribution.

B. ACOUSTIC SIZING

The capability of acoustic sizing of targets comes for the dual-beam technique. The spread of acoustic sizes is always greater than the spread of fish lengths, due to behavioral effects and the randomness of the scattering process (Figure 2). As a result, resolving acoustic modes from a bimodal population requires the mean lengths from the two modes to differ by a factor of about two (Ehrenberg, 1984). If acoustic modes can be resolved, population estimates by size class are feasible. The ability of the acoustic method to resolve acoustic modes can be enhanced by a technique called target tracking. This technique averages many measurements from each fish into a single target strength estimate for that individual. Figure 3 uses target tracking to reveal a bimodal target strength distribution from an acoustic survey over a mixed aggregation of threadfin shad and striped bass in Lake Mead. The large fish composed about 4% of the population by number.

C. RESOLUTION AND GROUND TRUTHING

Hydroacoustic sampling provides a high degree of resolution. The water column can be divided into depth strata of <1 m thickness. Data collection is continuous both vertically and along a transect. A knowledge of biological sampling theory is useful for determining location and spacing of transects. Spatial resolution between transects depends on transect spacing. This spacing is often as dependent on available boat time as on statistical considerations. If transect segments are considered replicates and are averaged together, variance estimates should be corrected for serial correlation. Transects are typically run perpendicular to distributional gradients, which in turn are often parallel to the shoreline. Since the hydroacoustic technique is nonobstructive, ground truthing is often difficult. Trawl catches are affected by physical structures and are often size selective and therefore may sample a different population than the acoustic equipment. Gill net catches may be affected by fish behavior or physical structure, resulting in selectivities. Ground truthing of acoustic data usually entails species identification by some sampling program. Under certain limited conditions, and with prior biological knowledge, species identification can be inferred from acoustic data.

D. QUALITY CONTROL

Hydroacoustic techniques are subject to the same quality control procedures as the other biological sciences, namely that of peer review. A community of scientists specializing in

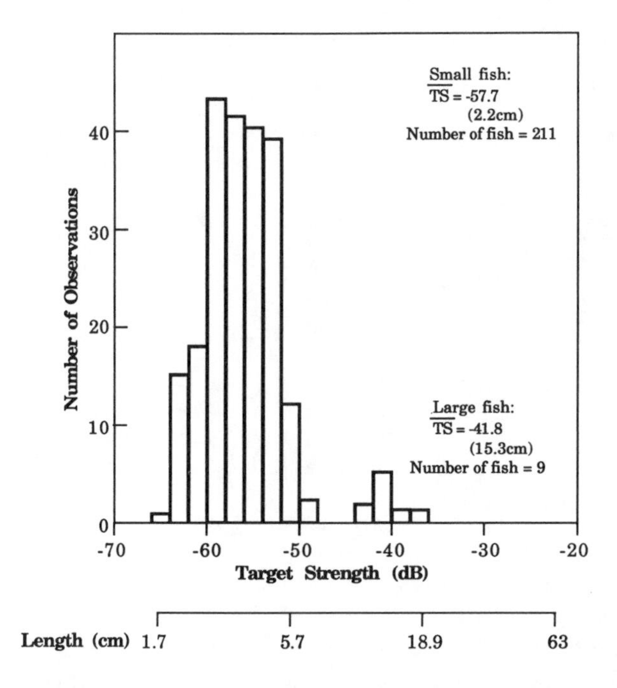

FIGURE 3. Target strength distribution of a mixed population.

hydroacoustic equipment and research exists, as well as a large body of publications, including the proceedings of several international symposia. Any valid scientific acoustic equipment manufacturer or scientist must be able to prove that the equipment and techniques used do conform to acoustic theory and sound biological sampling practices. Since hydroacoustics is regarded by many biologists as a discipline full of black boxes, it is very susceptible to unrealistic performance claims by "backyard" manufacturers and scientists. It is the duty of anyone that uses or is considering the use of hydroacoustic techniques to be sure the equipment and expertise that is utilized conforms to acceptable scientific standards.

E. ECONOMICS

The economics of using acoustic equipment differs from many other biological sampling techniques. The capital equipment costs at the beginning are quite high. With the design of powerful microprocessors, data analysis costs and time are being reduced. Once the capital costs are amortized, the technique is very cost-effective. In contrast, techniques such as mid-water trawling have high costs associated with data collection. These costs, associated with boat time and processing efforts, are continuous from year to year. The continuous nature of acoustic sampling makes more efficient use of available boat time. A complete, state-of-the-art hydroacoustic data collection and processing system now costs under $100,000.

F. GREAT LAKES USERS

Several research agencies have been or are currently using hydroacoustic techniques for assessment of mid-water forage and predator stocks in the Great Lakes. These agencies include the USFWS; National Fisheries Research Center-Great Lakes, Ann Arbor; the State University of New York at Oswego; the University of Wisconsin; the University of Maryland; and the Ontario Ministry of Natural Resources.

IV. APPLICABILITY OF HYDROACOUSTICS AS AN INVENTORY TOOL IN DIFFERENT GREAT LAKES HABITATS

The following classification of habitats is based partially on the sampling characteristics of the hydroacoustic technique. Habitats are defined as:

- Lake Environment
  - Nearshore (out to 5 m depth contour)
  - Offshore
    - Near Surface (upper 5 m)
    - Water Column (below 5 m)
    - Near Bottom (within 2 m of the bottom)
- Wetlands
- Rivers and Channels
  - Large Rivers
  - Small Rivers and Navigational Channels
  - Small Channels (up to 6 ft deep)

The strengths and weaknesses of hydroacoustic techniques are discussed for each of these environments.

A. LAKE ENVIRONMENT

1. Nearshore

Traditional mobile hydroacoustic surveys are not particularly well suited for assessing the nearshore environment. The sampling volume of the sound cone is relatively small due to the shallow depth. Many hazards to the boat and to navigation exist in this region. It is difficult to quantify boat avoidance. The presence of aquatic vegetation and gas bubbles may affect data quality. Traditional sampling techniques such as electrofishing, beach seining, or rotenone sampling may be viable methods in the nearshore environment.

Some modifications to the traditional vertical sounding method will improve performance in the nearshore environment. Mounting the transducer at or ahead of the bow of the boat minimizes avoidance, as does use of electric trolling motors. The transducer can also be mounted on a bottom or floating mount. This technique is called fixed-location. If the transducer is aimed up or down, information on acoustic size and vertical distribution is obtained. If the transducer is aimed horizontally, behavioral data such as direction of travel and fish velocity can be measured. Measurements of fish flux (numbers passing per unit time through a cross-sectional area) from a fixed-location acoustic system are analogous to fish density from a mobile system. The fixed-location technique obviously gives information only at a particular location. For fixed-location systems, large arrays of transducers may be deployed and subsampled to increase areal coverage. Correlating behavioral data from a fixed-location hydroacoustic system with samples from one of the traditional catching techniques can provide valuable insights into the ecology of the nearshore environment.

2. Offshore: Near Surface

The near surface environment of the Great Lakes, or the upper 5 m, is effectively sampled by hydroacoustics. Although the concerns on sampling volume and boat avoidance that were mentioned for the nearshore environment still apply, boat speed and direction are not as limited. Sampling power can be increased by using a faster pulse repetition rate. If the fish species of interest is surface oriented, the transducer can be mounted on a deep diving towed body and aimed up. This puts the greatest sampling power closest to the fish. Ground truthing for species identification is usually provided by surface tow netting. For some species, surface tow netting is not very effective because the fish sound when they sense the net approaching.

3. Offshore: Main Water Column

The hydroacoustic technique is ideally suited for the assessment of the pelagic zone of the Great Lakes. The acoustic size distribution determined by the dual-beam technique can be compared to the length frequency distribution from a mid-water trawl. Maps of fish and plankton density by depth can be constructed from transect data. Absolute biomass estimates of forage base fish or plankton provide inputs into modeling and management decisions. Although species identification is not currently possible using hydroacoustic techniques, a limited amount of ground-truth data applied to the distributional data from the acoustics allows inference of the distribution of various species. By using multiple acoustic frequencies, it is also possible to collect data on the particle size distribution of the plankton communities.

4. Offshore: Near Bottom

Since hydroacoustic techniques are somewhat limited in their ability to assess targets near boundaries, the near bottom region is considered separately. The minimum distance that the echo sounder can still detect a fish off the bottom is a function of the transmitted pulse width. Decreasing the pulse width decreases the resolution distance. Since the voltage outputs of the echo sounder are processed by automated microprocessors, the primary concern when assessing near bottom targets is recognizing and avoiding the echo from the bottom. The processors keep track of the position of the bottom echo. If its position changes gradually, the processors can "look closer to the bottom". Conditions that make the bottom echo position change rapidly include wave action and boat rolling, and rapid depth changes due to bottom topography. Other near bottom conditions that limit the effectiveness of using acoustic survey methods include standing timber and aquatic macrophytes.

B. WETLANDS

Wetlands are considered in this analysis as broad regions of shallow water depth with trees and aquatic vegetation. This environment mimics the nearshore lake environment from an acoustic sampling perspective, and the strengths and weaknesses already discussed for nearshore apply to the wetlands.

C. RIVERS AND CHANNELS

1. Large Rivers

The movement of fish in rivers and channels is quite often very dynamic. As a result, standard mobile survey techniques may give an accurate estimate of instantaneous density, but often do not provide a good measure of distribution. Population estimates and measure of habitat usage in rivers may be biased due to fish migration. Fixed-location techniques are usually better suited for studying fish in rivers. For large rivers, vertical sounding from surface- or bottom-mounted transducers at multiple sampling stations across a river gives estimates of horizontal and vertical distribution of fish. The angling transducer aims angles up- or downstream, and thus the direction of fish passage can be determined. For rivers that are deep enough and acoustically quiet enough, horizontal scanning from each bank is an effective method of monitoring. Irregular bottom contours and substrate, turbulence and air bubbles, and unpredictable fish behavior make acoustic riverine studies very complex. One promising technology is the Doppler river counter, which is basically a motion detector. This system uses the Doppler principle and filters out all signals that are stationary or moving downstream. It is also extremely effective in monitoring passage rates through restricted areas such as fish ladders.

2. Small Rivers and Navigational Channels

Although the principles discussed for large rivers also apply to small ones, hydroacoustic techniques are probably better suited for monitoring smaller rivers and navigational channels, primarily because the proportion of the river being "acoustically screened" is much higher. In many instances, the water flow is less turbulent, reducing the acoustic noise.

3. Small Channels

The principles and limitations discussed for the shallow water environments apply to small channels. Acoustic assessment may be limited by shallow water depth, gas bubble production, and aquatic macrophytes. Yet, in some locations complete acoustic coverage of a small channel is possible with the Doppler system or a standard research echo sounder. Assessment of aquatic macrophytes using hydroacoustic techniques is currently being evaluated.

V. SUMMARY

Hydroacoustic survey techniques are applicable in all Great Lakes habitats, but are particularly well suited to pelagic mapping and assessment and for monitoring rivers. Primary advantages include high sampling power and resolution, low operational costs, nonobtrusive nature, and automated processing and data storage. Limitations include reduced sampling power in shallow water, lack of species identification, and initial capital cost.

Acoustic techniques are accepted methodologies in the field of fisheries science for population mapping and estimation. Population estimates under ideal acoustic and distributional conditions can have confidence limits as low as 10 to 25% of the mean. Data are processed and stored in microcomputer disk files. These data can be downloaded to mainframes or database programs, or compared to satellite maps of environmental measures like surface temperature or chlorophyll.

Ground truthing primarily consists of methods for identifying species composition, since most catch methods have biases greater than the hydroacoustic technique. With the current advances in computer and electronic design, hydroacoustic assessment systems are becoming less expensive, more portable, more powerful, and simpler to operate.

REFERENCES

Burczynski, J. J. 1982. Introduction to the Use of Sonar Systems for Estimating Fish Biomass. Tech. Pap. No. 191, Rev. 1. FAO Fisheries, Rome.

Burczynski, J. J. and R. L. Johnson. 1986. Application of dual-beam survey techniques to limnetic populations of juvenile sockeye salmon (*Oncorhynchus nerka*). *Can. J. Fish. Aquat. Sci.* 43(9): 1776–1787.

Ehrenberg, J. E. 1983. A review of *in situ* target strength estimation techniques, in *Symp. Fish. Acoust., Selected Papers of the ICES/FAO Symposium on Fisheries Acoustics*. O. Nakken and S. C. Venema, Eds. FAO Fisheries Rep. No. 300. Food and Agriculture Organization, Rome. pp. 85–90.

Ehrenberg, J. E. 1984. Two applications for a dual-beam transducer in hydroacoustic fish assessment systems. *Proc. IEEE Congr. Eng. Ocean Environ.* 1:151–155.

Love, R. H. 1971. Dorsal-aspect target strength of an individual fish. *J. Acoust. Soc. Am.* 49:816–828.

McCartney, B. S. and A. R. Stubbs. 1971. Measurements of the acoustic target strengths of fish in dorsal aspect, including swimbladder resonance. *J. Sound Vib.* 15(3):397–420.

Nakken, O. and K. Olsen. 1977. Target strength measurements of fish. *Rapp. P.-V. Reun. Cons. Perm. Int. Explor. Mer.* 170:52–69.

PART C: USE OF SIDE-SCAN SONAR TO INVENTORY AND MAP GREAT LAKES BENTHIC HABITAT

T. A. Edsall

ABSTRACT

Recent advances in side-scan sonar technology have resulted in the production of relatively inexpensive survey-quality systems that are well suited for inventory and mapping of Great Lakes benthic habitat. These systems can yield detailed maps of the lake bed that closely resemble aerial photographs of land masses. Resolution of lake bed features smaller than 0.1 m is possible with some systems. Side-scan sonar data are readily processed in computer-linked GISs to produce substrate and bathymetric maps of any desired scale and related statistics describing, for example, the amount of lake bed covered by a particular substrate. A basis is given for estimating the cost of conducting side-scan sonar surveys of the Great Lakes benthic habitat.

I. INTRODUCTION

Sound was first used to examine the sea bed in the late 1920s and the first operable side-looking sonar was built in the late 1950s. The development of the side-looking sonar was a major technological breakthrough because it made possible the rapid, systematic study and large-scale mapping of surficial features of the substrate in marine and freshwater environments (Flemming, 1976). Side-scan sonar technology has developed rapidly and state-of-the-art systems are now capable of producing accurate, detailed records of the floor of a water body that can be readily assembled to provide geological mosaics closely resembling high quality aerial photographs of land masses (Clifford, 1979; EG&G, 1985; Klein, 1979, 1984).

Side-scan sonar has several diverse applications: hydrographic and geological surveys and mapping; underwater engineering surveys and inspection of underwater man-made structures; search for sunken objects, including vessels and aircraft; marine archaeology; study of underwater ice masses, water currents, and water turbulence (Russell-Cargill, 1982; EG&G, 1985); and uses by the military (Key, 1984, 1985). Side-scan sonar was also used extensively to inventory marine habitat on the outer continental shelf of the U.S. in connection with the sale of oil and gas extraction rights (Interstate Electronics Corp., 1979; Woodward-Clyde Consultants, 1979; Lissner, 1981; Rezak and Bright, 1981; Gettleson et al., 1982; Continental Shelf Associates, Inc., 1985), and to meet biological or geological objectives unrelated to hydrocarbon extraction in other marine areas (Boldrin et al., 1980; Jameson, 1981; Flemming and Martin, 1983). Benthic feeding grounds of gray whales (*Eschrichtius robustus*) were also mapped from side-scan sonar records that showed furrows made on the sea bed by feeding whales (Johnson et al., 1983; Nerini and Oliver, 1983). In fresh water, habitat-related applications of side-scan sonar include surveys and mapping of the spawning grounds of lake trout (*Salvelinus namaycush*) in the Finger Lakes of New York (Sly and Widmer, 1984) and in northern Lake Michigan (Edsall et al., 1989).

FIGURE 1. Side-scan sonar equipment used by the National Fisheries Research Center-Great Lakes to survey the lake bed on historical lake trout spawning grounds in the Great Lakes. The tape deck and microprocessor are in the rear and the towfish is in the foreground.

II. SIDE-SCAN SONAR TECHNOLOGY

A. PRINCIPLES OF OPERATION AND SYSTEM COMPONENTS

The National Fisheries Research Center-Great Lakes is conducting side-scan sonar surveys throughout the Great Lakes, using a survey-grade EG&G model 260 microprocessor, a Model 272-T 100 kHz towfish equipped with TVG, and a Model 360 digital tape deck (Figure 1).* This system, like other basic side-scan sonar systems, directs a fan-shaped beam of acoustic energy from the towfish to the lake bed in a plane normal to the track of the towfish (Figure 2). The beam impinges on the lake bed directly beneath the towfish and out to some preset distance on each side. The echo from the acoustic signal that reached the lake bed is received and transformed into an electrical signal by the towfish and transmitted to the microprocessor. There the signal is converted into a microprocessor (strip chart) record on which the surficial features of the surveyed swath of lake bed are shown in plain view (Figure 3). The signal received by the microprocessor also contains real-time water depth data along the towfish track; these data are recorded continuously in the profile section of the microprocessor record (Figure 3). The signal reaching the microprocessor is also transmitted to the tape deck for storage and retrieval, if needed, to produce duplicate microprocessor records.

B. FIELD SURVEY TECHNIQUES

Systematic mapping of the lake bed is the most rigorous habitat-related application of side-scan sonar. When side-scan sonar is used in this way, it is usually prudent to first conduct a reconnaissance survey of the area provisionally selected for mapping to help optimize the

* Mention of brand names or manufacturers does not imply U.S. Government endorsement of commercial products.

FIGURE 2. Surveying the lake bed with side-scan sonar.

FIGURE 3. Microprocessor record of Stannard Rock, Lake Superior. Rugged bedrock ridges near the peak of the shoal are clearly shown on the record. Water depths range from about 10 to 30 m.

mapping effort. The reconnaissance survey delineates shoal waters that must be avoided by the survey vessel and identifies portions of the area that do not need to be mapped. The survey is conducted by pulling the towfish at a constant speed behind the survey vessel over several transects that were selected to provide information on shoal areas and on the substrates along the boundaries of the area provisionally selected for mapping. The number and length of the reconnaissance transects needed is determined by the physiographic complexity and size of the area provisionally chosen for mapping. When the reconnaissance survey is completed and the area to be mapped is determined, Loran C coordinates are established bounding the area, and parallel mapping transects are laid out over that area. Transects are spaced to provide about 40% overlap of microprocessor records for adjacent transects to ensure that the lake bed within the mapping area is fully represented on the records. The towfish is then drawn over each mapping transect to produce the microprocessor records from which the mosaic map is created. Loran C vessel position coordinates are recorded in real time at intervals of 2 to 3 min on the margin of each microprocessor record to assist in orienting the record and to facilitate locating lake bed features.

Microprocessor records made during reconnaissance and mapping are examined in the field and used to determine where ground truth information should be collected to facilitate record interpretation and map preparation. Most of the ground truth information for the lake bed surveys we have conducted was obtained with a Benthos, Inc., Mini-Rover MK II remote-operated, submersible vehicle (Figure 4). This vehicle is equipped with a high-resolution, low-

FIGURE 4. Mini-Rover MK II remotely operated submersible. This is one of a class of small vehicles that can provide videotape and photographic records of underwater features.

light, color video camera that relays a video image of the lake bed to the survey vessel through a 300-m tether (umbilical); the MK II is also equipped with a 35-mm still camera. The MK II is deployed by an operator who maneuvers the vehicle with joystick controls and operates the cameras while monitoring the transmission from the video camera on a shipboard television screen. A permanent record of the video images is made by connecting a video cassette recorder to the shipboard monitor. An alphanumeric display of the depth at which the MK II is operating and the compass heading it is following are shown on the shipboard television screen, and the entire screen display is videotaped to provide a permanent record of lake bed features.

In 1985–1986, we also used a small, manned submersible equipped with video and still cameras to provide ground truth data for side-scan sonar surveys conducted on lake trout spawning grounds in Lakes Superior and Huron (Manny and Edsall, 1989). This submersible, the Johnson *Sea-Link II*, which carried two crew members and two scientists, was provided by the National Underseas Research Program of the National Oceanic and Oceanographic Agency (NOAA). The operating characteristics of the Sea-Link II and its support vessel were described by Askew (1985).

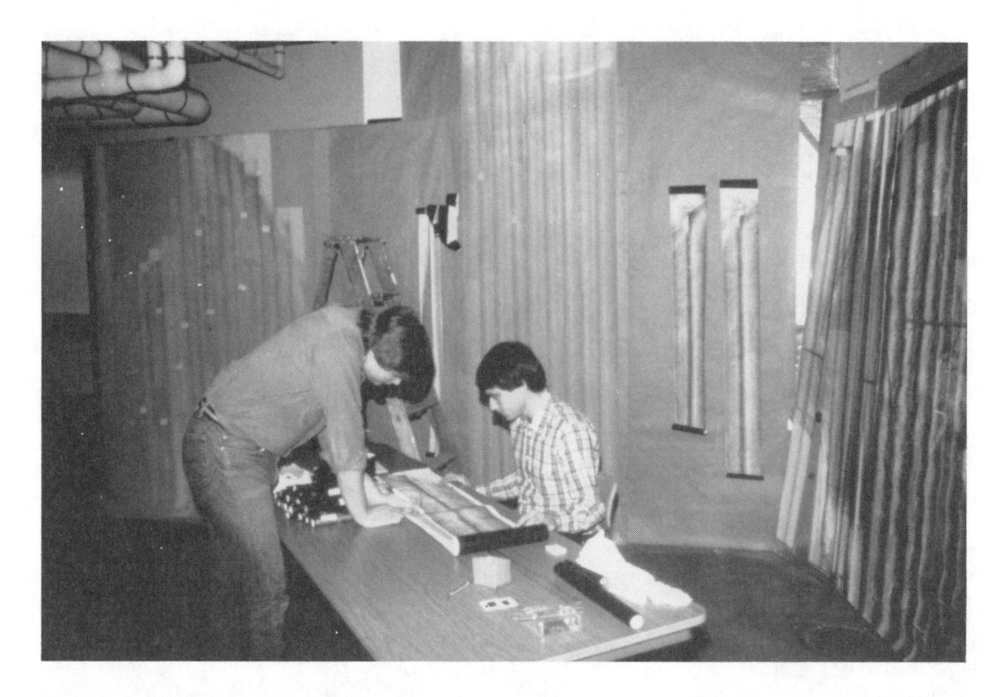

FIGURE 5A. Assembly of the side-scan sonar mosaic map of a lake trout spawning ground in Lake Huron. Individual microprocessor records are inspected and compared to ground truth information.

C. DATA PROCESSING AND PRODUCTS

Mosaic substrate and bathymetric maps are the major products of the side-scan sonar mapping surveys. Mosaic maps are created in the laboratory by arranging microprocessor records from adjacent transects side by side on a large map board and matching images of major substrate features that extend across adjacent records (Figure 5). This method of assembly produces an "uncontrolled" mosaic. An alternate procedure (Edge, 1986), which produces a "controlled" mosaic, assembles the records by matching the real-time vessel position annotations on each microprocessor record with the same positions on a preconstructed Cartesian grid of identical scale. Both methods produce similar results for surveys of small areas, but the controlled approach may produce more accurate maps for areas larger than about 20 km^2, where even small errors in navigation can impede the proper alignment of microprocessor records.

The mosaic is interpreted by inspecting the individual records that compose it and reviewing the ground truth information. The microprocessor records reflect both the material properties of substrate components and the topography of the lake bed. Darker images are produced by rock and gravel than by sand and mud. Steep slopes facing the towfish reflect the acoustic signal more strongly and produce darker images than the slopes that are more obliquely oriented. Individual rocks larger than about 0.25 to 0.5 m in diameter and other objects of sufficient size and acoustical density exhibit both material and topographic qualities on microprocessor records, showing a dark image on the side facing the towfish and a lighter "acoustic shadow" (Mazel, 1985) on the opposite side.

Mosaic maps are readily digitized, entered into, and retrieved from a computer-linked GIS database (see F. J. Horvath, this chapter) for display at a scale and in a format convenient to the user. Draft bathymetric maps prepared manually from real time water depth data displayed in the profile section of the microprocessor record (Figure 3) are also readily processed by computer. Substrate and depth information incorporated into the database are easily manipulated to produce various statistics, including the area of lake bed within a pair of depth contours, and the area of lake bed covered by a particular substrate.

FIGURE 5B. After the records are inspected they are attached to a backing to form the mosaic. The mosaic shown here covers an area of lake bed about 2.4 km long and 1.2 km wide. Note prominent bedrock ridges at the upper right-hand portion of the mosaic; sand is the substrate in the lighter areas on the right side of the reef.

Substrate maps produced by computer lack the detailed, photographic appearance of the mosaic, but are more easily used by the viewer who is untrained in the interpretation of side-scan sonar data. If a map is needed that retains the detail or appearance of the mosaic and is also portable, the mosaic can be cut into blocks of convenient size, photographed, and reproduced at the desired scale (Edge, 1986).

III. EVALUATION

A. SAMPLING POWER AND RELATED FACTORS

The rate at which the lake bed can be surveyed by side-scan sonar is determined by the basic sampling power of the system and the way in which effort is applied to meet survey objectives. The width of the strip or swath of lake bed represented on the microprocessor record (the swath width), the speed at which the survey vessel draws the towfish along the transect (the towing speed), and the resolution of lake bed features on the record are interrelated variables that can be manipulated to help meet study objectives. Generally, narrow swath widths and low towing speeds yield high resolution of lake bed features. The smallest objects that can be resolved on the records of our 100 kHz side-scan sonar system vary from about 0.1 m at a swath width of 50 m and a maximum towing speed of 8.3 km h^{-1}, to about 1.5 m at a swath width of 1200 m and a maximum towing speed of 23.5 km h^{-1}. Side-scan sonar systems are available that operate at 500 kHz and provide higher resolution at lower sampling power than the 100 kHz systems; 50 kHz systems are also available for situations requiring higher sampling power and lower resolution.

The maximum speed of the survey vessel may also be important in determining the rate at which mapping proceeds, particularly if the survey area is remote from port facilities that must be visited daily. Vessel speed is also an important consideration when all the mapping transects must be traversed in the same direction in order to minimize navigation error. The survey vessel must be run back across the survey area after each transect is mapped to position the vessel for mapping the next transect. For slow vessels this "runback" time can nearly equal the time required to make side-scan sonar records.

Weather, limnological conditions, and the bathymetric and physiographic complexity of the site to be mapped can also significantly affect the time required to complete a survey. Reconnaissance and mapping surveys are most easily conducted when wave heights are <1.5 m and no strong thermocline is interposed between the towfish and the lake bed. The absence of shoals or other impediments to the movement and safe operation of the survey vessel also facilitate the completion of field work. Mapping also proceeds more rapidly in areas that are not physiographically complex and therefore do not require extensive reconnaissance or ground truthing. Areas of uniform depth are more easily mapped than those with more complex bathymetry because changes in depth along the mapping transects may require repositioning the towfish in the water column to maintain it at the optimum height above the lake bed — a distance equal to about 10% of the swath width.

Examples of the sampling power of side-scan sonar were provided by two mapping surveys recently completed in the eastern portions of Lake Erie and Lake Ontario. These surveys were performed on historical lake trout spawning grounds by the National Fisheries Research Center-Great Lakes in collaboration with the New York Department of Environmental Conservation and the Great Lakes Fishery Commission. Both surveys were conducted under nearly ideal sea and weather conditions, with the side-scan sonar system shown in Figure 1, a 200-m swath width, and 120 m transect spacing (i.e., 40% overlap between records for adjacent transects). In Lake Erie, a noncomplex reef area of about 5 km^2, in water 13 to 26 m deep was mapped in about 10 h (mapping rate, 0.5 km^2 h^{-1}) from the New York Department of Environmental Conservation vessel *Argo*, which has a maximum speed of about 14.8 km h^{-1}. In Lake Ontario, the National Fisheries Research Center-Great Lakes vessel *Kaho*, with a maximum speed of about 18.5 km h^{-1}, was used to map a relatively complex reef area of about 7.6 km^2, in water 5 to 78 m deep, in 19 h (mapping rate, 0.4 km^2 h^{-1}). The completion of reconnaissance surveys and collection of ground truth information required an additional 4 h at the Lake Erie site and 12 h at the Lake Ontario site.

B. ECONOMICS

The cost of side-scan sonar equipment declined substantially within the 1980s and a survey-quality system suitable for mapping Great Lakes benthic habitat can now be purchased for about U.S. $50,000 to $70,000.

The cost of conducting a side-scan sonar survey is largely the cost of operating the survey vessel. Vessel costs can be negligible, if the survey is performed from a small boat with a crew of one or two persons, or as much as several thousand dollars per day if a large ship is used. The higher cost of the larger vessel may be offset significantly, however, if it carries more than one crew so that mapping can be conducted continuously. The time required to complete a side-scan sonar survey of a particular area can be estimated once the study objectives are defined and the survey vessel is chosen. About two thirds of the time spent collecting data on survey sites is usually needed to produce microprocessor records and the remaining one third is used to conduct reconnaissance surveys and collect ground truth data. Adverse weather and site-specific factors, including a physiographically complex lake bed, are variables that extend field operations and increase survey costs.

Costs to process field data and produce mosaic substrate maps and bathymetric maps are primarily salaries. A team of three trained persons can create the maps from microprocessor records (scale 1:1000) for 1 km^2 of Great Lakes lake bed in about 3 or 4 d. The mosaic and bathymetric maps prepared manually from the microprocessor records can be the final products of the side-scan sonar survey. However, 1:1000 scale maps of even relatively small (3 km^2) lake trout spawning grounds are too large for convenient handling, and computer or photographic processing of the mosaic and bathymetric maps is recommended to produce scaled-down copies. Color-coded substrate maps and corresponding black and white bathymetric maps can be produced at any desired scale from a computer-linked GIS at a cost of about \$300 km^{-2} of lake bed. The cost of reproducing mosaic maps photographically, using the method of Edge (1986), can be substantially lower than the cost of producing them by computer.

The cost of equipment needed to collect ground truth information for side-scan sonar mapping of benthic habitat in the Great Lakes can vary widely. Base prices for the less expensive remotely operated vehicles range from about \$25,000 to \$37,000, but optional accessories can exceed 50% of the base price for some systems (Busby Associates, Inc., 1986). Capital costs for manned submersibles vary widely, but exceed \$5 million for one of the Johnson *Sea-Link II* class and its highly specialized support vessel. The cost of deploying a remotely operated vehicle of the MK II class is essentially the salary of a skilled operator plus the cost of operating the survey vessel. The cost of operating the Johnson *Sea-Link II* and its support vessel in the Great Lakes in 1985 and 1986 was about \$13,000/d.

C. QUALITY CONTROL

The reliability of survey-quality side-scan sonar systems for mapping the floor of a water body has been extensively documented (e.g., Clifford, 1979; Klein, 1984). Survey-grade systems now produce images and data that accurately portray the surficial substrates and bathymetry. These systems have various built-in operating and calibration tests that can be easily performed in the field to ensure that the equipment is functioning properly.

Field methods applied in mapping surveys typically provide a preset 10 to 50% overlap of lake bed coverage between microprocessor records for adjacent transects. This overlap helps ensure that legible records are produced, thus facilitating mosaic preparation and interpretation. Videotape records, photographs, direct underwater observations, and grab samples of the substrate provide ground truth information that can aid in interpreting the mosaic and producing reliable substrate maps. Bathymetric maps prepared from the microprocessor record profile can be verified with records from the echosounder aboard the survey vessel.

Highly accurate and precise techniques and instrumentation, including survey-quality Loran-C, are available to determine the position of the survey vessel and reference map data to an established geographic coordinate system (see R. H. Brock, this chapter). Assembly of a distortion-free mosaic is facilitated by use of the controlled approach (Edge, 1986), or by the use of the uncontrolled approach followed by graphic and statistical evaluation of the map (Edsall et al., 1989). Data manipulation and map production are easily and reliably handled by computer-linked GISs that are now available to the Great Lakes research and resource management community.

D. EFFECTIVENESS OF SIDE-SCAN SONAR IN DIFFERENT HABITATS

Side-scan sonar can be used effectively to inventory and map surficial substrates in most benthic habitats in the Great Lakes and their connecting channels (Bryant, 1985; Edsall et al., 1989). The maximum operating depths of survey-grade systems exceed the maximum water depth in the Great Lakes, and the minimum practical operating depth of these systems may be 2 or 3 m. Operation in deep offshore waters may be impeded if the thermocline interferes with the acoustic signal. However, the towfish can be deployed below the thermocline, or these areas can be surveyed when the water is not thermally stratified.

Side-scan sonar operation in shallow water can be accompanied by a reduction in sampling power and a loss of effectiveness in some situations. For example, steeply sloping walls in navigation channels, beds of submersed aquatic plants, and shoreline bulkheads may create exceptionally strong return signals that appear as "noise" on the microprocessor record and make interpretation difficult. In these situations, multiple passes on different vessel bearings may be needed to provide positive identification of the target and useful records for habitat inventory and mapping. Use of side-scan sonar is problematic in Great Lakes wetlands and shallow littoral waters populated with submersed plants because sampling power is reduced in shallow water and signals are blocked by submersed vegetation (Bryant, 1985). Other remote sensing techniques (see R. P. Bukata, this chapter; Hudson et al., 1986; Schloesser et al., 1985, 1988) probably offer greater potential for inventory and mapping in vegetated, shallow water habitat.

The maximum operating depth of the Mini-Rover MK II is about 300 m, which allows it to collect ground truth information in virtually all habitats in the Great Lakes. Operation of the MK II is more time consuming and difficult in deep water (>35 m) than in shallow water because the influence of water currents on maneuverability of the MK II increases as the length of the tether increases. However, when the MK II is used in deep water, it can be attached by its tether to a 15-kg "downweight" that is raised and lowered on a cable from the survey vessel. This facilitates spot placement of the MK II on the lake bed; however, it also limits the searching radius of the MK II to the length of the tether (usually 15 to 30 m) between the MK II and the point at which the tether is attached to the downweight.

The maximum operating depth of the *Sea-Link II* manned submersible is about 700 m and deep water operation poses no problem for this vehicle in the Great Lakes. The minimum operating depth of the submersible is basically the grounding depth of the support vessel.

IV. SUMMARY AND CONCLUSIONS

Recent developments in side-scan sonar technology have produced reliable, low-cost systems that can be used effectively to inventory and map the benthic habitat in the Great Lakes and their connecting channels. Detailed substrate and bathymetric maps useful to resource managers and others can be prepared manually from the side-scan sonar data. Computer-linked GISs now available to the Great Lakes research and management community can also be used to produce substrate and bathymetric maps and perform other manipulations of side-scan sonar data. Remotely operated submersibles that can produce videotape and photographic records of the floor of the water body are also available at low cost. These vehicles can be easily deployed to provide ground truth information needed to interpret side-scan sonar records of Great Lakes benthic habitat.

Research performed in the Great Lakes in 1985 and 1986 from a manned submersible clearly demonstrates that this type of specialized sampling platform can provide the opportunity to conduct habitat inventory and classification studies that would otherwise be impossible. However, the very high capital and operating costs of manned submersibles will probably preclude their routine use in side-scan sonar surveys to produce ground truth information that can be obtained at lower costs by other means.

ACKNOWLEDGMENTS

I thank C. L. Brown, G. W. Kennedy, T. P. Poe, R. T. Nester, and J. R. P. French, III, who contributed significantly to the development of an in-house capability at the National Fisheries Research Center-Great Lakes for the use of side-scan sonar and remotely operated vehicles to survey and map benthic habitat in the Great Lakes. Use of the Johnson *Sea-Link II* and the R/V *Seward Johnson* was supported by the National Undersea Research Program of NOAA at the University of Connecticut. The Great Lakes Fishery Commission and New York Department of Environment Conservation provided financial and professional support for studies that contributed to the development of this report. Contribution no. 787 of the National Fisheries Research Center — Great Lakes.

REFERENCES

Askew, T. M., 1985. Johnson Sea-Link Users Manual. Misc. Publ. No. 17. Harbor Branch Foundation. Ft. Pierce, FL.

Boldrin, A., S. Rabitti, and A. Stefanon. 1980. A Detailed Biogeological Survey of Two Rock Outcrops in the Northern Adriatic Sea. Paper presented at 6th Int. Sci. Symp. World Underwater Federation (CMAS), September 14, 1980, Edinburgh, UK.

Bryant, R. S. 1985. Side-scan sonar for hydrography. An evaluation by the Canadian Hydrologic Service. *Int. Hydrogr. Rev.* 52(1):43–56.

Busby Associates, Inc. 1986. Remotely Operated Vehicle Evaluation. Prepared for U. S. Department of Commerce, National Oceanographic and Atmospheric Agency, Office of Underseas Research, Rockville, MD.

Clifford, P. 1979. Real time sea floor mapping. *Sea Technol.* 20(5):22–24, 26.

Continental Shelf Associates, Inc. 1985. Southwest Florida Shelf Regional Biological Communities Survey, Marine Habitat Atlas. Vol. 2. An Addendum and Update of the Southwest Florida Shelf Ecosystems Study — Years 1 and 2, Marine Habitat Atlas. Tequesta, FL.

Edge, H. P. J. 1986. Use of side-scan sonar mosaicing technique in Brunei. *Hydrogr. J.* 42:31–26.

Edsall, T. A., T. P. Poe, R. T. Nester, and C. L. Brown. 1989. Side-scan sonar mapping of lake trout spawning habitat in northern Lake Michigan. *North Am. J. Fish. Manage.* 9(3):269–279.

EG&G. 1985. Model 260 Image Correction Side-Scan Sonar Instruction Manual. EG&G Environmental Equipment, Waltham, MA.

Flemming, B. W. 1976. Side-scan sonar: a practical guide. *Int. Hydrogr. Rev., Monaco* 53(1):65–92.

Flemming, B. W. and A. K. Martin. 1983. Nearshore Physiography and Sediment Distribution between Mossel Bay and Natures Valley. Paper presented at 5th Natl. Oceanogr. Symp., January 24, 1983. Grahamstown, South Africa.

Gettleson, D. A., R. M. Hammer, and R. E. Putt. 1982. Geophysical and Biological Sea Floor Mapping of Four Oil and Gas Lease Blocks in the South Atlantic Georgia Embayment. Paper presented at Oceans '82 Conf., September 20, 1982, Washington, D.C.

Hudson, P. L., B. M. Davis, S. J. Nichols, and C. M. Tomcko., 1986. Environmental Studies of Macrozoobenthos, Aquatic Macrophytes, and Juvenile Fish in the St. Clair-Detroit River System. Great Lakes Fish. Lab. Admin. Rep. 86-7. U.S. Fish and Wildlife Service.

Interstate Electronics Corp. 1979. Biological and Geological Reconnaissance and Characterization Survey of the Tanner and Cortes Banks. Vol. 1. Synthesis of Findings. Prepared for Bureau of Land Management, Washington, D.C.

Jameson, S. C. 1981. Key Largo Coral Reef National Marine Sanctuary Deepwater Resource Survey. Rep. No. NOAA-TR-CZ/SP-1; NOAA-81121406. National Oceanic and Atmospheric Administration Office of Coastal Zone Management. Washington, D.C.

Johnson, K. R., C. H. Nelson, and J. H. Barber, Jr. 1983. Assessment of Gray Whale Feeding Grounds and Sea Floor Interaction in the Northeastern Bering Sea. Geological Survey. Open-File Rep. 83-727. U.S. Department of the Interior, Washington, D.C.

Key, W. H., Jr. 1984. Alternate Uses of Hydrographic Data Taken with Side-Scan Sonar. Paper presented at Hydro '84 — Natl. Ocean Serv. Hydrogr. Conf. April 25–27, 1984. pp. 17–23.

Key, W. H. Jr. 1985. 20 meter coastal minehunter with mine sonar and remote control vehicle capabilities. *Def. Syst. Rev. Mil. Commun.* 3:1–10.

Klein, M. 1984. High-Resolution Seabed Mapping: New Developments. Paper presented at 16th Offshore Technol. Conf., May 7–9, 1984, Houston, TX. pp. 73–78.

Klein, M. 1979. New Capabilities for Side-Scan Sonar Systems. Paper presented at Oceans '79, September 17, 1979, San Diego, CA.

Lissner, A. L. 1981. Biological and geological reconnaissance survey of two offshore southern California banks. *Estuaries* 4(3):304.

Manny, B. A. and T. A. Edsall. 1989. Assessment of lake trout spawning habitat quality in central Lake Huron by submarine. *J. Great Lakes Res.* 15(1):164–173.

Mazel, C. 1985. Side-Scan Sonar Training Manual. Klein Associates, Inc. Salem, NH.

Nerini, M. K. and J. S. Oliver. 1983. Gray whales and the structure of Bering Sea benthos. *Oecologia* 59:224–225.

Rezak, R. and T. J. Bright. 1981. Northern Gulf of Mexico Topographic Features Study: Vol. 3. Rep. No. TR-81-2-T; BLM-YM-P/T-81-009-3331. New Orleans Outer Continental Shelf Office. Bureau of Land Management. New Orleans, Louisiana.

Russell-Cargill, W. G. A., Ed. 1982. Recent Developments in Side-Scan Sonar Techniques. Central Acoustics Laboratory, University of Cape Town, Cape Town, South Africa.

Schloesser, D. W., C. L. Brown, and B. A. Manny. 1988. Use of aerial photography to inventory aquatic vegetation. *J. Aerosp. Eng.* 1:142–150.

Schloesser, D. W., B. A. Manny, C. L. Brown, and E. Jaworski. 1985. Use of low-altitude aerial photography to identify submersed aquatic macrophytes, in *Proc. 10th Biennial Workshop on Color Aerial Photography in the Plant Sciences, May 21–24, 1985,* J. Evaritt Ed. University of Michigan, Ann Arbor. pp. 19–28.

Sly, P. G. and C. C. Widmer. 1984. Lake trout (*Salvelinus namaycush*) spawning habitat in Seneca Lake, New York. *J. Great Lakes Res.* 10(2):168–189.

Woodward-Clyde Consultants. 1979. Eastern Gulf of Mexico Marine Habitat Study. Vol. 3. Rep. No. BLM/YM/ES-79/19. Bureau of Land Management, Washington, D.C.

PART D: REMOTE SENSING AS AN AID TO AQUATIC HABITAT SURVEILLANCE

R. P. Bukata

ABSTRACT

This section considers the application of visible, near-infrared (IR), and thermal IR remote sensing to those habitats. The applications are considered in four categories: (1) feasible and simple, (2) feasible but more difficult, (3) feasible but still more difficult, and (4) presently not feasible.

I. INTRODUCTION

Remote sensing has been employed in various forms and degrees of complexity for many decades. Particularly useful recommended readings in this area include "Remote Sensing and its Application to Inland Fisheries and Aquaculture" by Cheney and Rabanal (1984), along with many of the specialized papers and documents referred to therein. This section discusses remote sensing within LANDSAT-type energies, *viz.*, visible, near-IR, and thermal portions of the electromagnetic spectrum, for which backlogs of both space- and aircraft-borne data exist in fair abundance. Omitted from consideration are ultraviolet (UV), microwave, stimulated particle emissions, and such specialty areas as side-scan radar and hydroacoustics. Bodies of open water and wetlands are considered aquatic habitats, and the surveillance of these habitats provides representation for any parameter that can be inferred from the remotely sensed data.

It must be recognized that "inferred" is the operative word in remote sensing interpretation. What the remote sensing device directly records are the spectro-optical emanations (either stimulated or reflected) from some combination of components (organic, inorganic, or both) of the biosphere. No parametric information per se is collected. Consequently, the spectro-optical information must be converted, through appropriate multidisciplinary modeling activities, into "inferred" estimates of the chemical, biological, or physical parameter being sought. Clearly, such inferences are not always reliable. It must be further recognized that data formats for remotely sensed optical data generally include photographic, digital, or both, which require interpretive methodologies ranging from photogrammetry to environmental systems modeling and attendant "instrumentation" ranging from the unaided human eye to complex electronic display systems capable of high-speed digital analyses.

II. APPLICATIONS

The following list of remote sensing applications for aquatic habitat surveillance may be presented if (1) accurate evaluations of the reliability and usability of the remotely sensed data can be performed, (2) both visual and machine processing skills are available, as well as environmental or optical modeling and scientific interpretation expertise, and (3) the remotely sensed data being used, whether in photographic or digital formats, are restricted to visible, near-IR, and thermal IR wavelengths.

A. FEASIBLE AND SIMPLE

1. Obtaining shoreline information of the "atlas" type, including shoreline geometries, beach reaches, natural and man-made shore protection devices, bare-soil mapping, etc. The precision of such information is directly proportional to the photogrammetric techniques employed.
2. Thermal mapping of water masses.
3. Measurement of relative temperature of solid surfaces.
4. Surface area estimates of standing water. These estimates can be readily made synoptic overviews.
5. Relatively unsophisticated surveys of land use and vegetation cover.

B. FEASIBLE, BUT MORE DIFFICULT

1. Classification of water masses based on uniqueness of spectral signatures.
2. Vegetation mapping and classification of vegetation into general wetland cover types and identification of some plant species.
3. Demarcation of boundaries between wetland types, and estimation of the areal extent of each type.
   - Relationships between wetland acreage and water levels.
   - Classification of open meadow soil according to permeability.
   - Detection of areas of severe plant stress.
4. Generation of topographic maps. Careful use of stereoscopic methodologies on remotely obtained images can be a powerful tool for the precise determination of basin slopes required for many modeling applications. However, it is recommended that ground surveys be the definitive method.

C. FEASIBLE, BUT STILL MORE DIFFICULT

1. Estimates of some organic water quality parameters. Models have been painstakingly developed which display various degrees of success in estimating water quality largely from the optical qualities of the water body being remotely monitored. The further the remote sensing device is removed from the water surface, the more the intervening atmosphere interferes with the analyses, and the less reliable become the estimates of water quality. Thus, digital data acquired from aircraft at low altitude are much more likely to provide reliable estimates of water quality than are those acquired from satellites.
2. Monitoring surface water color. This is readily accomplished, and can be related with reasonable reliability to near-surface sediment or mineral concentrations, with usually slightly less reliability to chlorophyll *a* concentrations and dissolved organic carbon concentrations, if reliable water quality models are available.
3. Delineating aquatic regimes possibly impacted by pollutants. Very convincing ground-truth data are required to irrefutably identify the pollutant.
4. Delineation of groundwater flow pathways. This can be readily accomplished from satellite altitudes for the Great Lakes and other such freshwater basins through consideration of the vigor of canopies of phreatophytic vegetation.
5. Estimates of vegetative vigor.
6. Measurement of crop canopy and leaf-index ratio as a means of estimating wetland biomass.
7. Measurement of advances and retreats in infestations of vegetation by insects and diseases. Identification of the insect or disease agent might be possible.

8. Damage impact studies of storms, erosion, and other destructive climatic forces on wetlands, flora, and fauna.
9. Delineation of wetland subdivisions.
10. Collecting, maintaining, and analyzing baseline data for the above parameters.

D. PRESENTLY NOT FEASIBLE

1. Identification and inventory of fish stocks.
2. Precise mathematical relationships among the plant and animal life cycle parameters and their basin environments.
3. Classification of the identity and nature of aquatic contaminants and pollutants.
4. Reliable predictions of fish stock yield or biomass as a function of the trophic state of the habitat.
5. Quantitative determination of the effects of atmospheric pollutants on vegetation vigor.
6. Distinguish naturally from cultured fish stocks.

III. CONCLUSION

The list of current application and limitations is inconclusive. It in no way exhausts the spectrum of remote sensing applications within the visible, reflected IR, and self-emissive IR regions of the electromagnetic spectrum that could contribute to a sensible classification and inventory of Great Lakes aquatic habitats.

REFERENCES

Bukata, R. P., W. S. Haras, and J. E. Bruton. 1975. The application of ERTS-1 digital data to water transport phenomena in the Point Pelee-Rondeau area. *Verh. Int. Ver. Limnol.* 168–178.

Bukata, R. P., J. E. Bruton, J. H. Jerome, A. G. Bobba, and G. P. Harris. 1976. The application of LANDSAT-1 digital data to a study of coastal hydrography. Proc. 3rd Can. Symp. Remote Sensing. G. E. Thompson, Ed. Canadian Aeronautics and Space Institutes, Ottawa. pp. 331–348.

Bukata, R. P., A. G. Bobba, J. E. Bruton, and J. H. Jerome. 1978. The application of apparent radiance data to the determination of groundwater flow pathways from satellite altitudes. *Can. J. Spectrosc.* 23:79–91.

Bukata, R. P., J. H. Jerome, J. E. Bruton, S. C. Jain, and H. H. Zwick. 1981. Optical water quality model of Lake Ontario. I. Determination of the optical cross sections of organic and inorganic particulates in Lake Ontario. *Appl. Opt.* 20:1696–1703.

Bukata, R. P., J. H. Jerome, J. E. Bruton, S. C. Jain, and H. H. Zwick. 1981. Optical water quality model of Lake Ontario. II. Determination of chlorophyll *a* and suspended mineral concentrations of natural waters from submersible and low altitude optical sensors. *Appl. Opt.* 20:1704–1714.

Bukata, R. P., J. E. Bruton, and J. H. Jerome. 1982. The futility of using remotely determined chlorophyll concentrations to infer acid stress in lakes. *Can. J. Remote Sensing* 8:38–41.

Bukata, R. P., J. E. Bruton, and J. H. Jerome. 1985. Application of Direct Measurements of Optical Parameters to the Estimation of Lake Water Quality Indicators. IWD Sci. Ser. No. 140.

Bukata, R. P., J. E. Bruton, J. H. Jerome, and W. S. Haras. 1988. A Mathematical Description of the Effects of Prolonged Water Level Fluctuations on the Areal Extent of Marshlands. Inland Warer Directorate, Scientific Series, no. 166 p. 47.

Bukata, R. P., J. E. Bruton, J. H. Jerome, and W. S. Haras. 1987. An evaluation of the impact of persistent water level changes on the areal extent of Georgian Bay/North Channel marshlands. Submitted to *Environ. Manage.*

Cheney, D. P. and H. R. Rabanal. 1984. Remote Sensing and its Application to Inland Fisheries and Aquaculture. Food and Agriculture Organization of the United Nations Fisheries Circ. 768.

Haras, W. S., R. P. Bukata, and K. K. Tsui. 1976. Methods for recording Great Lakes shoreline change. *Geosci. Can.* 3:174–184.

PART E: AN OVERVIEW OF SURVEY CONTROL AND MAPPING CONSIDERATIONS FOR THE CLASSIFICATION AND INVENTORY OF GREAT LAKES AQUATIC HABITATS

R. H. Brock

ABSTRACT

With the increasing use of GISs, more attention must be given to attaching reliable control information to collected data. Available survey control and mapping techniques are briefly reviewed with respect to accuracy and cost including Loran-C, Global Positioning System (GPS), electronic distance measurement (EDM), aerotriangulation, and digitized scanner imagery. Special note is made of the advantage of combining techniques.

I. INTRODUCTION

This section explores ways to satisfy survey control needs for mapping and navigation. Survey control needs have expanded greatly since the introduction of GISs. Each piece of information that is entered into a GIS should be accurately identified and related to position and time for maximum utility. Position should be referenced to a coordinate system which has known and reliable parameters. Control or position can be thought of as either basic or supplementary. Basic control is that control which has been previously established and verified while supplementary control is tied to the basic control and it is established as necessary. Survey control is established to different levels of accuracy, depending upon the purpose (Federal Geodetic Control Committee, 1975).

II. SURVEY CONTROL DATA

Typically, a control point is identified by position such as latitude, longitude, and elevation above sea level. Inventories of large aquatic areas, above and below the water surface, however, could require adding other reference information to data such as time of collection and spectral characteristics. Each of these data sets should carry a reliability label.

Position information should be related to a known datum. Recent data gaining wide acceptance are the North American Datum of 1983 and the World Geodetic System of 1984. Control information that is available or to be established may need to be transformed to conform to the datum of the specific GIS. Computer programs are generally available for common transformations. In performing transformations the user must know the datum of the existing coordinates as well as the datum for the new coordinates. Other important considerations are accuracy of the old and new coordinates, the number of common points to each datum, and the distribution of the common points. These considerations are particularly important when deciding upon the method of transformation (Hothem, 1987).

The needs for survey control data could fall into two general groups. The first would be that of providing control to produce accurate maps. In most cases this control does not need to be real time, but specific applications could require the establishment of real time control. A second need for control data would fall under the broad category of navigation. Generally the control data used for navigation would be established in real time.

III. CONTROL METHODS

The object is to review the most feasible methods for establishing survey control for mapping, navigation, or both. The brief discussions concentrate on geodetic accuracy and general convenience for each method.

A. LORAN-C

Loran-C can be used for two-dimensional positioning and if properly applied, can generate geodetic positions to ±0.25 nautical mile 95% of the time. The system is also capable of locating positions with much better accuracy (±25 m) if used in the repeatable accuracy mode and adjusted to known stations. This accuracy would vary for particular circumstances, but the potential is available. Loran-C is inexpensive and receivers with waypoint navigation start at about $1000. A general review of Loran-C accuracy is given in the National Marine Electronics Association (NMEA) News (Anonymous, 1985).

B. GLOBAL POSITIONING SYSTEM

GPS has been under development since the early 1970s. When fully operational, GPS will provide continuous three-dimensional worldwide coverage. Using standard positioning code and a single receiver, geodetic positions can be established to ±0.01 nautical miles at the 95% level. Typical GPS receivers cost $45,000 to $65,000 with the range of costs going from $20,000 to $140,000. It is felt that these costs may be $2000 or less by the year 2000. The use of GPS is somewhat more complicated than Loran-C and presently does not provide real time two-dimensional geodetic positions as conveniently as Loran-C (Eschenbach, 1985; Wells, 1986; Hothem, 1987; Pealer, 1985; Stansell, 1980; Beser and Parkinson, 1984; Scherrer, 1985). (Editor's note: as of November 1991 the price range for GPS receivers is US $1500 to $3800 with an accuracy of ±15 m.)

C. ELECTRONIC DISTANCE MEASUREMENT

In many cases survey control can be supplied with surveying instruments capable of electronic distance measurement (EDM). Procedures incorporating EDM may not be particularly convenient, but the normal result is high accuracy geodetic positioning in two dimensions. EDM systems can cost from $4000 to $20,000 and is used most conveniently to establish shoreline control, but it can be extended to offshore positions.

D. AEROTRIANGULATION

Aerotriangulation is a spatial triangulation procedure that combines metric aerial photography with geodetic position information to produce three-dimensional positions for all points imaged on stereophotography (Slama, 1980). Presently it is not a real time system. The strength of aerotriangulation is in its ability to provide detailed information on an infinite number of points covered by the photographs. If existing photography can be utilized, the procedure is convenient. The acquisition of new photography may be time consuming. The geodetic position accuracy of computed control points will depend upon the scale of the photography and other factors specific to a given set of photos. Generally horizontal point positioning can be accomplished to ±0.3 m times each 3000 m of flying height for the photography. Elevations can be established to ±0.3 m times each 3000 m of flying height.

E. DIGITIZED SCANNER IMAGERY

Satellite imagery cannot presently be used to establish survey control to provide the framework for a map or for real time navigation purposes. It can be extremely useful if combined with control data from other systems, and future satellite imagery services may provide same-day coverage for a particular area of interest. The resolution of satellite imagery currently ranges from 10 to 30 m.

Other types of lower altitude scanner imagery such as thermal imagery may be used as supplemental information. The task of providing accurate maps with this type of imagery is more difficult since scale and geometry vary over the image format.

IV. COMBINATIONS OF CONTROL METHODS

None of the methods discussed are faultless. Consequently, it is best to use combinations to accomplish a given task (Braisted et al., 1986).

A typical problem is one of entering an area with no control and producing a detailed planimetric map of features of interest. First, survey control must be established and second, the feature details must be filled in between control points. If the final map features must be reliable to a few meters, then the initial control must be established by either GPS or EDM. EDM is simple, but may not be possible due to intervisibility, etc. (Wells, 1986). Assuming the detailed features in this case are smaller than 1 m, then aerial photos taken with mapping cameras should be used to provide the detailed map. The producers used to fit the detailed features to the control points may be as simple as using an optical transfer instrument to fit the projected imagery to the plotted control. The final product can then be digitized and incorporated into a GIS or reproduced as hard copy imagery. Appropriate field checks should be incorporated into the process.

Another problem dealing with real time temperature mapping might utilize GPS or EDM to establish geodetic position control on buoys and then utilize these control points to calibrate Loran-C positions taken on the same control points. The resulting mathematical transformation would then be applied to the Loran-C output at the temperature sampling points to provide adjusted positions. It is also possible that extensive thermal scanner imagery of the same area at the same time could be calibrated with both BPS (position) and Loran-C (position, temperature) data to provide a more extensive and detailed temperature map.

Many combinations are possible but strict attention to accuracy and final error propagation is needed to ensure the usefulness of the final product.

V. SUMMARY

Several points have been made in this overview. It is important that habitat inventory studies incorporate control information and that the specific control information be compatible with the GIS or in a form that can be transformed in the GIS. If this is adhered to, the GIS provides excellent potential to combine separate studies for a particular focus. Control must be established with detailed attention to accuracy and geometry. Several methods of establishing control have been discussed and general accuracy and costs statements have been given. No one control method is presently faultless, therefore, the appropriate combination of control methods is both important and necessary. Attention to error propagation is critical when combining control methods; the necessary field checks should be made.

REFERENCES

Anonymous. 1985. Loran-C accuracy. *NMEA News J. Mar. Electron.* 12(1):10–11, 69–70.

Beser, J. and B. Parkinson. 1984. The application of NAVSTAR differential GPS in the civilian community, global positioning system. *Navigation, ION* 2:167–196.

Braisted, P., R. Eschenbach, and A. Tiwari. 1986. Combining Loran and GPS — the best of both worlds. Global positioning system. *Navigation, ION* 3:235–240.

Eschenbach, R. F. 1985. GPS, The coming revolution in marine navigation. *NMEA News J. Mar. Electron.* 12(1): 13, 66–68.

Federal Geodetic Control Committee. 1975. Specifications to Support Classification, Standards of Accuracy, and General Specifications of Geodetic Control Surveys, U.S. Department of Commerce, NOAA, NOS, Rockville, MD.

Hothem, L. 1987. Applied survey techniques using GPS, in *COURSE 121 NOTES*. Navigation Technology Seminars, Inc. Falls Church, VA.

Pealer, K. A. 1985. U. S. radionavigation policy. *NMEA News J. Mar. Electron.* 12(1):20, 71–74.

Scherrer, R. 1985. The WM GPS Primer, Wild Heerburgg. Geodesy Division, WM Satellite Survey Company. Heerburgg, Switzerland.

Slama, C. C. 1980. *Manual of Photogrammetry,* 4th ed. American Society of Photogrammetry. Falls Church, VA.

Stansell, T. A., Jr. 1980. Civil marine applications of the global positioning system. Global positioning system. *Navigation, ION* 1:132–140.

Wells, D. 1986. *Guide to GPS Positioning.* Canadian GPS Associates. Frederickton, NB, Canada.

PART F: THE USE OF A GEOGRAPHICAL INFORMATION SYSTEM TO INVENTORY, CLASSIFY, AND ANALYZE GREAT LAKES AQUATIC HABITATS

F. J. Horvath

ABSTRACT

The Michigan Department of Natural Resources is developing a computer-based GIS to manage and analyze spatially related geographically referenced data describing cultural features and natural resources on land, along the shorelines, and in the waters of the Great Lakes. This section describes Michigan's GIS and related programs and discusses the potential application of a GIS to manage Great Lakes aquatic habitat information.

I. INTRODUCTION

The Michigan Department of Natural Resources is mandated to protect, conserve, and allow for the prudent use of the State's natural resources. These resources include a major share of the Great Lakes' shorelands, bottomlands, and water. To address intense demands on, and competing uses of, these resources, and to take advantage of the explosion of new scientific knowledge, Michigan is developing a computer-based GIS.

The Michigan GIS is administered by two program units: the Michigan Resources Inventory System (MIRIS) to address land resources, and The Great Lakes Information System (GLIS) to address shoreline and whole-lake resources. Both units reside in the Land and Water Management Division and use the GIS, complemented by more traditional data management tools, to catalog, inventory, and analyze cultural and natural resource data.

A GIS is a computer system that manages and analyzes spatially related geographically referenced data. Natural resource data, inventoried and managed traditionally on maps, is perfectly amenable to a GIS. The GIS allows data "themes" to be selectively overlain on base maps where spatial relationships and distributions can be visualized.

The Michigan GIS is being used to inventory a wide range of natural and cultural features. This approach can be used with a scheme to classify aquatic habitats.

II. GIS DATA REQUIREMENTS

A GIS uses digital spatial data that consists of points, lines, and polygons (multi-sided shapes) associated with a geographical location. The data take the form of roads, lakes, rivers, soil associations, plant communities, and a wide range of cultural and natural features. The data are associated in a geographically relevant plane and are separated by type ("theme") into different computer files ("levels") which can be composited ("overlain") on each other to show their spatial relationships. The data can be shown at various scales and displayed on a computer screen or plotted as high-resolution maps.

Point data, such as water chemistry data or the locations of sampling sites, can be displayed on maps. Raw data can be analyzed using traditional data processing techniques and then statistics derived from these data can be plotted. By using computer software that extrapolates area distributions from point data, the distributions can be displayed as isopleths on maps.

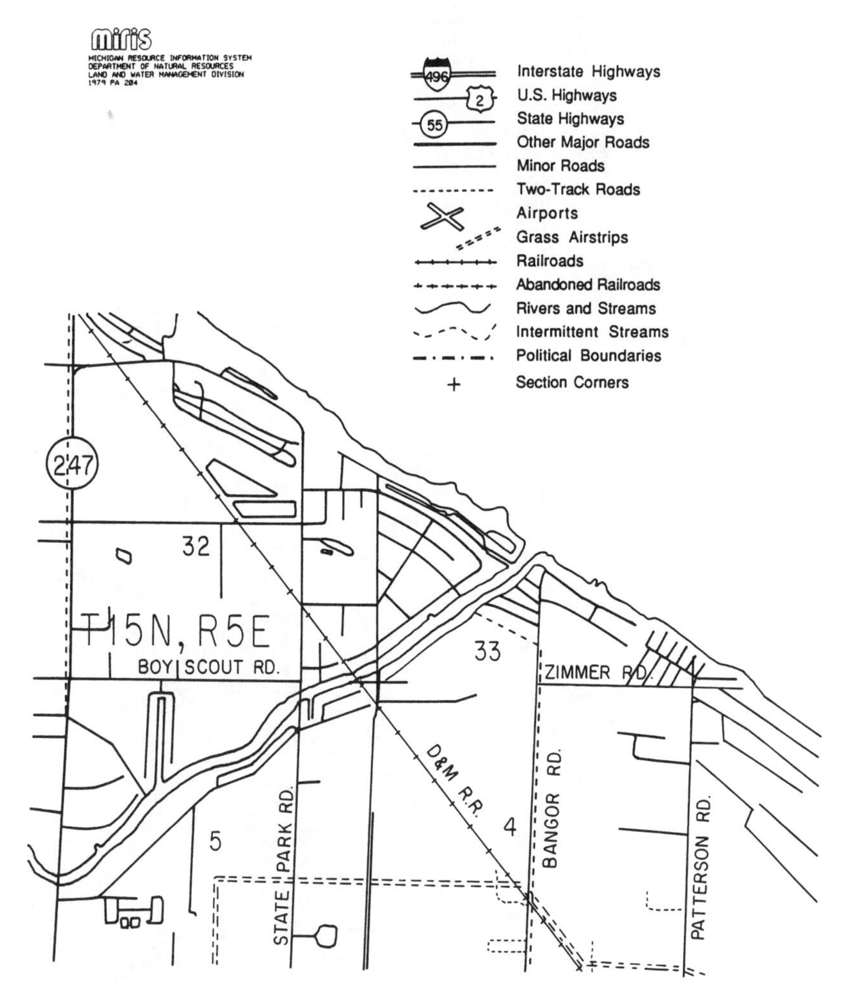

FIGURE 1. Base map of a typical Great Lakes shoreline produced by the Michigan GIS. The subjects shown in the legend (upper right) can be plotted separately or as composites. A composite of most subjects is shown here.

III. THE MICHIGAN GIS AND RELATED PROGRAMS

A. MIRIS AND GLIS

The Michigan GIS uses data derived from 1:24,000 scale (7.5 min) U.S.Geographical Survey (USGS) topographical maps as the functional base. Figure 1 is a portion of the Saginaw Bay (Lake Huron) shoreline showing features from the base map. The various levels of information that can be displayed separately or as composites are listed in Figure 1.

For the Great Lakes nearshore areas, the base maps are bathymetric maps at the 1-ft (30 cm) contour interval derived from National Ocean Service hydrographic surveys. Figure 2 is a portion of Saginaw Bay offshore of the area shown in Figure 1.

Table 1 lists the major natural and cultural themes being incorporated into the system. Data are classed either as area (e.g., fish spawning grounds) or point (e.g., water chemistry from a sampling site).

Point data such as the environmental quality data derived from monitoring and surveillance programs and locations of wastewater discharges are being assembled using traditional data processing techniques. Graphic overlays can be developed from these point data and displayed on base maps along with other themes to demonstrate their spatial relationships.

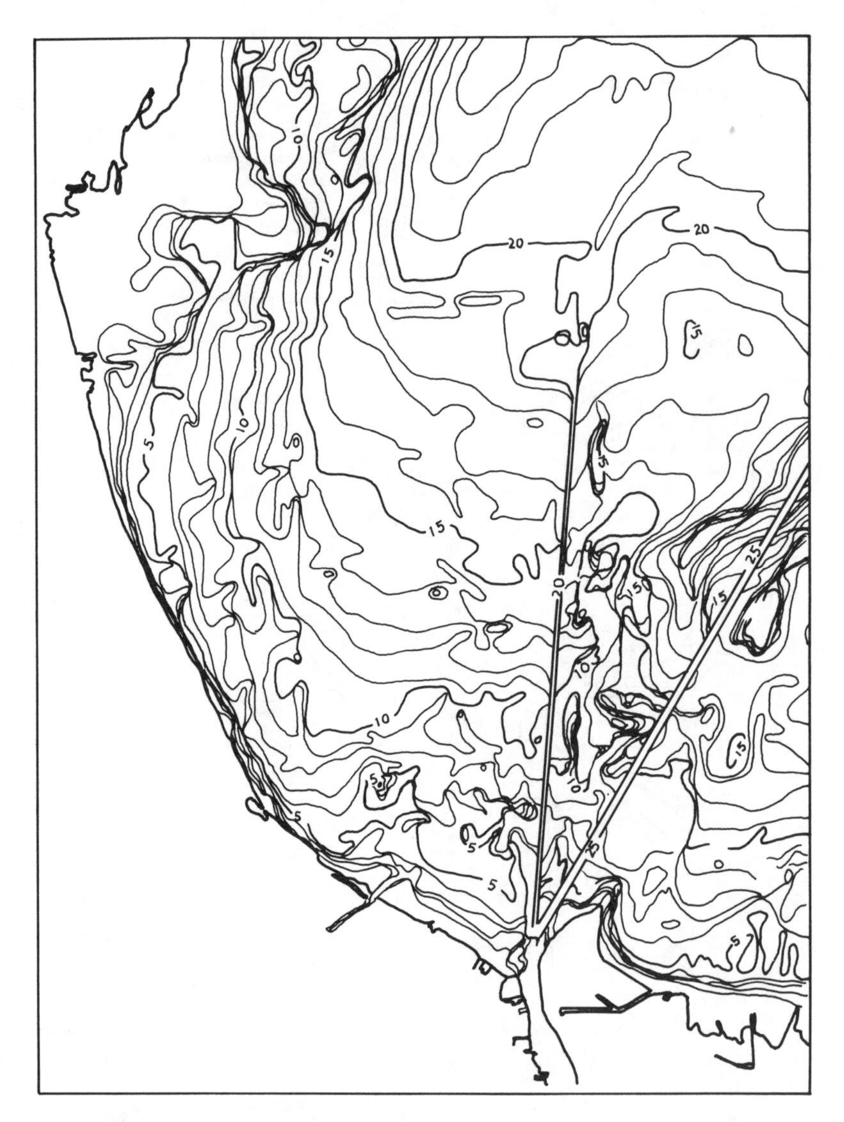

FIGURE 2. A portion of a bathymetric map of Saginaw Bay (Lake Huron) showing 1-ft depth contours. Map produced by Michigan GIS.

The Michigan GIS can produce reports on the area measurements ("acreage reports") of selected features and the area in which features co-occur.

B. THE MICHIGAN LAND USE INVENTORY

Mandated by Public Act 204 of 1979, the Michigan GIS is developing a major file on land use and cover. Source material is color infrared photography taken in 1978. The minimum size mapping unit is between 12.5 and 5 acres (5 and 2 ha), and the polygons are coded using a 64-part system described in MDNR (1976). Figure 3 presents the classes of land use and cover and shows the same section of Saginaw Bay shoreline as in Figures 1 and 2.

The land use and cover maps are updated regularly from new photographs, a process made possible and practical with the GIS. Land use and cover changes and trends can be rapidly detected and quantified by overlaying old and new files.

TABLE 1
Subjects ("Themes") that are Being Incorporated in the Michigan GIS for the Great Lakes

Land cover and use
Land ownership: owner: federal, state, local, municipal, private; size; easements and use restrictions; source of acquisition funds; boundaries; administering agency
Base map: Roads, highways, railroads, governmental boundaries, section corners, streams, lakes, drains, watershed boundaries
Topography
Bathymetry: 1-ft contour intervals nearshore; nautical chart bathymetry for open lake
Soils survey data
Sediment: type, particle-size distribution, depth of strata, chemical composition, including heavy metals and persistent toxic organic compounds
Effluent discharges: National Pollution Discharge Elimination System (NPDES) and Storage and Retrieval System (STORET) numbers, type (industrial, municipal), level of treatment, volume, composition
Water intakes: industrial or municipal, volume (consumptive vs. nonconsumptive), diversions out of basin
Navigation channels and harbors: commercial, recreational; bathymetry; capacity; use; dredging history; expansion potential
Population: demographics — distribution and composition
Energy facilities: storage, generation, processing, transfer; coal, oil, gas, nuclear, electricity; capacity; potential for spills; other impacts (e.g., impingement and entrainment); potential sites
Geological: bedrock and glacial formations; well locations and type; solution mining
Waste disposal sites: landfills and areas of known or suspected contamination; type and volume; score from Michigan Site Evaluation System; Act 307 and Superfund activities; known or suspected groundwater contamination; possible leachate to Great Lakes or tributaries
Critical habitats—by type and class: land cover, spawning and breeding areas, wetlands, recreational, sand dunes; habitats for threatened or endangered species
Hazardous areas: flood and high risk erosion
Legally restricted areas—by type and enabling legislation: wetlands, parks, natural rivers, historical and archaeological sites, submerged lands, designated environmental areas, underwater preserves, air quality zones
Limnology: Physical: thermal structure, currents, general water chemistry (chlorides, nutrients, solids, etc.), contaminants (pesticides, chlorinated organics, hydrocarbons, heavy metals, industrial chemicals)
Biological: plankton, macroinvertebrate, amphibian, reptile, fish distribution, diversity, and abundance; contaminants in fish and birds

C. BATHYMETRIC MAPS IN THE MICHIGAN GIS

Bathymetric maps showing 1-ft (30 cm) contour intervals are now being developed for Michigan's Great Lakes shoreline and nearshore areas. The area mapped will extend offshore to approximately the 40-ft (12 m) contour or to the edge of the wave-turbulence zone. The Michigan GIS is able to analyze the area between contours and is able to display each depth contour individually or all at the same time. Substrate of habitat maps can be displayed as overlays on bathymetric maps where the relationship between depth and substrate, or depth and habitat use, can be visualized and quantified. Similar relationships could be displayed from habitat classification data when the data become available.

IV. APPLICATION OF A GIS TO AQUATIC HABITATS

A technique of land use analysis similar to that used by the Michigan GIS could be adapted for application to aquatic habitats. Habitat type, abundance, and distribution, acquired from numerous techniques of remote sensing and ground survey, could be coded and incorporated into the GIS. Changes in distribution, size or class, and spatial relationships to other natural features or cultural intrusions could be analyzed. Application of this technology is feasible once the habitats are mapped and classified. Acquiring the basic data appears to be the biggest obstacle.

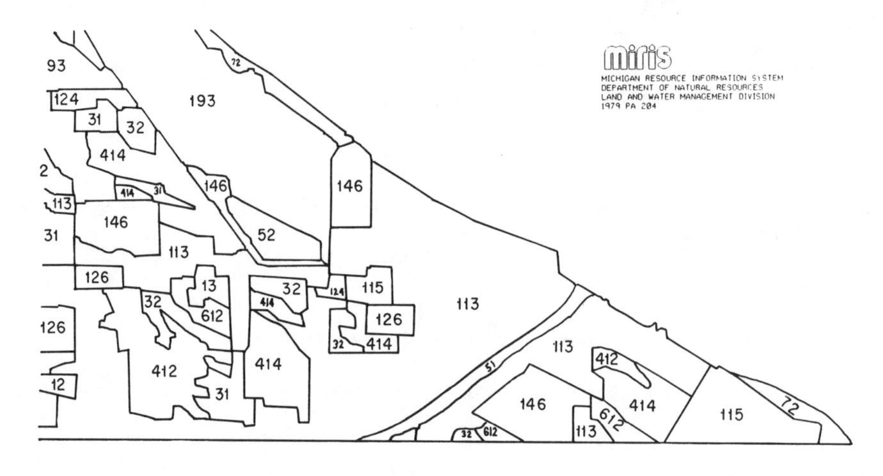

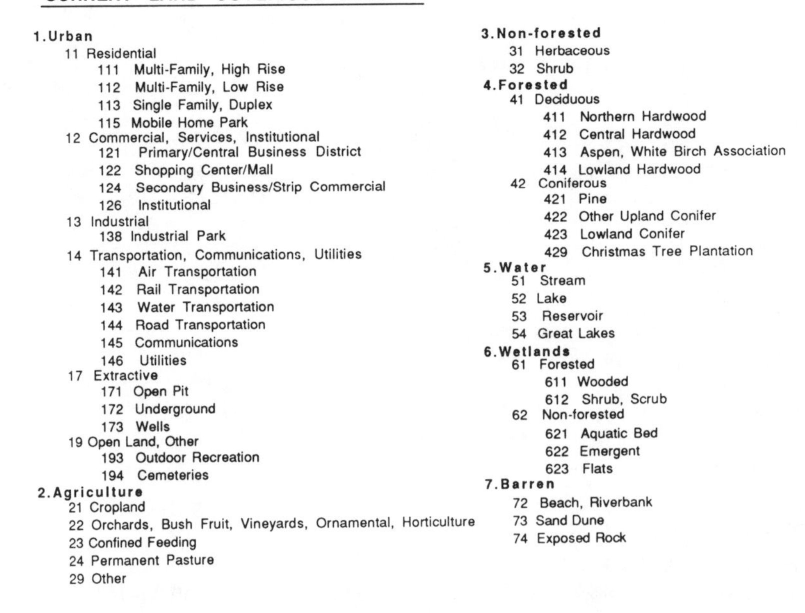

FIGURE 3. Land use and cover of a typical Great Lakes shoreline. Classification follows MDNR (1976). A classification scheme for aquatic habitats could be adapted for use in a GIS, as demonstrated above.

V. "EXPERT SYSTEMS" AND THE MICHIGAN GIS

Coulson et al. (1987) describes an "intelligent geographical information system" as an artificial intelligence system that relies upon the data files from a GIS as the "knowledge base". Various natural resource management decisions could be derived by linking data files with a set of rules or criteria. The Michigan GIS database could conceivably be used as a knowledge base in conjunction with a standard aquatic habitat classification system and a regional standard approach to habitat management.

REFERENCES

Coulson, R. N., L. J. Folse, and D. K. Loh. 1987. Artificial Intelligence and Natural Resource Management. *Science* 237(4812):262–267.

MDNR. 1976. Michigan Land Cover/Use Classification System. Developed by the Michigan Land Use Classification and Referencing Committee. Division of Land Resource Programs, Michigan Department of Natural Resources. Lansing. (July 1975, revised January 1976.)

INDEX

A

B

C

D

E

F

G

H

I

J

K

L

P

Q

R

Y

Z